Biological Systematics

T0074720

Biological Systematics

The state of the art

Alessandro Minelli

Dipartimento di Biologia
Universita' degli studi di Padova
Padova
Italy

CHAPMAN & HALL

London · Glasgow · Weinheim · New York · Tokyo · Melbourne · Madras

Published by Chapman & Hall, 2-6 Boundary Row, London SE1 8HN, UK

Chapman & Hall, 2-6 Boundary Row, London SE1 8HN, UK

Blackie Academic & Professional, Wester Cleddens Road, Bishopbriggs, Glasgow G64 2NZ, UK

Chapman & Hall GmbH, Pappelallee 3, 69469 Weinheim, Germany

Chapman & Hall USA., One Penn Plaza, 41st Floor, New York, NY10119, USA

Chapman & Hall Japan, ITP - Japan, Kyowa Building, 3F, 2-2-1 Hirakawacho, Chiyoda-ku, Tokyo 102, Japan

Chapman & Hall Australia, Thomas Nelson Australia, 102 Dodds Street, South Melbourne, Victoria 3205, Australia

Chapman & Hall India, R. Seshadri, 32 Second Main Road, CIT East, Madras 600 035, India

First edition 1993
First published in paperback 1994

© 1993 A. Minelli

Typeset in 10/12pt Palatino by EXPO Holdings Sdn. Bhd., Malaysia

ISBN 978-0-412-62620-3   ISBN 978-94-011-9643-7 (eBook)
DOI 10.1007/978-94-011-9643-7

Apart from any fair dealing for the purposes of research or private study, or criticism or review, as permitted under the UK Copyright Designs and Patents Act, 1988, this publication may not be reproduced, stored, or transmitted, in any form or by any means, without the prior permission in writing of the publishers, or in the case of reprographic reproduction only in accordance with the terms of the licences issued by the Copyright Licensing Agency in the UK, or in accordance with the terms of licences issued by the appropriate Reproduction Rights Organization outside the UK. Enquiries concerning reproduction outside the terms stated here should be sent to the publishers at the London address printed on this page.

The publisher makes no representation, express or implied, with regard to the accuracy of the information contained in this book and cannot accept any legal responsibility or liability for any errors or omissions that may be made.

A Catalogue record for this book is available from the British Library

Library of Congress Cataloging-in-Publication Data available

*For Federica
and Francesca*

Contents

Foreword

To some potential readers of this book the description of Biological Systematics as an art may seem outdated and frankly wrong. For most people art is subjective and unconstrained by universal laws. While one picture, play or poem may be internally consistent comparison between different art products is meaningless except by way of the individual artists. On the other hand modern Biological Systematics – particularly phenetics and cladistics – is offered as objective and ultimately governed by universal laws. This implies that classifications of different groups of organisms, being the products of systematics, should be comparable irrespective of authorship.

Throughout this book Minelli justifies his title by developing the theme that biological classifications are, in fact, very unequal in their expressions of the pattern and processes of the natural world. Specialists are imbibed with their own groups and tend to establish a consensus of what constitutes a species or a genus, or whether it should be desirable to recognize subspecies, cultivars etc. Ornithologists freely recognize subspecies and rarely do bird genera contain more than 10 species. On the other hand some coleopterists and botanists work with genera with over 1500 species. This asymmetry may reflect a biological reality; it may express a working practicality, or simply an historical artefact (older erected genera often contain more species). Rarely are these phenomena questioned. Yet, many of our analytical problems of biogeography, estimation of species numbers, evaluation of biodiversity or patterns of extinction depend on broad taxonomic assessments of comparability from fish to fowl, or flea to flowering plant.

Minelli surveys modern systematic methods and suggests that with the development of cladistics, computers and new molecular techniques, Biological Systematics of the nineties is advancing rapidly – and who would disagree? However, in his extended essay on the species (Chapter 4) he reiterates the many and varied ideas about what constitutes a species, the problems to which these lead and concludes that a species concept may not be appropriate for all organisms.

As a step towards describing the art in Biological Systematics, Minelli (Part II) surveys the major classifications of animals and plants to point out where the most rigorous ideas of interrelationships (systematization) are to be found. With the help of some 1450 references this provides the general reader with a rich source of material for discussion and further investigation.

Peter Forey
Natural History Museum
London, UK

Preface

Bad taxonomy, of which there has been plenty, persists. Unlike bad chemistry or bad physiology, of which there has probably been equally as much, it cannot be ignored; it must be undone and redone. Poor taxonomy is not only an ill unto itself; it is contagious, often with a very long incubation period ...One assumes that when [experimental biologists] state that they used 5 ml ethanol, they were not using 6 ml of methanol; and yet, if the experimental animal is wrongly identified, what are the grounds for such an assumption?

D.K. McE. Kevan (1973, p.1212)

'I write this book for fellow professionals at all stages of development': I borrow, for my book, these words from the preface to Niles Eldredge's *Evolutionary Dynamics* (Eldredge, 1989, p.viii).

What prompted me to write these pages, was the feeling that all main facets of biological systematics needed to be presented once and for all within the pages of one book. I was prepared to sacrifice depth of treatment to the advantage of broadness in coverage, because there are already plenty of specialized papers and books, where one can find out the details of a particular topic. Accordingly, this is not intended as a handbook or a reference book, but rather as an essay to read and to discuss. I hope that it will prove of some interest not only to fellow systematists, but also to other people, in the field of evolutionary biology or in biology at large, who are interested in getting a comprehensive idea of what is occurring today within the broad field of systematic biology. In addition, I hope that some parts at least will provide a refreshing read for biology teachers dealing with systematic biology in their courses.

Biological systematics has accomplished tremendous developments, not just since Aristotle or Linnaeus, but particularly in the last two or three decades. This is due, in part, to technical progress in two areas, particularly: molecular biology and computer techniques. Molecular biology has opened a new dimension to the comparative study of living organisms, whereas computers have allowed the development of powerful numerical techniques and their use for straightforward handling of large data matrices of (adequately coded) descriptive data.

These are only the most easily perceived among the many facets of the recent progress in biological systematics. Still more important, in my view, is the profound theoretical rethinking of the very foundations of biological systematics we have witnessed in the recent literature: a very deeply felt and still open debate, the outcome of which, as we shall see in the following pages, is a first, definitive entrance of sound phylogenetic thinking into biological systematics.

Despite these huge developments, many traditional barriers still hamper the rapid and fruitful circulation of ideas. It is perhaps not by chance, that for several years cladistics could not expand within botany, whereas it was already strong in zoology and palaeontology; still more, even in these fields it did not circulate too much, for several years, outside specialists' fields (insects, especially Diptera; Recent and fossil 'fishes') cultivated by some leading figures such as Hennig, Brundin and Griffiths, or Nelson, Patterson and Wiley.

Do botanists really know what zoologists are doing, and *vice versa*? Consider the following sentences. 'Zoologists have not attempted a definition of species based exclusively on crossability' (Raven, 1986, p.17); and 'botanists have historically restricted the movement (so common in animal-oriented evolutionary biology of the past 50 years) to recognize 'species' primarily as reproductive communities' (Eldredge, 1989, p.92). Who is right? Both pictures are very probably inaccurate and reflect excessive specialization in biological systematics, as in other scientific areas.

Hence, the idea to write a book like this. I am well aware of the risks I run, when adventuring in many fields traditionally remote from my zoological background, but I run them deliberately. Surely, all individual topics could be dealt with better than I can do here, but the main message I try to convey with these pages is that most systematists need, more often than not, to look much further than they usually do, confined as they are within their narrow specialistic fields.

A few examples of the topics to be developed in the following pages are given here. First, species concepts (and taxonomic practices) developed by students of birds need to be compared with those developed by students of flies, ciliates, or flowering plants.

Second, fossil taxa cannot be simply classified along with Recent ones, without adequate discussion, not only of the possible meaning of 'species' in palaeontology, but also of the role of fossils in phylogenetic reconstruction, as well as of the opportunity, and feasibility, of developing a diachronic, rather than synchronic, classification.

Third, nomenclatural problems will not be resolved as long as zoologists, botanists and bacteriologists continue to work with their separate Codes, or to revise them without trying to overcome the many puzzles posed, for instance, by the 'ambiregnal' nomenclature of protists.

The book is divided into three parts. Part One deals with general issues, both theoretical and methodological, beginning with a discussion of the very nature and aims of biological systematics (Chapter 1), followed by a survey of concepts and procedures in the analysis of phylogenetic relationships of living organisms (Chapter 2). A fairly detailed discussion (Chapter 3) is devoted to the rapidly developing field of molecular systematics, whereas Chapter 4 reviews the matters concerning species, speciation and infraspecific taxa. The last chapter of this first part deals with the resources and media in systematics, especially with matters concerning literature and nomenclature.

Part Two is devoted to a survey of the state of the art. First, how many species do we know, and how many species do we expect to discover (Chapter 6). What follows (Chapter 7) is a short account of the major advances in the systematic knowledge within the major taxa of the living world (but it should be very clear that this book does not aim at all at a comprehensive, encyclopaedic coverage of all major taxa, such as in the well known *Synopsis and Classification of Living Beings* edited by Sybil P. Parker (1982)). Chapter 8 summarizes different experiences with daily taxonomic work, as witnessed by many colleagues I have interviewed, on matters such as recognition of genera, species and subspecies, as currently treated in individual groups. The same chapter also briefly touches on recent technical and organizational improvements of systematic work. The last two chapters of Part Two are devoted to less usual questions: to an analysis of the unequal distribution of taxonomic diversity within the living world (Chapter 9) and to the taxonomic treatment of domesticated animals and cultivated plants (Chapter 10).

Part Three is very short (Chapter 11 only) and deals with several problematic areas that I wish to call attention to, when looking towards the future.

I would like to acknowledge my enormous debt to the many authors, of the past as well as of the present, whose works have stimulated my views on biological systematics. To remember them all would necessitate duplicating the references given at the end of the book. Therefore, here I quote only the names of numerous friends and colleagues who have helped me directly, kindly answering my specific requests for information for this book: E.F. Anderson, W.R. Anderson, I. Andrássy, R. Argano, H. Aspöck, F. Athias-Binche, P. Audisio, D.F. Austin, B. Baccetti, G. Baldizzone, E. Balletto, M. Balsamo, S.C.H. Barrett, S. Barbagallo, L. Bartolozzi, F. Bernini, R. Bertolani, R.G. Beutel, M. Biondi, F. Boero, E.L. Bousfield, H. Brailowsky, K. Bremer, R. Brinkhurst, S. Brookes, K. Brown Jr., D. Burkhardt, J.M. Carpenter, S. Carter, F. Cassola, D. Clayton, J. Coffey Swab, E. Colonnelli, J.O. Corliss, J.L. Crane, P.S. Cranston, M. Daccordi, G. Dahlgren, D. Danielopol, D.A.L. Davies, W. Dietrich, R.H. Disney, R.P. Essen, K. Fauchald, J.H. Frank, G. Fryer, P.A. Fryxell, R.J. Gagné, G. Gardini, G.A.P. Gibson, R. Gibson, A. Giordani Soika, F. Giusti, J. Golding, H. Goulet, W.F. Grant, R. Grolle, J.P. Haenni, C.L. Häuser, S.A. Hasan, R.L. Hauke, E. Heiss, I.D. Hodkinson, J.-L. d'Hondt, D. Isely, M. Javaheri, C. Jeffrey, L.A.S. Johnson, S.B. Jones, Z. Kabata, D.H. Kavanaugh, I. Kerzhner, K.C. Kim, N.A. Kormilev, H. Koyama, N.P. Kristensen, M. La Greca, N. Levine, K.G. McKenzie, F. Margaritora, J. Martens, B.R. Maslin, A. Messina, C.D. Michener, K. Mikkola, C. Monniot, L. Mound, L. Munari, E.S. Nielsen, P.L. Nimis, B. Nordenstam, M. Olmi, P. Omodeo, G. Osella, P. Paiero, P. Passerin d'Entreves, B. Pejler, T.D. Pennington, J. Péricart, G.L. Pesce, E. Petitpierre, M.N. Philipson, W.R. Philipson, R.E.G. Pichi Sermolli, D.J.G. Podlech, R. Poggi, V.

Puthz, T. Racheli, S. Ragusa, K. Rahn, G. Relini, C. Ricci, E.S. Ross, R. Rouch, R. Rozkosny, S. Ruffo, A. Ruttner-Kolisko, O.A. Saether, S. Sakai, M. Sarà, S.C. Schell, G.D. Schmidt, H.K. Schminke, M. Schmitt, R.T. Schuh, R. Shiel, M.R. Siddiqi, J.A. Slater, A. Smetana, C.N. Smithers, M. Solinas, W. Sterrer, T.F. Stuessy, P. Sundberg, E. Sylvén, R. Tarjan, I. Tavares, M. Thayer, R.F. Thorne, P. Tongiorgi, G. Underwood, C. Utzeri, G. Viggiani, A. Vigna Taglianti, V. Vomero, H.-P. Wagner, R. Wallace, H.E. Weber, G.C. Williams, T. Yamasaki, Th. Zanoni, A. Zullini and R. zur Strassen. Previous drafts were very carefully read and constructively criticized by Chris Humphries and Peter Forey, who deserve my most sincere gratitude. Without the help of all these people, this book could never have been written. However, they are not responsible for the views I have expressed in these pages. My debt towards my friends and colleagues at the Natural History Museum (London) is even larger, however, beginning with Laurence Mound giving the first critical look at my manuscript three years ago, and ending with Peter Forey kindly writing the Foreword.

Finally I would like to acknowledge the Consiglio Nazionale delle Ricerche and the Italian Ministero della Pubblica Istruzione, grants from which have allowed me to work on systematics for all these years; and the editorial staff of Chapman & Hall for rare patience with my manuscript and for skilful editing of the same.

The last word of acknowledgement is for my wife Pia and our daughters, for their continuous stimulation and varied help during the many months I have dedicated to writing this book, rather than to them.

A. Minelli
Padova, August 1992

PART ONE
Problems and Methods

... the stability [often seen] as the advantage of at least some evolutionary classifications is largely attributable to the built-in resistance of such systems to change and the poor resolution of cladistic relationships that they often intentionally reflect. The supposed didactic advantages of stable classifications will prove illusory in the long run to the extent that the classifications fail to advance biological theory.

P.F. Stevens (1986, p.329)

Systems and classifications

1

If the goal of systematics is to depict relationships accurately, then any tradition that interferes with this goal should be abandoned.

K. de Queiroz and M.J. Donoghue (1988, p.332)

1.1 SYSTEMATICS AND TAXONOMY

There is no general agreement as to the definition of systematics, and of related words, such as taxonomy, biological system and classification. Capricious use of these and related terms is a very old plague of our discipline. Even in 1848, A. de Jussieu lamented the mixed use of the terms, methode and système. He finished his critical review of the literature by abandoning any hope of restoring the terms to what he believed to be their original, legitimate sense. Possibly matters are worse today, because of the widespread influence of some ill-inspired definitions, such as Simpson's (1961) well-known ones:

'Systematics is the scientific study of the kinds and diversity of organisms and of any and all relationships among them' (Simpson, 1961, p.7).

'Taxonomy is the theoretical study of classification, including its bases, principles, procedures, and rules' (Simpson, 1961, p.11).

'Zoological classification is the ordering of animals into groups (or sets) on the basis of their relationships, that is, of associations by contiguity, similarity, or both' (Simpson, 1961, p.9).

What Simpson referred to as systematics could be described today as comparative biology (Nelson, 1970; Wiley, 1981; Rieppel, 1988a; see also Brooks and McLennan, 1991; Harvey and Pagel, 1991). The field of systematics, as currently understood, is narrower (Wiley, 1981), although not limited only to producing classifications and identification keys. Still more unfortunate was Simpson's definition of taxonomy, although many authors agree with him in recognizing this discipline as a subset of systematics. At variance with Simpson's definition, which emphasized theoretical or methodological problems as the core of taxonomy, this discipline is largely seen as the practical branch of systematics. Indeed, Bisby (1984), Abbott, Bisby and Rogers (1985) and Hawksworth and Bisby (1988) regard the products of taxonomy as 'a taxonomic information system comprising classification, nomenclature, descriptions, and identification aids', whereas systematics 'is the taxonomy plus the biological interrelations – breeding systems and genetics, phylogeny and evolutionary processes, biogeography, and synecology' (Hawksworth and Bisby, 1988, p.9). However, as Hawksworth and Bisby (1988, p.11) observe, with the increasingly wide array of fields currently investigated by taxonomists to substantiate classificatory arrangements 'the separation in practice between taxonomy and systematics as defined above has become obsolete.' Hawksworth and Bisby's (1988) paper reviews further definitions of systematics.

Several authors, such as Ax (1984, 1988), regard systematics and taxonomy as synonymous. It has become fashionable to recognize comparative biology as a major field of biology, conceptually broader than systematics, as this science is understood here. I shall use the two terms, systematics and taxonomy, more or less interchangeably. What matters is not the dispute about the confines of academic provinces but identifying targets worth considering and problems worth solving. Then developing strategies for attacking them, improving our knowledge and communicating the results of our endeavours.

There are several approaches to biological systematics (man's oldest profession, according to Hedgpeth, 1961). Different views from those expounded here as well as useful additional insights are given by Cain (1954), Simpson (1961), Davis and Heywood (1963), Mayr (1963, 1969), Hennig (1966, 1982), Sneath and Sokal (1973), Ross (1974), Endler (1977), Eldredge and Cracraft (1980), Grant (1981), Nelson and Platnick (1981), Wiley (1981), Ax (1984, 1988), Briggs and Walters (1984), Ridley (1986), Matile *et al.* (1987), Rieppel (1988a), Stace (1989), Hillis and Moritz (1990), Stuessy (1990) and Mayr and Ashlock (1991).

1.2 CLASSIFICATION VERSUS SYSTEM

A classification is generally regarded as the primary product of systematic effort, although hardly two students agree as to their concepts of classification. We shall see later how difficult it is to compare characters or to get agreement for ranking taxa. Very few voices, however, have been raised against the identification of the primary aim of taxonomic work with the production of classifications.

It was probably Griffiths (1974, 1976) who first recognized that a classification may not be the sole, or even the best way of representing the outcome of systematic work. Accordingly, he drew a distinction between classification, the arrangement of entities in classes, and systematization, the ordering of entities into systems. Hennig (1975), Ax (1984, 1988), Gosliner and Ghiselin (1984), de Queiroz (1988), de Queiroz and Donoghue (1988) and Minelli (1991a) may be cited among the few explicit supporters of this view.

All entities appearing in a biological classification (such as species, genera and families) are generally called taxa (singular: taxon), as distinct from the taxonomic categories, (i.e. species, genus, family, order, class, phylum and kingdom) to which we refer the individual taxa, according to the rank we recognize for them. Accordingly, man, dog and horse are examples of species; Rosaceae and Palmae are examples of families; and Echinodermata and Mollusca are examples of phyla.

A classification is a hierarchical arrangement of objects within classes. Two objects are placed in the same class because they share one or more attributes that are the defining properties of the class. In this sense, two individual men are members of a species-level class called *Homo sapiens*. At higher levels, a family such as Canidae or Rosaceae is also a class, the members of which are genera; and a genus such as *Canis*, or *Rosa* is a class, the members of which are species. Some systematists also regard species as classes, the members of which are individual organisms, but this point is much disputed (cf. Chapter 4).

By contrast with a classification, a system is a representation of the living world as a whole. Its structure is the result of some natural process through which the parts of the system are related. Organisms are related through two different processes: interbreeding and common descent. Interbreeding has often been advocated as the prime criterion to identify species (which, in this sense, can also be regarded as wholes, rather than classes; cf. Chapter 4). Common descent unites all Recent and extinct living beings into a single tree, the structure of which we try to reflect in our phylogenetic system. In this view, the individual taxa are the branches of the phylogenetic tree, rather than hierarchically arranged subsets of an all-containing class of living beings; they are the parts of a whole, more or less orderly nested within larger parts of the same whole.

Suggesting a system, rather than a classification, as the more important outcome of scientific research, may sound useless, irrelevant, or even absurd. I think that it is simply an unconventional, but legitimate, view whenever we develop biological systematics within a phylogenetic framework.

1.3 BIOLOGICAL CLASSIFICATIONS, FROM ANDREA CESALPINO TO THE NEW SYSTEMATICS

To classify means arranging objects into groups. Linnaeus' classification of animals and plants was a hierarchy, as are all classifications in zoological and botanical textbooks. A classification is explicitly presupposed to have a hierarchical structure by the provisions of the International Codes of Nomenclature, with some possible exceptions at the infraspecific level (cf. Chapter 5).

In principle, a hierarchical classification could be developed either in a descending (downward) or an ascending (upward) way. Downward classification was developed by Andrea Cesalpino (*De Plantis*, 1583) by applying the Aristotelian *principium divisionis* to the universal set of plant species known to him. A first choice between two opposite traits (e.g. trees and herbs) allows partitioning of the universal set into smaller subsets. The procedure is then repeated until a satisfactory level of discrimination has been reached. This is the method used in identification keys, where couplets of contrasting traits are presented for choice, allowing recognition of increasingly smaller sets of species. As an operational tool to facilitate identification, downward classification is very useful, and rightly used. However, the artificiality of this method is evident. Size and circumscription of the groups, as well as the arrangement of species within the key depend upon the order we follow in introducing the different characters in the discriminating couplets.

Things are different in upward classification. John Ray (1690) may have been the first author to suggest (in his *Synopsis methodica stirpium Britannicarum*, 1690) that a classification should be built by putting together individual species into increasingly larger units (in modern terms: genera, then families, orders, classes etc.). Although the original method was often obscured by a widespread pragmatic concept of classification (presumably following Cesalpino's footsteps), Ray's concept of an upward classification gained increasing popularity. Its popularity generally paralleled a re-formulation of classificatory procedures in evolutionary terms.

When building an upward classification, it is necessary to identify the elementary units to be classified and a hierarchical structure which allows recognition of several, increasingly more inclusive ranks. Moreover, we need criteria for grouping the elementary units into classes as well as the lower, less inclusive classes into higher, more inclusive ones, and, finally, criteria for giving them specified ranks. There are several further problems, such as that of a suitable nomenclature, but those are less central to the argument I am developing here.

Mayr (1982b) and Knight (1981) give preliminary accounts of the history of systematics.

In the system developed by Linnaeus, the elementary units of classification are species. Their existence is an expression of God's will, and their relationships mirror a divine design. The naturalist's job is to trace this design from the evidence he can find in nature. As for ranks, matters can be settled by the principle later formalized by Georges Cuvier (1799, 1805): genera are distinguished according to trivial differences in organization, whereas orders and classes are separated by differences in the organs most important to the animal. The highest level in classification separates animals with wholly different, non-comparable body plans. In this way, Cuvier could recognize, within the animal kingdom, four irreducible embranchements (types, phyla): Radiata, Articulata, Mollusca and Vertebrata. With Cuvier, the taxonomic ranks are determined by idealistic morphology. The embranchements are therefore defined in terms of differences in the nervous system, as the (ideally) most important trait of anatomical organization.

The same criterion for ranking taxa according to differences in organization was advocated by Agassiz (1857) and other nineteenth century authors, but it seems to be still alive, judging from the following sentence: 'evolutionary lines of algal morphological progress represent the classes. The single grades of algal advance in a class represent a lower taxonomic rank, the order, which is thus defined morphologically by the level of plant body construction.' (Fott, 1974, p.449).

It is difficult not to see an idealistic vein in this approach. It soon became clear that biological systematics needed different theoretical foundations.

The advent of the evolutionary theory led to the reinterpretation of species and the relationships between them by reference to natural phenomena such as interbreeding and common descent. Common descent (with modifications) explains the hierarchical pattern of similarities exhibited by different species, as recognized in purely descriptive terms by the pre-evolutionary systematists. Roughly speaking, similarities may be understood as proofs of relatedness in the phylogenetic tree. Living things can be classed into genera, families and so on, according to some pattern of affinities, from which we can infer their genealogical relationships.

Soon after the publication of Darwin's (1859) *Origin of Species*, the well-known, aesthetically attractive, drawings of phylogenetic trees by Ernst Haeckel appeared (e.g. Haeckel, 1866, 1874). The way seemed to be open for a theoretical re-foundation and an operational re-formulation of biological classification. In fact, the enthusiastic beginnings laid down by Haeckel failed to revolutionize our science. It was not until several decades later that the Darwinian and Haeckelian heritage was welded together and extensively received by systematists.

1.4 EVOLUTIONARY SYSTEMATICS

One of the most problematic areas revolved around the concept(s) of species. A merging of different research traditions, in systematics as well as in palaeontology and genetics, prompted the development of what is currently known as 'The New Systematics', the major achievements of which are mirrored in a book of the same name (Huxley, 1940) as well as in Rensch (1929), Mayr (1942, 1963), Stebbins (1950) and Simpson (1961). See also Mayr and Provine (1980) for a historical account. Later developments of these debates on the species, centring mainly around the so-called biological species concept, are considered in Chapter 4.

For a long time, these developments, in what Mayr (1982b) calls micro-taxonomy, or the science of species, ran quite independently from the course of macrotaxonomy, the science of classification. Theoretical comments ranged from a denial that supraspecific taxa were natural entities, to vague statements that a phylogenetic classification is the more natural. How do we reconstruct phylogeny? How do we represent it in a formal classification? For many years, the most comprehensive theoretical essays on comparative biology (e.g. Remane, 1952), as well as the otherwise useful textbooks on biological systematics (e.g. Simpson, 1945, 1961; Mayr, Linsey and Usinger, 1953; Mayr, 1969) did not advance substantially the theoretical background of systematics.

During the 1940s and 1950s, systematics polarized around a few leading figures, especially Ernst Mayr and George Gaylord Simpson. Later, with the development of different approaches to biological systematics, it become fashionable to identify a School of Evolutionary Systematics corresponding to the views of Mayr, Simpson and others. We owe a large debt to these scientists, especially to Mayr, whose work has inspired generations of systematists and evolutionary biologists.

At the same time, one cannot simply ignore the conceptual progress of the last decades, which cannot be simply integrated within the programme of the Evolutionary Systematics school, but requires a different approach to several important issues, especially regarding macrotaxonomy.

1.5 NUMERICAL TAXONOMY

A first comprehensive effort to develop a new theoretical and practical approach to biological systematics was accomplished by Robert R. Sokal and P.H.A. Sneath, as illustrated in their 1963 book, *Principles of Numerical Taxonomy*. Sneath and Sokal (1973) gives a more detailed account. Sokal and Sneath's numerical taxonomy developed from an operational attitude, by explicitly refusing to incorporate into taxonomy dubiously retrievable phylogenetic information; objects are compared 'at face value', regarding as many characters as possible. To underline their methods compared with other, more or less explicitly phylogenetic, or evolutionary approaches, Sokal and Sneath's philosophy has been defined as phenetic, i.e. directly dependent on the overall similarity of the characters (the phenotypes).

In phenetics, therefore, there is no place for homology, or for history-dependent concepts, such as ancestry, or evolutionary change. Also, theory-laden biological terms, such as species, are not used in phenetics. The branching diagrams found in works on phenetics are not phylogenies of species, but simple dendrograms clustering together operational taxonomic units (OTUs). The criteria for clustering are equally provided by criteria other than biological.

A phenetic analysis begins with the construction of a data matrix with suitably coded data for *n* characters in *m* OTUs. From this data matrix, a distance matrix is calculated, which gives a measure of overall similarity (calculated over all *n* characters) between each individual pair of OTUs.This is similar to a matrix giving the distances, by road or by aeroplane, for a given set of localities. The distance matrix is used to group the OTUs using a suitable algorithm, to give the branching diagram of similarity or phenetic relatedness, i.e. the dendrogram. Formal groups (i.e. more or less inclusive clusters), are arbitrarily obtained by cutting the dendrogram at one or more conventional levels of similarity.

This work with matrices and branching diagrams cannot be developed by hand. It requires the use of several algorithms and computer programs have been developed to carry out the calculations. The technical advantages of numerical techniques have proved to be a more lasting contribution to systematics. Phenetics, as a system of biological systematics excluding phylogeny and homology, has progressively declined. The claim that phenetics had an objectivity not achievable within an evolutionary framework, was an interesting challenge, but an unsubstantiated one.

1.6 HENNIG'S PHYLOGENETIC SYSTEMATICS

Sokal and Sneath's work prompted a major debate on the foundations and the methods of biological systematics. The whole scenario changed rapidly, however, with the publication of Willi Hennig's *Phylogenetic Systematics* (1966), a translation by Davies and Zangerl into English of an updated version of a textbook previously published in German some 15 years before (Hennig, 1950). The history of the way Hennig was received by the scientific community has been written by Dupuis (1979, 1984). See also Hull (1988) for a very extensive account of the last 25 years in the history of biological systematics, written by this philosopher of science as a base for his speculations on the sociology of science. For a discussion of Hull's view of systematics in these years see Donoghue (1990), Farris (1990a) and Oldroyd (1990).

Hennig stressed the absolute need for a thorough and consistent use of phylogenetic information in building biological classifications. He described a few precise rules for reconstructing phylogeny as well as for translating phylogenetic trees into classificatory schemes.

1.7 CONTRASTING SYSTEMATIC SCHOOLS

Hennig's work initiated a polarization of fronts. On one side, Nelson (1970, 1972, 1974), Brundin (1972) and Patterson (1982) were quick to accept his suggestions; partly because of the consistency and logic of his approach, and partly through dissatisfaction with the uncertainty of traditional approaches. Other students rejected Hennig's approach, or dismissed its importance, regarding his contribution as a rewording of old concepts and procedures in new, more formal terms. However, the most explicit and persistent reproaches to Hennig (as expressed, for instance, by Mayr (1974) and by Charig (1982); but see Hennig's (1974) reply to Mayr) do not concern his rules for reconstructing phylogeny, but his contention that the phylogenetic information (i.e. the branching order of taxa) should be retained as the sole criterion for recognizing and ranking taxa. These difficulties are still raised, with much emphasis, in a textbook by Mayr and Ashlock (1991).

Thus, between the late 1960s and 1970s, the attitudes of most students towards the principles and methods of systematic biology polarized around the leading figures of the three different 'schools'. It became fashionable to speak of Hennig's Phylogenetic Systematics or Cladistics, of Sokal and Sneath's Numerical Taxonomy, and of Mayr and Simpson's Evolutionary Systematics. The inconvenience of some of these terms has been often lamented, for example by Charig (1982). However, these terms have been firmly established in the specialized literature and there is no point in changing them here. The debates between these schools have often been acrimonious, as expressed in the pages of books and journals, particularly in *Systematic Zoology* (Hull, 1988). The new ideas also led to the foundation of new societies or research groups. In 1967, the first Numerical Taxonomy Conference was held in Lawrence, Kansas. These conferences, born out of the Society of Systematic Zoology, initially incorporated all numerical approaches. Eventually, however, the burgeoning cladistic school grouped around a new Willi Hennig Society, founded in 1981. This group also developed the journal, *Cladistics*, first issued in 1985.

Despite the clear-cut labels so often applied to these matters, the opinions held by the partisans of all three groups have varied, both individually and with time. Within cladistics, for example, the so-called 'pattern cladistics', or 'transformed cladistics' (Platnick, 1980) developed; an approach apparently convergent towards phenetics, in its refusal to advocate the evolutionary theory when constructing tree diagrams (cladograms) from the analysis of the patterns of character distribution among the organisms under study. Pattern cladists have also increasingly turned to using numerical techniques, thus giving rise to what is often described as numerical phyletics, numerical phylogenetics or numerical cladistics.

The relationships between pattern and process still remain the focus of most workers, despite the apparent lack of adequate theoretical foundations. Rieppel

(1991, p.94) accordingly advocated 'a causal theory of representation that translates observed 'group' membership into conjectured phylogenetic relationship, [because] homology relates to common ancestry as classification relates to systematization'.

The study of phenetics now seems to be dead in most circles, whereas the debate between cladists and evolutionary systematists is still lively.

However, it is quite difficult to identify a set of qualifying principles common to all students assigned to a given school by their opponents. Evolutionary systematists, in particular, are often recognized in strictly negative terms, as students refusing to accept either the phenetic approach of Sokal and Sneath or the strictly phylogenetic classification advocated by Hennig. With the passage of time, Hennig's suggestions for the reconstruction of phylogeny have progressively become incorporated within the mind and language of the adherents of the evolutionary systematic school. However, strong contrasts remain, for example, regarding the acceptance of the so-called paraphyletic taxa (cf. section 2.6), which are rejected by cladists but not by the followers of evolutionary systematics.

Radical attitudes continue to be expressed on many occasions. However, several of the leading representatives of the different schools have begun to acknowledge that a compromise is possible, and not too far away (Patterson, 1988). That is possible, but may be excessively optimistic.

Mayr and Ashlock (1991) have used a lot of ink to clarify the differences between evolutionary and cladistic systematics, to conclude, as expected, with a desperate defence of the first approach. An outline of the most important differences between the two schools follows. A few technical terms employed by necessity in the following lines are explained in more detail in Chapter 2.

The aim of cladistics is to construct cladograms (branching diagrams depicting the genealogical relationships between species or other suitable terminal taxa). The aim of evolutionary biology is to construct phylograms (branching trees incorporating, as well as cladistic information, a phenetic measure of morphological 'advance' of the individual branches). Comparisons of organisms, within the framework of evolutionary systematics, take all homologous features into account for grouping and ranking taxa (however, see section 2.3, on homology), whereas cladistics selectively searches for orderly sets of synapomorphies (i.e. shared derived (advanced) characters, as opposed to symplesiomorphies, or shared ancestral characters).

Accordingly, cladistics only accepts monophyletic taxa (holophyletic, in Mayr and Ashlock's terminology), these are 'complete' phylogenetic units, comprised of all the taxa issued from a given hypothesized ancestor, irrespective of their degree of divergence. Evolutionary systematics, on the other hand, in addition to phylogenetic information also takes phenetic information into account. Accordingly, this school does not reject the so-called paraphyletic groups (i.e. those comprised of species issued from a common ancestor, but deprived of a 'more advanced' subgroup, which is recognized as an independent taxon of the same rank).

This point is important, not just for systematics, but for evolutionary biology at large. To give an example, consider the enormous palaeontological literature on macroevolutionary trends, rates of evolution, mass extinction and related matters.

The data for the ensuing speculation is provided by tedious tabulation of events, such as first occurrence, temporal range or extinction of taxa (mostly orders, families or genera). Patterson and Smith (1987) and Smith and Patterson (1988) have elegantly demonstrated the extremely dangerous artefacts deriving from systematically including paraphyletic taxa in these tabulations. Paraphyletic taxa seem to be extinct when they are survived by a more advanced subgroup, which is not treated as part of the same taxon. Many putative extinctions are thus nothing else than 'taxonomic extinctions'.

Within cladistics, there is no criterion other than relative inclusiveness, for establishing the rank of a taxon, whereas evolutionary systematics works within a pre-defined (Linnaean) ranking framework and each taxon is ranked according to the degree of difference it exhibits in respect to other taxa.

1.8 TOWARDS A NATURAL SYSTEM OF LIVING ORGANISMS

I wonder whether there is anything more than an 'overall similarity' between the search for a natural system encompassing the whole diversity of living beings and medieval efforts to achieve a universal knowledge of the world. To develop a system encompassing every existing entity may seem like getting a 'clavis universalis', a key to the knowledge of everything. An instrument for such knowledge is, by necessity, a universal language. It is possibly not by chance that some outstanding systematists of the past have been associated with programmes to reform language or writing. John Ray acted as counsellor to John Wilkins, when the latter worked at his project requiring a classification of 'all possible' concepts and things. However, Ray gave up the project as soon as he recognized that the principle of tripartite division, according to which Wilkins was working for his *Essay towards a Real Character and Philosophical Language* (cf. Raven, 1942), failed to match the progression of nature. One century later, Michel Adanson tried to reform the writing of French language, as one can see while browsing through his *Familles des Plantes* (1763).

Mayr (1982b) summarized several possible meanings for the term natural system. (1) Natural as reflecting the essence of the things so classified, as in Linnaeus. (2) Natural because rational, and rational because it mirrors God's design, as in Agassiz (1857). (3) Natural because of its broad empirical, not merely utilitarian, character. (4) Natural because it mirrors relatedness (i.e. phylogeny). One could add Gilmour's (1940) concept of natural because it explains most. The first two meanings are (or, at least, should be) foreign to the current philosophy of science, and the third, although germane to a modern school of thought, such as phenetics, also sounds inadequate and undefined. Thus, we are left with the fourth phylogenetic sense of the term. No wonder that it has become the flag for cladists, who are apt to criticize as unnatural every classificatory statement contrasting with a reconstructed phylogeny.

These points have been developed by Ridley (1986) in his defence of phylogenetic systematics, as the only school aiming to construct a natural classification, although he argues, in the same book, that 'transformed cladism' is a method degenerating towards phenetics. However, at variance with transformed cladism, phenetics has no concept of homology. It is based entirely on overall similarity, or isology. Isologous characters are simply converted into a distance matrix and then into a phenogram. Therefore, there is no way of applying any criteria of choice, like parsimony or maximum likelihood (cf. Chapter 2), to approach a more natural classification, as in cladistics, whether it be Hennigian or transformed.

Phylogenetic systematists have argued strongly that taxa must be strictly monophyletic (i.e. that they must comprise only the descendants from a given ancestor, without excluding any of them, even if they have become very specialized). In a sense, these monophyletic taxa (or monophyla *sensu* Ax, 1984) are 'real supraspecific natural bodies, in so far as they correspond to closed phylogenetic communities' (Ax, 1988, p.64; author's translation). In a sense, therefore, monophyletic taxa are individuals and that is probably true also of species (cf. Chapter 4). Insofar as they are individuals (not classes) they do not have members, only parts. In the same sense, a branch is a part (not a member) of a tree and a smaller branch is a part both of the bigger branch and of the whole tree. These relationships are evidently hierarchical, like those between taxa in a classification, but the units they involve are definitely not classes, but parts of a whole. These parts can (and should) be arranged to form a system, whereas to arrange them in a classification would mean to disregard their ontological status as individuals.

Working for a systematization, instead of for a classification, has important consequences for the ranking of taxa. Ax (1988, p.58) develops this point with his usual clarity and strength: 'Linnaean categories are useless to identify the hierarchical levels of the phylogenetic system. In addition, categories are painfully unsuitable to represent the reality of living nature. As a result of my arguments I resolutely propose to completely reject the ballast of category naming. With the rejection of categories we lose but nothing' (author's translation).

Accepting the view that systematics (according to its very name) should proceed with building up a phylogenetic system of monophyla, to be understood as (ontological) individuals, means that the eagle and the sparrow are not instances of a monophylum Aves (the birds), but parts thereof. Similarly, Carabidae is a part of Coleoptera; Coleoptera a part of Pterygota, and so on. In my view, this courageous and clear-cut choice recognizes the correct way to develop biological systematics as a science. Of course, all our problems are not solved by simply renaming, as monophyla, the old taxa of the current classification; nor are they solved by simply putting out of our minds the familiar names (and concepts, if any) of genus, tribe, family, order, and so on. Problems are both theoretical and operational and I firmly believe that they should be careful discussed, before trying any major revolution.

I will mention a few of these problems; let me begin with the species concept. De Queiroz and Donoghue (1988) have clearly argued how hopelessly contradictory it is to define species in terms of monophyly, especially if we require, at the same time, to define species in terms of reproductive cohesion (interbreeding) (cf. Chapters 2 and 4). How do we match this state of affairs with our contention that all natural taxa should be monophyletic? Further problems are of a more operational nature. For instance, how many levels of nested monophyla are to be recognized and named within the system? Naming all possible monophyla runs the risk of producing a plethora of names which only satisfies some kind of baptismal fever.

As for nomenclature, rejecting the Linnaean categories means that we cannot rely any more on the International Codes of Nomenclature. Ax (1988, p.57) even suggests avoiding the terms 'generic name' and 'specific name' or 'epithet', and replacing them with 'first name' and 'second name', respectively. But is it reasonable to fill papers and books with new names for monophyla, without reaching a consensus on how to develop a 'systematic' nomenclature and without devising a set of rules for reconciling the old ('classificatory') nomenclature and the new ('systematic')?

It has been suggested that more than one formal naming system can be applied to a single branching diagram depicting phylogenetic relationships (a cladogram, cf. section 2.7). Therefore, 'as words are used more frequently for communication than diagrams, schemes, or graphs, the 'general reference system of biology' remains somewhere in utopia. The 'phylogenetic system' is rather more synonymous with a cladogram than with a written classification' (Tassy, 1988, p.43). The point is, that a system is not a classification.

All these problems (and many more) should be of direct concern to students of biological systematics. In addition, there are the problems (and the needs) of 'consumers' of systematic knowledge, such as ecologists, horticulturists and experimental biologists. It has often been argued, that divorcing scientific systematics from the production of a commonly used classification is unavoidable. Although uncomfortable to many of us, this route is perhaps the only practical one (Minelli, 1991a). De Queiroz (1988, p.246) has stressed the point that both classification and systematization have 'central and complementary roles in science' and both should be adequately developed.

Users of systematic information need a species name (possibly, a stable one; cf. Chapter 5) for every organism in which they are interested. They cannot have any concern for semispecies, or superspecies, or for claims that many living organisms are not parts of species, for example unisexual forms (Dobzhansky, 1937; Mayr, 1969; Hull, 1981; Ghiselin, 1987; cf. Chapter 4). As for infraspecific taxa, their requirements often contrast with the current trend (cf. section 8.4) to reduce the traditional burden of names for races, varieties and so on.

There are also problems with the so-called higher taxa. However, these are probably less difficult to resolve, even when they are very conspicuous. How easy is it to persuade people to abandon traditional taxa, finally unmasked as not natural, when they are so universally recognized, such as Turbellaria, Polychaeta, Reptilia and Dicotyledones? Nothing is impossible and today few seriously believe that worms, invertebrates, fishes and algae are natural groups to be recognized as taxa in a formal classification as they were in the past. In the near future, it is possible that turbellarians, polychaetes, reptiles and dicots will be used only as vernacular terms. I hope so, in so far as we are correct, in denying that they are natural (monophyletic, cf. section 2.6) phylogenetic units.

To sum up, 'The equation of systematics with classification is one of the ancient and petrified notions we have inherited from the pre-evolutionary period, and that we ought now to abandon. The problems of systematics now and in the future are not the problems of classification, they are the problems of historical reconstruction and historical representation: the reconstruction and representation of the evolutionary past.' (O'Hara, 1992, p.136).

Some steps in comparative biology

2

A natural system is something we discover, not something we create.

M.T. Ghiselin (1987, p.130)

2.1 CHARACTERS AS 'SYMPTOMS' FOR RECOGNIZING TAXA

'Epistemologically, a natural system is based on whatever scientific evidence legitimizes it. Ontologically, a natural system is based upon the objective reality to which it corresponds.' Accordingly, 'a scientific classification is based upon an understanding of the objects classified, and ... any characters are therefore epiphenomenal.' (Both quotations from Ghiselin, 1987, p.132). In this sense, characters should be regarded as 'symptoms' of an existing order we are trying to discover and to translate into systems (or classifications).

Characters are the basis of our investigations, but we cannot leave a causal analysis of the diversity of living organisms outside our research programme. The problem is whether the reconstruction of phylogeny depends on specifying a set of hypotheses about how evolution works, or, as contended by pattern cladists, evolutionary theory should only be considered after one has constructed, by 'theory-free' character analysis, tree diagrams and a classification. 'Corroborated theories of relationships (cladograms) can be used as part of the data array to test predictions of evolutionary patterns stemming from theories of the evolutionary process.' (Eldredge and Cracraft, 1980, p.325).

An in-depth discussion of this issue would be out of place in this book, but to build a classification on the simple belief that evolution has occurred and that organisms are therefore linked in a more or less tree-like way is not enough, if we want to get a truly phylogenetic, natural classification or a phylogenetic system. Evolution does not simply mean splitting lineages (cladogenesis, speciation). It also means adaptation, and constraints. When deciding which state is primitive and which is advanced between, say, the absence or presence of wings, we cannot content ourselves with pattern analysis of presences and absences in a character matrix. We also need to know something about the functional value of the wing, when present, and the possible adaptive significance of its absence. Still more, we wish to know something about the development of the wing, its genetic control, and the network of causal relationships of wing ontogeny with other developmental processes. Of course, all this information is seldom available, but that does not mean it is (or could be) extraneous to our endeavours in reconstructing phylogeny.

For a critical analysis of the theoretical foundations of pattern cladistics, I suggest reading Scott-Ram's (1990) book, rather than Ridley's (1986) account, which is much more readable, but not always reliable.

2.2 CHARACTERS FOR CHOICE

I do not intend to discuss here the concept of 'character' and related matters; these are discussed, for example, by Wiley (1981), Ghiselin (1984a) and Rieppel (1988a). We could accept as operationally useful the traditional distinction between character and character state, as when distinguishing, for example, white and red, as character states of the character 'colour of flowers'. However, a better account was developed along cladistic lines by Pimentel and Riggins (1987). They begin their analysis of the nature of cladistic data by operationally defining a character as 'a feature of organism that can be evaluated as a variable with two or more mutually exclusive and ordered states' and continue by explaining that character, in their usage, 'corresponds to a morphocline (Maslin, 1952, p.201) and its multistate variable representation to a transformation series'. Pimentel and Riggins (1987, p.202) regard character states as 'characters with a more restrictive conditional phrase', or as Eldredge and Cracraft (1980) put it, 'character and character state merely signify different hierarchical levels'.

As for the role of characters in systematics and phylogenetic reconstruction, it is clear that all characters are not equally informative. To take an extreme example, a character uniformly distributed without variation among all the species we are comparing cannot give us any information about their interrelationships. In more general terms, the importance of character distribution among the organisms under comparison is certainly of prime interest in deciding on their usefulness for reconstructing phylogeny and building a classification.

It has been suggested that there are *a priori* criteria of choice for identifying characters of use in systematic work. Duncan and Stuessy (1984, p.49), for instance, warn that 'it is important ... that care and attention be given to the selection of characters, that their states be described effectively, that the homology of these states be assessed, that these states be placed in a logically acceptable character-state network, and that the evolutionary polarities (primitive versus derived conditions) of these states be determined'.

Pimentel and Riggins (1987) contend that continuous variables cannot provide evidence on phylogeny. However, other systematists, such as Donoghue and Sanderson (1992, p.359) 'are suspicious of assertions to the effect that some forms of data are useless'. What is true, is that continuous characters are much more insidious to code adequately, without excessively biasing the results (Farris, 1990b; Stevens, 1991).

These questions lead us to the problem of character weighting. Should all characters be treated as if they were equally important in systematic analyses? If not, how differently have they to be weighted?

Weighting was contrary to phenetic principles, because it seemed to introduce a subjective bias into an objective technique, but today the issue of weighting is hotly debated in more articulate terms for example by Farris (1969, 1983), Neff (1986), Wheeler (1986), Sober (1986), Bryant (1989) and Albert and Mishler (1992).

In principle, one could differentiate between weighting *a priori* and weighting *a posteriori*. Weighting *a priori* means that we decide to make different use of different data before gathering them or, at least, before looking at the matrix where they are collected. An extreme instance of weighting *a priori* is the intentional neglect of a given type of data. We could also consider weighting differentially because we know, for example, the different probability of a given molecular change (more probable, therefore less informative, hence low weight), or the partial interdependence of some character (less independent, therefore less informative, hence low weight).

Weighting *a posteriori* means weighting according to the distribution of characters within the data matrix. For example, by lowering the weight of the characters that prove to be less informative within a given set of taxa, for example because of multiple, independent occurrence in different lineages. Some type of weighting *a posteriori* is nearly universal, in taxonomic analysis, whereas the rationale and utility of *a priori* weighting is still debated. I am in favour of some weighting, in so far as the weighting procedures do not introduce more noise than they are expected to filter out.

Mutual dependence versus independence of characters should be of major concern for all students, whatever their philosophical outlook, but we are very seldom informed as to the true state of affairs. Only in a few cases do we know the genetic control of the characters we study and the pleiotropic effects of the genes involved. Unfortunately, some redundancy is generally difficult to avoid in listing characters, owing to the impossibility of completely identifying their subtle genetic or epigenetic interdependence.

Many students are inclined to discard the traits which they believe are most strongly subject to selective forces. This may be a way of discarding evidence disagreeable to them, which does not agree with their favourite phylogenies.

However, it is widely believed that strongly selected characters vary too easily to retain much useful phylogenetic information. In particular, they are prone to parallelism. This was underlined by Cronquist (1969, p.188) as 'especially true in the angiosperms, in which the evolutionary barriers between different ecologic niches are frequently minimal ... The characters used to distinguish the families and orders of angiosperms are in large measure things which are difficult to relate to adaptation and survival value: hypogyny, perigyny, and epigyny; polypetaly, sympetaly, and apetaly; apocarpy and syncarpy; placentation; numbers of floral parts of each kind; the presence or absence of endosperm or perisperm in the seed; the sequence of development of the stamens, and the like'. In other terms, and with relation to insects, rather than plants, Gisin (1967, p.118) affirmed that adaptive characters identify what Huxley (1958) had called grades, the product of anagenesis (Gisin's 'évolution quantique'), whereas non-adaptive characters identify clades (the branches of the phylogenetic tree, as produced by cladogenesis). The concern for the consequences of strong selection on the different kinds of characters is probably one major reason for the widespread use of allozymic (electrophoretic) data in recent systematic studies (cf. section 2.4). Things are not so simple, however. As Endler (1986, pp.158–9) points out, although it is often relatively easy to demonstrate the effects of natural selection on morphological characters, it is not so easy for allozymes. The current notion that they are influenced little by natural selection is often a suggestion, rather than positive knowledge. These questions involve some of the most debated issues of evolutionary biology, such as the contrast between the selectionist and neutralist viewpoints. We cannot deal with these themes in detail here. However, a few aspects will be developed in Chapter 3, when discussing molecular clocks.

2.3 HOMOLOGY

Homology is probably the most important concept in comparative biology. It has been treated in different ways, however, and more than one concept of homology is probably defensible. New insights or special necessities may suggest something better than our current concepts. We should not forget that the concept of homology is of pre-evolutionary origin and it would not be useful to stick to Owen's (1843) definition of homology for reasons of priority. Detailed discussions of the concept of homology, together with ample historical references, have been given by Patterson (1982) and Rieppel (1988a).

Some students, however, reject the validity of homology. In this respect, an interesting argument was developed by Goodwin and Trainor (1983). They say that homologizing limb elements involves an artificial atomism, but individual skeletal elements do not have a developmental identity of their own, the whole limb being an integrated developmental field (cf. Goodwin, 1984). However, Shubin and Alberch (1986, p.375) correctly observe that, 'if Goodwin and Trainor (1983) took their approach to its logical conclusion, they could not discuss limb development and evolution in the first place, since this requires them to use the concept of homology at another level of the taxonomic hierarchy – to homologize a structure called a limb in the first place. They implicitly accept a certain degree of compartmentalization by assuming that the limb is an independent embryonic field (*sensu* Waddington (1957) and Wolpert (1969)). The degree of developmental compartmentalization corresponds to the level at which homologies can be resolved. We suggest that limb development can be further decomposed as a sequence of branching and segmentation events. This allows us to establish more specific homologies.'

A related argument is developed by Sattler (1984, pp.391–2). He suggests that instead of trying to establish structural correspondences between organisms, we may make comparisons in terms of transformational processes. As a consequence, according to Sattler, the term 'homology' may be dropped. Instead we should refer to 'similarity', 'structural correspondence,' or 'structural relationship'.

The first question should be, what should be compared? Every comparison, be it in terms of homology or otherwise, cannot begin without an initial willingness to compare two objects (Eldredge and Cracraft, 1980; Rieppel, 1988a). Why are we interested in discussing the possible homology between the wing of a bird and the forearm of a man, while we never try to compare the wing of a bird with the caudal fin of a fish? Our preliminary hypotheses of homology are generally guided by general similarity, including previous knowledge about the organisms to be compared and their various relatives, including other hypotheses of homology retained from previous studies of the same organisms. The soundness of previous work will strongly affect the meaningfulness of new statements and the validity of new arguments. Similarity is not a reliable criterion for homology; nevertheless, it often initiates the study of a disciplined comparison. Therefore, homology will be sought between 'comparable' objects. In this sense, there are some elements of truth in the definition of homology given by Nelson (1970, p.378): 'Features assumed to be homologous may be assumed, simply, to be comparable, and when compared may be expected to result in an hypothesis of ancestral conditions'. But, such a notion is just the beginning of our efforts to circumscribe the limits of our investigation of homologies.

The concept of homology developed in comparative morphology. Before being given its modern name by Owen (1843), it had already been applied in a more or less conscious way by several comparative anatomists, especially by Etienne Geoffroy Saint-Hilaire (1830). According to Geoffroy, the bewildering diversity of animal forms conceals the true unity of a structural plan, the existence of which allows us to compare the single parts of different animals. Every bone in a dog has correspondence with a given bone in a horse, in a monkey or in a bat. Many correspondences can be traced beyond the limits of a taxonomic class, for example a mammal with a fish and a frog with a bird. Geoffroy also tried to identify the structural correspondences between animals belonging to different phyla (or embranchements, to use the term coined by Cuvier). In Geoffroy's terminology, the corresponding parts in different organisms were called analogues (a very different meaning of the term we now use).

When introducing the term homology, Owen substantially retained Geoffroy's concept of equivalent parts (analogues), but substituted an ideal type of biological form (the archetype) for the indeterminate transformational series devised by the French author. In a historical perspective, we can understand this philosophical shift towards idealism. At the same time, we must be careful to avoid this notion when 'translating' Owen's concept into modern language and fitting it as far as possible within an evolutionary framework.

There is no doubt that an established homology is often more interesting when the organisms compared are distantly related. However, one could question whether it should be useful and/or justifiable to abandon the concept of homology when comparing two very closely related organisms. Is it legitimate to speak of homology between comparable structures of two individuals belonging to the same species? Ax (1989c), for example, argues that it is not. Furthermore, is it legitimate to speak of homology between parts of the same organism? These questions are answered in different ways by different students. I defend the most comprehensive concept of homology, allowing intraspecific and intra-individual comparisons (Minelli, 1991c; Minelli and Peruffo, 1991; Haszprunar, 1992b).

Serres observed (1827, pp.52–3) that the different parts of an animal are the repetition of each other. As aptly summarized by Rieppel (1988a, p.141), Serres relied, in part, on previous studies by Bordeu (1751), Vicq d'Azyr (1786) and Duméril (1806). They had recognized, respectively, the equivalence of the two sides (the right and the left) of the body of a vertebrate; the equivalence of the anterior and posterior appendages of a vertebrate; and the possible relationship between the bones of the skull and the vertebrae (an idea developed by Goethe). The equivalence between symmetrically or serially arranged parts of the body has subsequently been called homonomy or serial homology.

Van Valen (1982, p.307) observes that the homologies occurring within a given individual are not limited to serially repeated units, like the segments of an earthworm or the teeth of a mammal. On the contrary, most homologies within individuals are nonserial. It is difficult to deny the homologies between two hairs or two red blood cells. These homologies are not exciting discoveries in themselves, but a wide-ranging concept of homology may be more than of anecdotal interest. First, it allows the concept of homology to be extended to higher plants, colonial hydrozoans and bryozoans; those organisms consisting of modules (colonoids *sensu* Van Valen (1978)). It also allows the recognition of a link between the traditional, morphological meaning of homology and the more recent use of the term as practised by the molecular biologists, when considering homologous proteins or nucleotide sequences. Second, this concept of homology allows us to define homology, by reference to the common informational background underlying the features we regard as homologous. It is easy to see that this relationship applies to inter- and intraspecific comparisons, as well as to inter- and intra-individual, and to morphological and molecular comparisons. By informational background I do not only mean genetic, because other kinds of information are involved in morphogenesis. Owing to the scanty knowledge we have of these non-genetic sources of information, I shall simply call them epigenetic (cf. Løvtrup, 1974; Roth, 1982; Van Valen, 1982; Minelli and Peruffo, 1991; Haszprunar, 1992b).

My definition of homology is essentially the same as that of Van Valen (1982): Homology is the relationship between two structures or features traceable (in a more or less important and more or less readable way) to a common informational background. Using such a definition, every homology should be regarded as relative, conditional and eventually quantifiable. Everything depends on our ability (or interest) in specifying the historical relationship between the alternative versions of this 'common informational background'. We need to assess what the two structures have in common and the historical (or developmental) relationship between the informational background of the structures we compare.

In this context we can appreciate Morris and Cobabe's (1991) claim that information on biosynthesis should be required before assessing the homology between molecules. This does not necessarily mean that notions of homology in morphology and molecular biology are essentially the same (De Pinna, 1992). All too often simple sequence similarities between macromolecules of more or less distantly related organisms are called homologies, despite the strong criticism raised against this practice by an authoritative group of molecular biologists (Reeck *et al.*, 1987).

This is probably the best place to introduce Wagner's (1989a,b) distinction of historical and biological concepts of homology. Historical concepts (those of more direct concern for the systematists) develop by reference to the historical continuity of descent from a common ancestor, whereas biological concepts develop by reference to some mechanism, which may be developmental, rather than to genealogy.

I shall concentrate on those aspects of homology that are of concern to a systematist, beginning with a few definitions of the term. I follow here Patterson's (1982) elegant subdivision of classical, evolutionary, phenetic, cladistic and 'utilitarian' definitions.

Let us compare the classical definition given by Owen (1843, p.374), according to which a homologue is 'the same organ under every variety of form and function', with Mayr's (1969, p.85) evolutionary definition: 'homologous features (or states of features) in two or more organisms are those that can be traced back to the same feature (or state) in the common ancestor of those organisms'. According to Ax (1984, p.167; author's translation) Mayr's definition is the 'factually and logically widest possible meaning [of homology] within the context of evolutionary theory'. Bock's (1977, p.881) definition ('features (or conditions of a feature) in two or more organisms are homologous if they stem phylogenetically from the same feature (or the same condition of the feature) in the immediate common ancestor of these organisms') may look identical, but it is not, because of the qualification 'immediate common ancestor' called in to substantiate the relationship.

For an example of a phenetic definition of homology, Jardine (1967, p.130) stated that homology is nothing more than 'the relation between parts which occupy corresponding relative positions in comparable stages of the life-histories of two organisms', whereas a 'utilitarian' definition (generously so-called by Patterson (1982)) is that of Blackwelder (1967, p.141): 'Certain structures are homologous, such as the wing of a bird, the foreleg of a horse, and the arm of a man, and these can be compared usefully with each other'. Nothing less, and nothing more.

Let me also provide some cladistic definitions: 'Different characters that are to be regarded as transformation of the same original character are generally called homologous' (Hennig, 1966, p.93). 'Homology is the study of the monophyly of structures' (Hecht and Edwards, 1977, p.6).

In the last few years, most debates on homology have developed around evolutionary and cladistic concepts, whereas phenetic and utilitarian concepts have lost ground for theory-minded systematists.

Let us now look more closely at the issue of homology as developed within cladistics, beginning with Hennig's *Phylogenetic Systematics* (1966). Hennig introduced the now widely applied distinction between plesiomorphy and apomorphy. We can define the plesiomorphic and the apomorphic character states as relatively primitive and relatively derived states, respectively, in what Hennig (1966) and Wiley (1981) call an evolutionary transformation series. For instance, at one point during the evolution of the hexapod arthropods the wing came into existence. Absence of a wing is thus the plesiomorphic condition, whereas the presence of a wing is the corresponding apomorphic state. The problem is how to establish the polarity (plesiomorphic to apomorphic state) of an evolutionary change that we cannot directly observe, but only infer from the comparative evidence at hand.

A synapomorphy is an apomorphy shared by two or more species in respect to the alternative, plesiomorphic feature occurring in other, less closely related forms. Whenever two species, A and B, share a common apomorphy a', while a related species C exhibits the corresponding plesiomorphy a, we can hypothesize that the phylogenetic lineage common to A and B diverged from the lineage of C before the first occurrence of the apomorphy a'. In other terms, we can hypothesize an ancestor X common to A and B, but not to C. This phylogenetic relationship involving species A, B and C exemplifies the three-taxon statements, the elementary units of phylogenetic systematics.

What happens if we take into account a fourth, less closely related species, D? In that case, a different synapomorphy b', uniting A, B and C, but not D (exhibiting instead the plesiomorphy b) allows the identification of a more basal split between the A plus B plus C lineage and the D lineage. Accordingly, we can hypothesize Y, a common ancestor to A, B and C, but not to D; and so on.

Symplesiomorphy is the joint possession of ancestral, rather than of derived, traits. Symplesiomorphies are homologies, but are devoid of any phylogenetic information content. For example, let us think of a group of 10 species, two of them sharing an apomorphy c', the remaining eight species all exhibiting the corresponding plesiomorphy c. The synapomorphy c' allows the identification of a phylogenetic branch (a clade) within the phylogenetic tree of the whole group, but what about the structure of the remainder of the tree? We cannot simply put together the other eight species in a group characterized by the symplesiomorphy c, because some of them may be more closely related to the synapomorphic clade than to other species outside it. Therefore synapomorphies allow us to recognize natural groups, with their own origin, identity and history, whereas symplesiomorphies only define 'incomplete' groups, deprived of their more derived offshoots, the subgroup with the synapomorphy c'.

The terms, apomorphy and plesiomorphy, are strictly context dependent. In the preceding example, b' was the synapomorphy for A plus B plus C, compared with D, but the same character, b', reduces to a symplesiomorphy within the A plus B plus C group. In a sense, both apomorphies and plesiomorphies are homologies, but at different relative levels of generality (Patterson, 1982).

These concepts underline how relative and precarious the concepts of homology are. In the words of Roth (1984), we must be prepared to acknowledge that there are no 'perfect' instances of homology. Besides the trivial case of complete identity between the structures we are comparing (isology), there will always be differences, in the genetic and/or epigenetic background of the traits we are homologizing. How we place the line between homology and non-homology will always be a matter of subjectivity.

2.4 HOMOPLASY

Major difficulties with the reconstruction of phylogenetic relationships derive from the widespread occurrence of what comparative morphologists have long described as parallelisms and convergences. There is also the possibility of reversal of character states, for example of a change from b' to b following a previous change from b to b'. These events are the cause of similarities that are neither synapomorphies nor symplesiomorphies, but homoplasies. Wiley (1981, p.12) defines them as 'characters that display structural (and thus ontogenetic) similarities but are thought to have originated independently of each other, either from two different preexisting characters or from a single preexisting character at two different times or in two different species. Put simply, homoplasy exists if the two taxa showing the character have a common ancestor that does not have the character'.

Gauld and Mound (1982, pp.73–4), for instance, found that 'certain Ichneumonidae and fungus-feeding Phlaeothripidae appear to have undergone frequent reversal or parallelism (homoplasy) of characters during evolution so that extant species present almost every imaginable permutation and combination of characters. Recognition of monophyletic genera in such groups is difficult. Large monophyletically defined genera are often not monophyletic, whilst small genera need to be defined by a large and invariable character-suite'. For such groups, where a high degree of homoplasy is present, Gauld and Mound (1982) suggest that polythetic genera should be recognized. These are taxa which lack characters exclusive to them and common to all included subtaxa, but are identifiable by a set of characters of which a large subset (not always the same) is present in each subtaxon (cf. Sneath, 1962). Polythetic taxa were recognized in eighteenth century taxonomy; Mayr (1982b) quotes a definition from Vicq d'Azyr (1786) which says that 'a group may be perfectly natural, and yet have not a single character common to all species that compose it'. Polythetic taxa are unacceptable to most cladists (see comments by Hall, 1991), but I agree with Gauld and Mound's (1982) contention that 'polythetic genera can be holophyletic [i.e. natural; cf. section 2.6] groups and are not merely phenetic assemblages'. Homoplastic similarity is often regarded as devoid of phylogenetic interest. However, that is not always the case. Parallelisms reflect a type of 'underlying similarity' that we can evaluate when reconstructing phylogeny.

However, one is obviously more ready to accept a phylogenetic tree which implies less homoplasy than otherwise. Several indexes have been proposed, to measure the amount of homoplasy in a given phylogenetic tree. In the following lines I follow Farris' (1989) lucid account (alternative formulations were given by Archie (1989); see also Farris (1990b) and Archie (1991)). The consistency index c was introduced by Kluge and Farris (1969), whereas the retention index r and the rescaled consistency index rc were both suggested by Farris (1989). Their simple formulations are:

$$c = \frac{m}{s} \text{ with } s = m + h$$

$$r = \frac{g - s}{g - m}$$

$$rc = r \cdot c$$

where: m is the minimum amount of change that the character may show on any tree (it is a function of the number of possible states and, for numerical characters, it may be regarded as the range of the character); s is the amount of change in the character required in the least homoplastic way, by the considered tree; h is the fraction of change that must be attributed to homoplasy; g is the greatest amount of change that the character may require on any tree (that is the greatest possible value of s).

Several numerical methods developed as tools for the reconstruction of phylogeny (cf. section 2.9), rank the possible trees according to the decreasing values of the consistency index associated with each tree. The least homoplastic tree is generally taken to be the best estimate of phylogeny, within the limits set by the data available and the numerical method employed.

However, Goloboff (1991a,b) claims that lower values of consistency index or retention index do not necessarily imply that the data provide less information for choosing between alternative trees. Although the consistency index behaves better when comparing homoplasy between different data sets, the retention index works better when comparing the fit of different trees for the same data set. According to Goloboff, what matters more than homoplasy is decisiveness. This is the information content in the data matrix which allows one to choose between competing trees, because they differ more in the number of steps they require.

According to Donoghue and Sanderson (1992), homoplastic characters might be misleading, but it is important to see how they interact with the other characters in the data set. The degree of homoplasy increases with the number of taxa involved (Sanderson and Donoghue, 1989). The consistency index does not appear to depend on the number of characters used in the analysis or on the taxonomic rank of the terminal taxa in the tree, be they species or higher taxa (Sanderson and Donoghue, 1989). What is critical is not the overall amount of homoplasy, but the way it is distributed in the data matrix (Jansen et al., 1990). Confidence in a tree can be low even if homoplasy is low, if the clades are only supported by one synapomorphy each.

For several years it was contended that flowering plants exhibited a much higher level of homoplasy than animals, but that is demonstrably not true. There seems to be no difference in the levels of homoplasy between molecular and morphological data (Sanderson and Donoghue, 1989; Donoghue and Sanderson, 1992).

2.5 CHARACTER CODING

The results of phylogenetic analyses are sensitive to the way they are scored, or coded, in a matrix, before evaluating them, either by hand or with the aid of a computer program. I refer the reader to the specialistic literature, particularly to the many papers in *Cladistics* and *Systematic Zoology*.

A recent example may explain what is at issue. Bremer (1991) tried to score the results of restriction site analysis of chloroplast DNA of 33 taxa of the plant family Rubiaceae in four different ways. The data sets were analysed with the same algorithm (Wagner parsimony) and resulted in four different cladograms. The problem is to decide which cladogram represents the phylogenetic relationships in the most reliable way.

Problems with the identification of character states have been very carefully studied by Stevens (1991). His impressive analysis demonstrates that most of the character states relating to quantitative characters, which have been used in species-level cladistic analyses, are ambiguously coded, because ingroup and outgroup variation is seldom taken into account. Stevens stresses that outgroup variation may compromise the character states recognized in the ingroup, or reverse the polarity they have been assigned. He also says (1991, p.553) that 'character states often seem to be delimited in conjunction with developing ideas of the phylogeny, rather than in a step prior to a phylogenetic analysis'. One cannot but argue that 'explicit justification for the delimitation of character states should be given as a matter of course in all phylogenetic studies'.

Characters are easier to score, when they can be reasonably regarded as two-state characters (e.g. presence versus absence; hairy versus glabrous). Hauser and Presch (1991) wonder whether multistate characters would be better treated as ordered (additive) characters, or as unordered (non-additive) ones. Ordering is obviously theory-laden because, in accepting a given hypothesis of order, alternative hypotheses are discarded. They have examined 27 data sets, from cladistic analyses in the literature, to test the idea that hypotheses of character state order are more informative than hypotheses of non-order, thus facilitating the choice among alternative trees. Their results disprove this hypothesis and support Stevens' (1991) concern for the bias introduced when coding characters according to the researchers' favourite, though often unconfessed, *a priori* hypothesis of phylogenetic relationships. Problems of choice between different transformations for the same character set are also dealt with by Mickevich and Lipscomb (1991) and Lipscomb (1992), in terms of a correspondence between degrees of homology and proximity of states in the transformation series. Once more, the danger of circularity is likely in these matters.

Theoretically more trivial, but operationally important, are the problems of coding missing entries. The reader is referred to Nixon and Davis (1991) and Platnick, Griswold and Coddington (1992).

2.6 MONOPHYLY, PARAPHYLY, POLYPHYLY

Assuming that the phylogenetic relationships of all living beings can be represented by a dichotomously branching tree (which is not true in the cases of reticulate evolution), the most elementary statements about phylogeny then take the form of three taxa statements (cf. Hennig, 1966; see also Eldredge and Cracraft, 1980; Nelson and Platnick, 1981; Wiley, 1981; Rieppel, 1988a; Funk and Brooks, 1990).

B shares a common ancestor with C not shared by A (Figure 2.1). A, B and C are the terminal taxa of our analysis, either species or higher taxa. Diagrams, such as Figure 2.1, do not specify the identity of the ancestors common to B and C, or to A, B and C, but only indicate the order of branching in the phylogenetic history of the group. B and C, the two taxa representing the two branches emerging from a common (though unknown) ancestor are defined as sister-groups (Hennig, 1966), or adelphotaxa (Ax, 1984). Analogously, taxon A is the sister-group or adelphotaxon of B plus C.

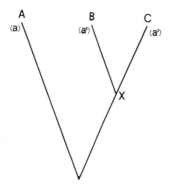

Figure 2.1 Phylogenetic relationships of three taxa (A, B and C). B and C are sister-taxa (adelphotaxa), sharing a common putative ancestor X. B plus C (together with their common putative ancestor X) form the sister-group of A. The character a is present in state a in taxon A and in state a' in taxa B and C.

We can now introduce the concepts of monophyly, paraphyly and polyphyly (for definitions and discussions see Hennig, 1965, 1966; Mayr, 1969; Ashlock, 1971, 1972, 1974, 1979, 1984; Nelson 1971, 1973; Farris, 1974, 1991; Platnick, 1977a, 1980; Holmes, 1980; Lorenzen, 1981; Wiley, 1981; Charig, 1982; Ax, 1984; Dubois, 1986, 1988; Oosterbroek, 1987). The terms monophyly and polyphyly were introduced by Haeckel (1866) more than 100 years ago. Paraphyly was introduced by Hennig (1965) and a related term, holophyly, by Ashlock (1971).

Hennig (1966, p.146) distinguished between monophyletic, paraphyletic and polyphyletic groups, according to the type of similarity serving as the basis for grouping together two or more taxa: 'In systems where ... groupings are based on the simple principle of morphological 'similarity', there arise group formations ... that may be distinguished as monophyletic (if the similarity is synapomorphy), paraphyletic (if the similarity is based on symplesiomorphy), and polyphyletic groups (if the similarity is due to convergence)'.

These concepts could be redefined in terms of genealogy, instead of characters. Accordingly, Nelson (1971, pp.471–2) redefined monophyly as 'a quality of a group including all species, or groups of species, assumed to be descendants of a hypothetical ancestral species. The members of such a group are all interrelated, form a sister-group system, and include all species of that system. Such groups may be considered complete sister-group systems'. Nelson contrasted monophyly with non-monophyly, 'a quality of a group not including all species, or groups of species, assumed to be descendant of a hypothetical ancestral species'. He then distinguished two types of non-monophyletic groups: paraphyletic 'an incomplete sister-group system lacking one species (or monophyletic sister-group)'; and polyphyletic 'an incomplete sister-group lacking two or more species (or monophyletic species groups) that together do not form a monophyletic group'. Nelson's definitions proved attractive to many students, in their reference to genealogy rather than to characters, but they led to a proliferation of alternative and often misleading definitions.

A few authors felt that origins and composition were both involved in these matters. Ashlock (1971) tried to clarify them, by introducing an additional term (holophyly), in order to separate genealogical from compositional aspects. His holophyletic taxa are the same as Hennig's monophyletic ones, but Ashlock's monophyletic taxa include both Hennig's monophyletic (his holophyletic) and paraphyletic taxa. This use has been followed by a few authors (e.g. Charig, 1982; Ghiselin, 1984b; Stuessy, 1990; Mayr and Ashlock, 1991). For a while, I also favoured this approach (Minelli, 1991b), but now I feel it more reasonable to return to Hennig's (1966) concepts. I have been convinced by Farris' (1991) analysis, the conclusion of which is (p.304): 'Monophyly can be defined (though not recognized) without reference to character evidence only because monophyletic groups have real and independent historical existence. Paraphyletic and polyphyletic groups have no such existence; they are nothing but the characters by which they are delimited. Without characters, paraphyly and polyphyly mean nothing. It is the reality of monophyletic groups that ultimately distinguishes phylogenetic systematics from syncretistic taxonomies'.

Let us read a few more lines from Hennig's (1966) book: 'While polyphyletic groups, when recognized as such, are today probably no longer permissable in any system, the distinction between paraphyletic and monophyletic groups is still often overlooked. The paraphyletic groups (much as the polyphyletic ones) are distinguished from the monophyletic ones essentially by the fact that they have no independent history and thus possess neither reality nor individuality'. Thus, Hennig clearly saw that a battle was to be developed with traditional taxonomists for banishing paraphyletic taxa altogether. In his criticism of cladistics, Mayr (1974) defended paraphyletic taxa in strong terms. His arguments are reiterated in Mayr and Ashlock (1991), as well as in Stuessy (1990), but this attitude is not compatible with the development of a sound and consequent phylogenetic system of living organisms.

In less abstract terms, within a sound and consequent phylogenetic system there is no place for the paraphyletic taxa that most of us have 'known' since schooldays. Reptilia (the reptiles) and Aves (the birds) are recognized as two classes in most conventional vertebrate classifications. However, according to the current understanding of phylogenetic relationships, birds are a specialized group of Archosauria, which is a group of reptiles, also comprising crocodiles, dinosaurs and various other extinct forms. Therefore, in keeping with the definitions given above, the class Reptilia is paraphyletic if it excludes birds (and possibly also mammals). To avoid paraphyletic groups, the taxon Reptilia could be enlarged to include birds (as for example, Gauthier, Kluge and Rowe, 1988a,b). Alternatively, Reptilia should be removed as the name of the taxon and the larger natural group comprising birds (and mammals, if supported by phylogenetic reconstructions) together with the conventional 'reptiles', could be given a different name. In so far as the underlying reconstruction of phylogeny holds true (cf. section 7.24), there seems to be no alternative to these procedures. In a natural system, there is no simultaneous place for two taxa of equal rank, bearing the traditional names of Reptilia and Aves.

The same case could be made for monocots and dicots in the flowering plants. It seems to be fairly well established, that monocots represent a specialized offshoot of dicots, originating from within magnoliids (e.g. Huber, 1982). As far as this phylogenetic reconstruction can be trusted, Dicotyledoneae (Dicotyledonopsida, Magnoliopsida) becomes paraphyletic and should be dismantled, or enlarged to include the monocots.

2.7 DETERMINING CHARACTER POLARITY

The problem is how to establish character polarity. Three main criteria are generally discussed in cladistic literature (e.g. Eldredge and Cracraft, 1980; Wiley, 1981): outgroup comparison; palaeontological or stratigraphical correlation; and ontogenetic correlation.

In cladistics, an outgroup is a taxon external to the taxa involved in a problem of phylogenetic reconstruction (i.e. taxa A, B and C in a classic three-taxon state-ment), but closely related to them. Ideally, the outgroup is the sister-group of the group comprising all the taxa under examination. In many cases, the actual sister-group of the group we are studying is not known with certainty but this difficulty is not essential. One or two reasonable candidates for sister-group relationships with the group under study can be taken into account. Let us consider, then, a suitable outgroup D, together with the three taxa A, B and C we began with. Let us further assume that we have identified, within the group (A,B,C), one or more characters, each present in two alternative states. For example, character a is present in a state a in A and in a state a' in B and C. Which state is apomorphic, a, or a'? According to the criterion of outgroup comparison, the decision rests on the recognition of the state present in the outgroup. To quote Wiley (1981, p.139), 'given two characters that are homologues and found within a single monophyletic group, the character that is also found in the sister group is the plesiomorphic character whereas the character found only within the monophyletic group is the apomorphic character' (Figure 2.2).

Accordingly, if we find a in the outgroup D, we can regard a' as an apomorphy within the ingroup A plus B plus C; more precisely, as the synapomorphy defining a clade B plus C. In the opposite case (outgroup D with a') we are left with an uninformative symplesiomorphy a' for B and C, together with an autapomorphy a

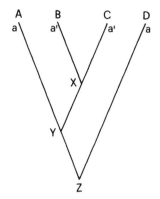

Figure 2.2 Outgroup comparison. To determine the polarity of the character a within the taxon comprising A, B and C, a related group ('outgroup') D is also taken into consideration. Within A plus B plus C, a is present in two different character states (a and a'). State a, which is also present in the outgroup D, is regarded as plesiomorphic (primitive), whereas a' is regarded as apomorphic (derived). The transition from a to a' is postulated to have occurred between Y (the common ancestor of A, B and C) and X (the common ancestor of B and C).

for A. Therefore to resolve the cladistic relationship within our group, we would be obliged to look for characters other than a.

Outgroup comparison is not always enough to provide a firm knowledge of character polarity. Limits to the workability of this method are experienced whenever a phylogenetic analysis is carried out involving more than one character and three taxa. Often a consequent use of the outgroup criterion leads to contradictory results, for example when recognizing a mosaic distribution of plesiomorphic and apomorphic character states among the taxa we are studying. However, there are approximate solutions to this problem. Maddison, Donoghue and Maddison (1984) have developed several useful rules to indicate the extent to which different resolutions of outgroup relationships influence our assessment of character polarity. Moreover, the remaining doubts or incongruencies are often reduced (but not always cancelled) by recourse to criteria other than outgroup comparison.

The palaeontological or stratigraphical criteria rest on the assumption that primitive (plesiomorphic) traits are exhibited by more ancient forms, and advanced (apomorphic) ones by their more recent relatives. Common sense seems to support this method, but it is not as reliable as it might seem (Eldredge and Cracraft, 1980). Our palaeontological record may hold the remains of older but more specialized forms (possibly belonging to fully extinct branches of the phylogenetic tree), compared with younger (or even Recent) taxa, belonging to less specialized branches retaining many plesiomorphic traits. There is no way of avoiding this difficulty, although, as suggested by Harper (1976), it is less dangerous in those groups where palaeontological documentation is more complete or more 'continuous' in time. However, this applies only to a small number of groups (Schaeffer *et al.*, 1972).

Modern discussions of the other criterion, ontogeny, have revived a re-examination of comparative studies introduced into science more than a century ago. Within an historical perspective, the work of von Baer (1792–1876) and Haeckel (1834–1919), and even Serres, Meckel and Agassiz have been re-examined. For the historical aspects of these developments the reader is referred to Gould (1977), Mayr (1982b) and Rieppel (1988a). Haeckel's biogenetic law is more broadly known, even among non-specialists, than von Baer's embryological rules. Von Baer's principles are summarized by de Beer (1958, p.3):

1   In development from the egg the general characters appear before the special characters;
2   From the more general characters the less general and finally the special characters are developed;
3   During its development an animal departs more and more from the form of other animals;
4   The young stages in the development of an animal are not like the adult stages of other animals lower down on the scale, but are like the young stages of those animals.

These statements have been defined by Gould (1977, p.56) as 'the most important words in the history of embryology'. Von Baer attained these astonishingly straight-forward results during the 1820s and they appear in his 1828 treatise: *Entwicklungs-geschichte der Thiere: Beobachtung und Reflexion*. The general statement in Haeckel's biogenetic rule 'ontogeny recapitulates phylogeny' is much more synthetic. There is one important difference between the two concepts. According to Haeckel, juvenile stages of descendant forms correspond to the adult forms of their ancestors, whereas von Baer directly compared equivalent developmental stages of different animals. To understand their formulations better, it is useful to remember that Haeckel's statement was coined in evolutionary, phylogenetic terms, whereas von Baer's laws were independent from evolutionary thinking. They are, however, amenable to interesting interpretations from an evolutionary point of view. Haeckel's recapitulation rule cannot be taken seriously as a guide to phylogenetic reconstruction (cf. de Beer, 1958). In a sense, von Baer's laws are more cognate to the cladistic ontogenetic criterion for establishing polarity (Rieppel, 1988a).

The ontogenetic criterion was formally introduced by Nelson (1978, p.327) in the following terms: 'given an ontogenetic character transformation, from a charac-ter observed to be more general to a character observed to be less general, the more general character is primitive and the less general advanced'. This criterion has been supported by several authors (e.g. Wheeler, 1990). However, Rieppel (1979) rejected Nelson's arguments, mainly because the comparison of ontogenetic stages simply follows the same criteria as outgroup comparison. Other authors have raised queries against the adoption of the ontogenetic criterion. Kluge (1988, p.73), for instance, points to numerous putative synapomorphies that would fail a test of ontogenetic similarity. One is reminded, here, of the widespread feeling that features sharing a substantial degree of homology may be based on different genes (Sander, 1989; Wagner, 1989a,b; Haszprunar, 1992b). However, some of these instances may be the effect of genetic piracy (Roth, 1988, 1991). This describes the recent involvement in a given, well-buffered developmental pathway of genes previously foreign to it. At some time, the new genes take over the control of some key developmental step, previously under different genetic control, but no major change ensues in the resulting phenotype, hence the possibility of recognizing homologies despite the renewed genetic background.

According to Mabee (1989), ontogeny provides useful data to evaluate for phylogenetic reconstructions, but the ontogenetic criterion is not an accurate predictor of character polarity. Ontogenetic transformations may also be considered as characters (e.g. de Queiroz, 1985; Kluge and Strauss, 1985; Creighton and Strauss, 1986; Kluge, 1988; Mishler, 1988). This concept is not new, for example Danser (1950) claimed that systematists should classify life cycles, rather than single stages, such as adults and juveniles.

Outgroup comparisons, and the additional criteria of palaeontological and ontogenetic evidence, more or less exhaust the weaponry of the cladist, although biogeographical evidence can also be used. There are, however, less direct but useful clues, for example what Saether calls underlying synapomorphies (Saether, 1977) (homoiologies *sensu* Hennig (1966, pp.117–28), as well as to the functional and developmental–mechanical criteria. All these criteria are rejected by strict pattern cladists.

The inherited tendency to develop parallel similarities has been repeatedly advocated as a clue to relationships (Wernham, 1912:390; Crampton, 1929; Hennig, 1966, 1970, 1972; Brundin, 1976; Saether, 1977, 1979a,b,c, 1983, 1986; Lehtinen, 1979; Cantino, 1982; Møller Andersen, 1982; Gosliner and Ghiselin, 1984; Sluys, 1989). A passage from Hennig's book (1966, p.118) has often been cited in this context: 'It can be no accident that so conspicuous and peculiar a character as stalked eyes occasionally occurs only in a relatively small group of about 5000 species (in which it apparently arose repeatedly) among the approximately 65 000 species of living Diptera'. In a sense, the parallel development of similar features is a type of litmus paper, helping to reveal an otherwise hidden similarity. Parallelism should be differentiated from other kinds of relationships (those Saether (1983) calls 'objective synapomorphies' and 'subjective synapomorphies'). To dismiss it as non-informative would arbitrarily reduce our chances of reconstructing the past.

Functional criteria for evaluating character polarity are similarly rejected by many cladists, as extraneous to the pattern analysis of a matrix of character states. In fact, functional analysis involves knowledge and judgements extraneous to cladistic schemes of character evaluation. However, that does not mean that they cannot contribute to checking our reconstructions of phylogeny. They may also be used to verify the functional and adaptive value of the character states postulated as present in the hypothetical ancestor Y of two taxa A and B.

In some traditional, pre-cladistic reconstructions of phylogeny it was customary to see hypothetical ancestors and intermediates discussed against hypothetical physiological and ecological scenarios. Many reconstructions were quite fanciful and Hennig's exhortation for more straightforward methods has certainly been important. However, to plead for accuracy in reconstructing phylogeny does not mean to search for lifeless abstractions or to prefer numerically likely but functionally impossible ancestors. I think that the evolutionary transitions postulated by our phylogenetic reconstructions must be checked, as far as possible, for plausibility in adaptive as well as in developmental terms.

As for developmental–mechanical arguments, consider the views of Maynard Smith *et al.* (1985) on developmental constraints. Without adequate knowledge of the obligate pathways that evolutionary changes seem to follow, owing to intrinsic (although historically acquired) constraints of developmental processes, our evaluations of character polarity may be seriously wrong. Nevertheless, our current understanding of development and its control devices does not enable us to infer phylogeny in the light of developmental mechanics only.

2.8 CLADOGRAMS AND TREES

Branching diagrams are increasingly common in the systematic literature. Haeckel delineated them literally with roots, stems and branches, but modern authors have contented themselves with simple, geometric diagrams, expressing only topological relationships between taxa.

The differences between one tree-like diagram and another are not simply matters of aesthetics. Several conceptually different tree-like diagrams are possible and one should check carefully, whether these are phylogenetic trees, cladograms or something else. It is also important to note that definitions and use of terms are often inconsistent. This is particularly true when one compares papers written by representatives of different schools (cf. Cracraft, 1979; Eldredge, 1979; Eldredge and Cracraft, 1980; Wiley, 1981).

Eldredge (1979, pp.167–8) defines a cladogram as 'a branching diagram depicting the pattern of shared similarities thought to be evolutionary novelties ('synapomorphies') among a series of taxa', whereas a phylogenetic tree is 'a diagram (not necessarily branching) depicting the actual pattern of ancestry and descent among a series of taxa'. The tree-like diagrams in Figures 2.1 and 2.2 are examples of phylogenetic trees, because they fully resolve the genealogical relationships between a set of taxa (A,B,C,D) and postulated ancestors (X,Y,Z).

Cladistic theory and methodology deals with the construction of cladograms and most investigations make no attempt to move from cladograms to phylogenetic trees. However, I think that climbing up this step is worthy of effort. Using the words of O'Hara (1992, p.135), it is true that 'If we are to understand the true nature of the evolutionary past we must adopt 'tree thinking'', but that means we must 'develop new and creative ways, both narrative and non-narrative, of telling the history of life'. I feel that in this conceptual and methodological area we need more insight and courage than is required in many other areas of biological systematics.

The topology of phylogenetic trees has been studied by comparing the results of phylogenetic analyses with theoretical models. Guyer and Slowinski (1991), for example, found that most trees of small size have a predictable, non-random topology. The major problem, however, is how to find adequate null models against which to test the data. Another problem is how to proceed, when the character evidence suggests alternative, contrasting cladograms, as happens in most cases. There are many reasons for obtaining different cladograms. Contradictions exist because of homoplasy, or because we do not have enough data to resolve a particular node. For example, some character polarities might have been wrongly evaluated, or a given character state might have been coded as a homology when in fact it had independently evolved on more than one occasion or a character might have reversed on a tree.

These problems have always faced systematists either when reconstructing phylogenies or arranging species in a phenetic classification. Individual solutions vary but most include establishing priority rules for weighting some characters. Many students are dissatisfied with a subjective, qualitative approach to these problems, and have developed a series of criteria and techniques for evaluating, in quantitative terms, the distribution matrices of character states (Saether, 1986). These approaches, variously named quantitative phyletics, numerical phyletics, numerical cladistics, or even neocladistics (Cartmill, 1981), have been developed through appropriate numerical methods, supported by computer packages.

2.9 NUMERICAL METHODS FOR THE RECONSTRUCTION OF PHYLOGENY

A very useful survey of the numerical procedures currently employed in phylogeny reconstruction has been provided by Swofford and Olsen (1990). They view these algorithms as tools to obtain a 'best estimate' of a theory of evolutionary relationships, based on the premise that we only have incomplete information about organisms. In principle, there are two ways to accomplish this goal, either by using a specific algorithm for constructing the best tree, or by applying a criterion for choosing between alternative trees. The first procedure is faster to perform than the second, but does not provide a ranking criterion for suboptimal trees. The second procedure provides ranking for all trees under comparison, but is much slower to perform and cannot give exact results for matrices with more than 12 taxa, because of computational difficulties.

Swofford and Olsen (1990) identify four types of methods for estimating phylogenies:

- methods based on pairwise distances;
- parsimony methods;
- Lake's method of invariants;
- maximum likelihood phylogenies.

2.9.1 DISTANCE BASED METHODS

These are generally regarded as phenetic, because distance data do not acknowledge character polarity. Data deriving from some experimental methods, as DNA–DNA hybridization, can only be analysed by distance-based methods, because they are generated in the form of pairwise similarities. Accordingly, these methods are widely used by molecular systematists, whereas most cladists regard them with suspicion (e.g., Carpenter, 1992).

2.9.2 PARSIMONY METHODS

For a more detailed account of parsimony methods see Camin and Sokal (1965); Farris (1969, 1970, 1973, 1977, 1983); Kluge and Farris (1969); Farris, Kluge and Eckhardt (1970); Kluge (1984); Sober (1988). In general terms, parsimony methods select the shortest tree(s) (i.e. the tree(s) requiring the least number of character changes) to explain the data matrix. It is important to distinguish between the optimality criterion underlying a given parsimony method from the actual algorithm used to implement it. Optimality criteria are hypotheses on how evolution proceeds.

Fitch and Wagner parsimony methods are the simplest of all, because they assume that the changes in character states are free from every constraint (Fitch) or only minimally constrained (Wagner). The Wagner parsimony method was formalized by Kluge and Farris (1969) and Farris (1970); it applies to binary, multistate and continuous characters, measured on some appropriate scale. The Fitch parsimony method (Fitch, 1971) is a generalization of the Wagner method. It was devised to allow the use of unordered multistate characters, such as those deriving from amino acid or nucleotide sequences. Fitch's method allows any state to change directly into any other state, without following a transformation through some intermediate states, as assumed in the Wagner parsimony method.

The Dollo parsimony method was introduced by Farris (1977). It requires that each character originates once on the tree and allows homoplasy to take only the form of reversal to a more ancestral state; parallel or convergent evolution of the same derived state is not allowed. It has been suggested that this method is most appropriate for restriction-site data (cf. Chapter 3) (DeBry and Slade, 1985), but see Swofford and Olsen (1990) for a contrary view.

The Fitch, Wagner and Dollo parsimony methods allow for the reversibility of changes, a condition excluded in other parsimony methods. The Camin–Sokal parsimony method assumes irreversibility of evolutionary changes but does allow parallelism and convergence; accordingly, it does not seem applicable to molecular data, but may be used for morphological data.

An interesting, but computationally very expensive method of generalized parsimony assigns a 'cost' to every state change. Recent developments in parsimony methods are discussed by D.R. Maddison (1991) and W.P. Maddison (1991).

In addition to the operational aspects of parsimony, there has been much debate over the philosophical meaning and justification for this approach. I subscribe to the following sentences from Gosliner and Ghiselin (1984, pp.259–60): 'Cladists (Hennig, 1966; Kluge and Farris, 1969; Eldredge and Cracraft, 1980; Wiley, 1981; Lauder, 1982) invoke the criterion of parsimony to justify their posit that parallelism is secondary to divergence ... It may be more parsimonious to assume that closely related organisms will respond to similar selection pressures in much the same way, producing many identical responses ... Parsimony, however, is a poorly understood principle of inference, and one that has long been controversial. Parsimony is not a proposition about the material universe. Its applicability does not depend upon 'nature being parsimonious' (Ghiselin, 1966b). This point has also been developed by Farris (1983) and Sober (1983), with whom we are basically in agreement ... One does not violate parsimony when one has good reason to invoke convergence'. Similar arguments are developed by Kluge (1984). Sober (1988) has devoted a whole book to a deep analysis of parsimony.

Problems occur when a single data set yields two or more equally parsimonious trees, or when two different kinds of data (e.g. molecules and morphology), when separately analysed, yield two imperfectly congruent trees for the same set of taxa. Two different strategies have been proposed, to obtain the 'best' trees in these difficult cases. The first strategy is to obtain a consensus tree (Adams, 1972), by computing a tree which represents only the cladistic information shared by all rival trees. This procedure has been criticized as a solution to the problem of mismatch between different data sets (Barrett, Donoghue and Sober, 1991). First, consensus trees are often poorly resolved (i.e. many nodes are multibranched). Second, if a consensus tree is derived from a tree based on morphology and another tree based on molecules, molecular data are likely to be much more numerous than morphological data, resulting in unequal weighting of the constituent characters (Cracraft and Mindell, 1989). Third, consensus trees often require more character-state changes than those required by any of the separate trees: in that case, the consensus tree may misrepresent the pattern of character evolution (Miyamoto, 1985). Fourth, there are several possible methods for constructing a consensus tree and a choice between them is arbitrary (Kluge, 1989). Owing to these difficulties, several authors reject the use of consensus trees and take an alternative way of handling different data sets related to the same taxa. This procedure consists of analysing the total evidence at hand as a single data set (Kluge, 1989). Barrett, Donoghue and Sober (1991) have demonstrated that a consensus tree, in addition to all the weaknesses listed above, can also positively contradict the most parsimonious tree obtained from the pooled data. A demonstration of the advantages of using total evidence rather than consensus trees is given by Vane-Wright, Schulz and Boppré (1992), in their cladistic analysis of the butterfly genus *Amauris*.

2.9.3 LAKE'S METHOD OF INVARIANTS

This method was developed (Lake, 1987) as an improvement on the parsimony methods applied to comparisons of nucleotide sequences, with provisions for reducing the effects of homoplasy.

2.9.4 MAXIMUM LIKELIHOOD METHODS

These methods (e.g. Felsenstein, 1973, 1978a, 1979, 1981a) evaluate the likelihood that a given tree will yield the observed data (e.g. the observed molecular sequences) to end up with adopting the tree with the highest likelihood. It requires the specification of a given evolutionary process responsible for the change in character state. In maximum likelihood analyses of molecular data, frequently used models are the Jukes–Cantor model, the Kimura two-parameter model and the generalized two-parameter model; they differ in the assumptions made regarding the frequency of occurrence of the four nucleotides and the frequency of the different types of nucleotide substitutions.

Swofford and Olsen (1990) remark on the close relationship between finding the likelihood of a tree and finding the cost of the tree under the general parsimony criterion. As for the procedures for searching for the optimal trees (the second main strategy for reconstructing phylogeny, in Swofford and Olsen's (1990) dichotomy), exact algorithms are possible for a few taxa only. Felsenstein (1978b) calculated that for 20 taxa there are more than 2×10^{20} possible (unrooted) trees. Accordingly, exhaustive search methods have been given up for the sophisticated branch-and-bound methods (Hendy and Penny, 1982) or other more practicable but approximate heuristic methods. Questions concerning the reliability of the trees inferred by these methods have been dealt with by Felsenstein (1988b).

Finally, the so-called bootstrap and jackknife methods (cf. Mueller and Ayala, 1982; Felsenstein, 1985a; Sanderson, 1989) have been used to estimate the robustness of the results of these analyses. Both methods involve 'resampling' from the original data set. In the bootstrap method 'data points are sampled randomly, with replacement from the original data set until a new data set containing the original number of observations is obtained. Thus, some data points will not be included at all in a given bootstrap replication, others will be included once, and still others twice or more. For each replication, the statistic of interest is computed. ... The jackknife ..., on the other hand, resamples the original data set by dropping k data points at a time and recomputing the estimate from the remaining $n-k$ observations'. (Swofford and Olsen, 1990, p.491).

In addition to these methods, we should mention the character compatibility method, (e.g. Le Quesne, 1969, 1972, 1982; Estabrook, 1972, 1983; Estabrook, Johnson and McMorris, 1975, 1976a, 1976b; Felsenstein, 1981b) which searches for the largest 'clique' (the largest set of characters that can evolve without homoplasy on a given tree). This method is criticized by Swofford and Olsen (1990) who point out its underlying dependence on an unrealistic model, according to which a character that has been excluded from the largest clique cannot continue to convey any useful phylogenetic information.

Felsenstein (1988a) provides an extensive discussion of the relative merits and ranges of applicability of the different quantitative approaches to phylogenetic inference. To conclude the discussion of this topic, let me quote some passages of his review. 'Exactly which of the various parsimony methods (Camin–Sokal parsimony, Wagner parsimony, Dollo parsimony, or polymorphism parsimony) or compatibility methods were recommended depended on the relative rates of different kinds of events; forward changes, reversion, gain and loss of polymorphism, and misinterpretation of the data. ... When rates of evolution are not small, there seems to be no general correspondence between maximum likelihood and parsimony (or compatibility) methods. ... Cavender (1978) and Felsenstein (1978a) investigated the statistical properties of parsimony and compatibility methods in a simple evolutionary model with four species. ... When we assumed that evolutionary rates varied greatly enough among lineages, it turned out to be more probable that a shared derived state was the result of parallel changes in two branches of the tree that had high evolutionary rates, than that they were the result of a single change in a branch having low evolutionary rate. In these cases parsimony and compatibility methods fail to have the statistical property of consistency ... When the rates of change in lineages are low, the requisite inequality of rates becomes greater and greater for parsimony and compatibility to misbehave. This counter-example to the use of parsimony and compatibility has no force if evolutionary rates are sufficiently equal among lineages.'

2.10 ANCESTORS

When a cladogram is converted into a phylogenetic tree, the problem of ancestors arises. Nelson and Platnick (1984) and Rieppel (1988a) argue that the notion of monophyly is incompatible with a notion of ancestry. Rieppel (1988a, p.153) identifies two kinds of ancestry, Haeckelian and von Baerian: 'Haeckelian groups are actual ancestors and descendants respectively, and thus by definition privative or paraphyletic. As compared to the ancestor, the descendant is derived by the addition of an evolutionary novelty ... As a consequence the ancestor must be a privative or paraphyletic group with respect to its descendant. Von Baerian groups on the other hand are types, logical conceptions of similarity providing a guide to common ancestry. ... The real organisms sit out on the branches, and cannot be ancestral one to another'. Von Baer's concept of ancestry developed within an idealistic, pre-evolutionary view, but it finds a place also in the minds of pattern cladists.

Although ancestors are or have been actual organisms, rather than abstractions, they cannot be recovered with cladistic methods. A method of overcoming this difficulty has been suggested by a few authors. Eldredge and Cracraft (1980), for example, relax the universal requirement of monophyly for the species taxa that are the stem-species (ancestors) of a monophyletic group. However, Platnick (1977b), Willmann (1983) and Ax (1984) regard species monophyly as inappropriate to discussion. In their opinion, the concepts of monophyly, paraphyly etc. only apply to groups of species; that is, to supraspecific taxa.

De Queiroz and Donoghue (1988) reject these positions and explore alternative methods of defining species, which could be used for stem-species as well as for species representing terminal taxa of the cladistic analysis. Unfortunately, they failed to obtain a satisfactory solution. In strictly formal terms, there seems to be no way of escaping Rieppel's (1988a, p.156) dilemma: 'If we admit privative [para-phyletic] species, we objectify the species taxon as an open system and we obtain evolutionary lineages. If we close the species taxon by its objectification (i.e. individualization) on the basis of some uniquely shared character(s) [i.e. if we identify species taxa as monophyletic units on the basis of one or more apomorphies], species can no longer serve as actual ancestors unless reversal for those diagnostic characters is assumed'. However, are we justified in discussing the diachronic behaviour of species in so much detail, whenever we are assured that the concept of species has any meaning in a geological, instead of an ecological, time dimension (cf. Chapter 4)?

2.11 FOSSILS AND CLADISTIC ANALYSIS

There has been a great deal of discussion regarding how to treat fossils in systematics. Besides the particular problems raised by fossils regarding the delimitation of species taxa (cf. Chapter 4), many authors (e.g. Hennig, 1966, 1969; Crowson, 1970; Løvtrup, 1977, 1985; Patterson and Rosen, 1977; Jefferies, 1979, 1986; Rieppel, 1979; Patterson, 1981a,b, 1982; Rosen *et al.*, 1981; Fortey and Jefferies, 1982; Gardiner, 1982; Ax, 1985b, 1987, 1989b; Willmann, 1985; Forey, 1986; Craske and Jefferies, 1989; Loconte, 1990) have contended that extinct taxa cannot be accommodated within the same system, or classification, as Recent taxa, and in the same way. Their positions range from excluding fossils from the system of Recent organisms (e.g. Crowson, 1970; Løvtrup, 1977, 1985), to simply introducing special ranks to accommodate fossils within the same system (e.g. Patterson, 1981a; Gardiner, 1982; Ax, 1985b, 1987, 1989b; Forey, 1986).

The opposite view has been defended by others, especially by Gauthier, Kluge and Rowe (1988a,b), Donoghue *et al.* (1989) and Kluge (1990), who not only consider that fossils should be treated in the same way as Recent taxa but also that they can influence the classification of modern groups. This last point has been strongly rejected by other authors, especially by Gardiner (1982) and Løvtrup (1985), according to which fossils cannot overturn a relationship established on Recent forms only. In my view, their arguments have been convincingly rebutted by Gauthier, Kluge and Rowe (1988b) with their re-analysis of the relationships of major groups of amniotes. The first step in their study was reconstructing a phylogeny strictly based on Recent taxa only. This resulted in the following pattern of sister-group relationships: (lepidosaurs (turtles (mammals (birds, crocodiles)))). However, adding the fossil evidence resulted in a different, most parsimonious cladogram: (mammals (turtles (lepidosaurs (birds, crocodiles)))), closer to the traditional view, as well as in closer agreement with the stratigraphic record. These results were expected, but when they tried to identify the fossil taxa most responsible for changing the phylogenetic pattern, they concluded that the changes were determined by ingroup fossils, not by outgroup fossils, particularly by some synapsid reptiles exhibiting a particular combination of primitive and advanced traits. In my view, this analysis supports both the need, for systematists, to evaluate all types of available evidence, without excluding *a priori* some types of data, and the need for a careful analysis of intra-group variability before coding characters as they are present in one or a few 'representative(s)' of each terminal taxon. In seed plants, Doyle and Donoghue (1987, p.63) found that fossil information does not radically change the relationships established among extant taxa, but 'it provides unique evidence on the sequence of events and possible adaptive factors involved in the origin of groups. Both contributions are especially important when there has been extensive homoplasy and/or when living groups are isolated from each other by large gaps, as is often true of higher taxa'. Besides remarking that 'higher taxa' are isolated from each other by large gaps, I feel that Doyle and Donoghue have aptly underlined, in the passage above, the need for sensitivity to the adaptive context of the evolutionary changes, a sensitivity too often lacking in modern phylogenetic analyses (Ghiselin, 1984b).

For fossil vertebrates, Norell and Novacek (1992) found fairly good agreement between the age of a group, as established by the stratigraphic record, and the rank usually attributed to the corresponding clade in phylogenetic classifications. This result looks attractive but, as we shall question later, what legitimizes our ranks?

2.12 GROUPING AND RANKING

How do we arrange the monophyletic groups within a classification or system? Recent developments in phylogenetic systematics have revealed several short-comings in the traditional Linnaean hierarchy. The number of levels in the hierarchy is a problem. A fully resolved phylogeny of any given group would re-quire a very large number of hierarchically nested taxa, to accommodate all monophyletic taxa. Two recent arrangements of higher taxa of Plathelminthes and Chelicerates (cf. Appendices 7 and 11, respectively) exemplify this trend well. It is easy to understand that an exceedingly complex hierarchy would proliferate the number of names, unless we adopt a more expedient procedure.

An interesting proposal aimed at overcoming this difficulty, put forward by Nelson (1972, 1974) and subsequently developed by Wiley (1981) and others, is the so-called phyletic sequencing convention. Wiley (1981, p.209) formally enunci-ates it as follows: 'Taxa forming an asymmetrical part of a phylogenetic tree may be placed at the same categorical rank ... and sequenced in phylogenetic order of origin. When an un-annotated list of taxa is encountered in a phylogenetic classification, it is understood that the list forms a completely dichotomous se-quence, with the first taxon listed being the sister-group of all subsequent taxa [conventionally ranked at the same level] and so on down the list'. This convention reduces the number of ranks recognized and taxa named. Consequently, the Linnaean hierarchy could still work, despite the large number of hierarchically nested taxa we have identified. Additionally, the branching order of the individual lineages, although not immediately obvious from the rank and names of taxa, can be fully retrieved. As an example, rearranging the higher taxa of the Chelicerata as recognized by the system of Weygoldt and Paulus (1979) (cf. Appendix 11 and Figure 2.3) according to the sequencing convention, we obtain the following.

Aglaspida
Xiphosura
Eurypterida
Scorpiones
Lipoctena
  Megaloperculata
    Uropygi

Amblypygi
Araneae
Apulmonata
  Palpigradi
  Haplocnemata
    Chelonethi
    Solifugae
  Cryptoperculata
    Opiliones
    Ricinulei
    Acari

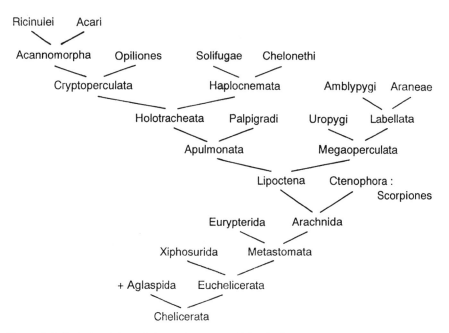

Figure 2.3 Relationships of the major taxa within chelicerate arthropods, according to Weygoldt and Paulus (1979).

From this exercise, it is easy to see four hierarchical levels, instead of ten, which adequately reproduce the phylogenetic branches, as identified by Weygoldt and Paulus.

There are problems with the sequential method, however. One must consider Tassy's (1988, p.51) remark, that 'strictly speaking, information is stored at the nodes. As some nodes are [left] unnamed, some of the information cannot be retrieved from the classification'. In fact, the names of taxa in a classification adopting Nelson's sequencing convention become strongly context dependent. Citing a name without full reference to a table such as that in the previous lines would obviously reduce its 'cladistic information content'.

Another interesting convention was proposed by Patterson and Rosen (1977) for extinct fossil groups. When classified together with Recent species or groups of species, fossil species or monophyletic groups of species are classified as plesions (i.e. as taxa of variable, indeterminate rank). This proposal was an attempt to overcome a difficulty perceived by Hennig (1969), who had suggested that all those fossil species within a group, that are more primitive than the most primitive Recent species of the same group should be identified as Stammgruppe (i.e. stem group). It is clear that such stem groups are likely to be paraphyletic, if treated in the same way as Recent taxa. For detailed, convincing arguments in favour of the plesion convention see Patterson and Rosen (1977) and Wiley (1981) although these two definitions of plesion do not overlap completely (Tassy, 1988).

Nelson's sequencing convention has been developed by Haszprunar (1986) in the form of a 'clado-evolutionary classification'. In addition to the monophyletic taxa, Haszprunar also accepts what he defines as 'orthophyletic taxa'. These are paraphyletic taxa with only one emerging (and taxonomically excluded) lineage. Haszprunar accepts Nelson's sequencing convention, but also tries to accommodate recognized and named taxa within the conventional Linnaean categories (cf. Appendix 10). His efforts may help biologists to accept some types of phylogenetic classifications and the sequencing convention in particular. However, I think that the way is already open towards the development of a genuine phylogenetic system, definitively free from its formal dependence on the traditional, Linnaean categories.

The existence of Linnaean categories represents a second problem in classification. When we speak of families, for example in birds, gastropods and ascomycetes, are we referring to meaningful comparable entities? Most people suppose that scientific terms are used in a definite and consistent way. People do not fear that words such as molecule, acceleration and chloroplast each refer to a lot of different and incomparable objects. However, that is what generally occurs, when we say, for instance, that there are more families of Recent birds than there are families of Recent amphibians.

One of the most detailed discussions of comparability of taxonomic categories (ranks), as applied to different phyla, is given in Van Valen (1973), where several possible standards of comparison between, for example, families of mammals and families of insects, are listed. They include the following:

- number of species or genera included;
- number of individuals, or biomass, or energy flow, represented at any one time by the numbers of each family;
- phenotypic diversity;
- genotypic diversity;
- ease of hybridization among taxa;
- phylogenies, i.e. time of divergence, or average number of branchings, or other measure of phylogenetic relationship;
- 'adaptive diversity'.

However, all these criteria were then rightly discarded as unsuitable. Instead van Valen developed a 'metataxinomic' criterion, based on the structure of the classification itself. Van Valen argued that taxa ranked at the intermediate levels between the ranks of species and class should be 'proportionately' spaced, so as to determine a geometric series, the terms of which are the average numbers of species taxa respectively included within genera, families, orders and classes. Therefore, in a larger class the contained orders, families and genera will also be larger, on average, than the corresponding taxa in a smaller class. The artificiality of this proposal is evident (Levinton, 1988).

A different proposal was put forward by Hennig (1966), to overcome the problem of non-equivalence of rank. He suggested that the upper rank limit of the group be fixed according to the time of origin of the group itself. This, in turn, would be decided either by the earliest fossil of that group or by the age of its sister-group. Hennig divided Earth history into six time intervals, each corresponding to a Linnean rank. However, this proposal to equate ranks with age found no following.

Reality versus artificiality of supraspecific taxa has been discussed by many authors. La Greca (1987) convincingly rejected the use of subgenus. Dubois (1982) tried to redefine the genus on reproductive criteria. I think the main critical issue, within systematics, is relative hierarchy, not those biological criteria which are required when dealing with an evolutionary interpretation of the biological diversity around us. Before deciding on a definition of genus, one should look at a figure in Allmon (1992), where the variation in geological time (Cretaceous to Present) of the generic diversity of a group of gastropods (Turritellidae) is separately plotted according to three different classifications: the resulting trends in taxonomic diversity are not comparable.

The semantic burden caused by the Linnaean tradition is generally underestimated. 'Abandoning the notion of taxonomic rank would go a long way toward eliminating sequences of contemporary taxa, because sequences are most often created out of collections of taxa of equal rank. ... And while taxa may be referred to as ancestral or descendant, or as older or younger, taxa should never be referred to as primitive or derived or higher or lower (de Queiroz, 1988, p.252).' (O'Hara, 1992, p.154). I ask the benevolent reader to forgive me whenever I have allowed the fingers on the computer keyboard to be guided by tradition more than by reason.

2.13 PHYLOGENY VERSUS ADAPTATION

A sound reconstruction of phylogenetic relationships is the most suitable background against which to discuss the evolution of adaptations. This topic has been developed in several papers (e.g. Cheverud, Dow and Leutenegger, 1985; Felsenstein, 1985b; Schaefer and Lauder, 1986; Coddington, 1988, 1990; Pagel and Harvey, 1988; Bell, 1989; Larson, 1989; Maddison, 1990; Baum and Larson, 1991; Harvey and Purvis, 1991; see also Bock (1982b), for a different theoretical approach) and also in book form (Brooks and McLennan, 1991; Harvey and Pagel, 1991).

One can dispute the precise terms in which synapomorphies are related to what in evolutionary biology are known as key adaptations. Coddington (1988) defines adaptation as apomorphic function promoted by natural selection, compared with plesiomorphic function. Adaptation, he stresses, is a relational term, and hypotheses of adaptation which fail to specify plesiomorphic and apomorphic states are, accordingly, weak. This important area pertains to evolutionary biology, rather than to systematics. The reader is referred to the literature just cited for more details.

Biochemical and molecular systematics

3

Clearly, a method that is independent of fossils and morphology, not prone to convergence, and that can measure the genealogical distances between living species should improve the reconstruction of avian phylogeny. Since we are trying to measure degrees of genetic relationship, it makes sense to look for help on the genetic material itself.

C.G. Sibley and J.E. Ahlquist (1987, p.97)

The last two decades or so have witnessed an enormous development in biochemical and molecular systematics. The approaches in this field have been diverse, both in theory and in the operational aspects of gathering and handling data. A broad distinction should be made between the systematic studies of (a) the different classes of micromolecules or chemical phenotypes and (b) macromolecules which are more or less close to the core of hereditary information, i.e. the (genotypic) DNA sequence versus the (phenotypic) RNA and protein sequences.

3.1 MICROMOLECULES

In botany, the systematic study of a large array of molecules has often been the by-product of techniques developed in response to economic demands, for example in rubber technology, pharmacy and cosmetics. Considerable systematic information is now available for several classes of compounds, such as flavonoids, terpenoids, alkaloids, glucosinolates and iridoids (Hegnauer, 1962–1973; Gibbs, 1974; Giannasi and Crawford, 1986).

For flavonoids (Giannasi, 1979) there has been a shift from studies of intraspecific variation towards comparative surveys, from the interspecific to the supraordinal level. The reader is referred to reviews by Niemann (1988) for gymnosperms and Giannasi (1988) and Williams and Harborne (1988) for angiosperms. As a consequence, many interesting patterns have emerged. For instance, the presence of flavonoids in many families of monocotyledons similar to those of the Magnoliidae gives support to the theory that monocotyledons are an offshoot of magnoliidean dicotyledons (cf. section 7.26). Flavonoid evidence also supports the definitive recognition of the artificial character of Sympetalae, previously based on fused corolla tubes. At the species level, flavonoids often provide a good means of detecting hybrids. In some instances, flavonoid inheritance seems to be very simply controlled and hybrids possess both parental complements (a pattern of inheritance also found in terpenoids and other compounds). In other cases, as in some *Rhododendron* (e.g. King, 1977a,b) or *Phlox* (e.g. Levy and Levin, 1974, 1975), the situations are not so simple, partly because of polyploidy. A problem with the use of flavonoids in phylogenetic reconstruction has recently been raised by Bohm and Chan (1992). They argue that the scattered occurrence of a subclass of these molecules (the B-ring deoxyflavonoids) among dicotyledons is perhaps the consequence of convergent adaptive changes, rather than an expression of common descent.

Terpenoids have been extensively studied in Coniferales and in a few angiosperm families (Umbelliferae, Labiatae), where they are widespread. As in the case of flavonoids, terpenoids have sometimes been used to detect hybridization, e.g. in *Pinus, Juniperus* and *Eucalyptus*. In other genera, they have also been useful at the species level (*Abies*) and in groups of higher rank (*Eucalyptus*). Some chemical results are surprising, for example, Giannasi and Crawford (1986) regard as phylogenetically important the taxonomic distribution of sesquiterpenes (together with acetylenes and some phenolic derivatives) between Umbelliferae and Compositae (Heywood, Harborne and Turner, 1977; Seaman, 1982). Sesquiterpene chemistry also provides many suggestions for a more natural arrangement of genera within the Compositae.

The alkaloids are similarly important in systematic studies of plant families, but they are less accessible for study. Their significance also seems to be less easy to analyse than other classes of metabolites. This is because of the structural transformations they often undergo within living plants, while transported from the organs of production (e.g. the roots) to the other parts of the plant. This fact, while representing a possible source of error whenever these endogenous chemical changes are taken into account, nevertheless suggests the possibility of using whole biosynthetic pathways as taxonomic characters. On a larger taxonomic scale, biosynthetic pathways are sometimes considered to provide important clues to the relationships between the major groups of living organisms. At a lower taxonomic level, they may also help to resolve some problems of plant systematics.

Of the other types of micromolecules of chemosystematic interest, I shall only mention iridoids and monoterpene lactones and their derivatives. The distribution of these compounds is of major interest in the rearrangement of the families traditionally placed within the flowering plants orders Rubiales, Caprifoliales, Scrophulariales, Cornales and others.

Chemical data have even greater significance in the study of lichens, in some groups of algae and in bacteria. Chemical tests for lichen identification were first developed by Nylander (1866) and have been used routinely for more than 50 years. Chemical tests permit fine taxonomic discrimination. Six different 'chemical races' or chemotypes have been identified in Europe within the *Ramalina siliquosa* complex, whereas, on morphological grounds, two 'species' can be recognized, both with the same chemical 'races' (for a review see Culberson, Culberson and Johnson, 1977).

Within the scope of this book I can mention only a few additional interesting approaches, which are discussed at length elsewhere (e.g. Giannasi and Crawford, 1986). Interesting chemical studies include comparative studies of phytoalexin induction (especially for the study of Leguminosae) and the analysis of amino acids in the nectar (e.g. for *Aquilegia, Penstemon* and *Geranium*).

Small-molecule chemosystematics is much less developed in the animal kingdom than for plants, however Løvtrup (1977) discusses an unconventional approach to the study of the relationships of vertebrates. There is an increasingly large body of evidence which demonstrates the enormous potential of these data for animal systematics. Sex pheromones are particularly interesting at the species level, because of their high specificity. In a few cases, pheromone analysis has been the first clue in the discovery of sibling species (cf. Chapter 4). Similarly, toxic and repellent substances are interesting, at various taxonomic levels. They exhibit specificity at the tribal and familial levels in many groups of beetles and bugs (cf. Dettner (1987) and Aldrich (1988) for up to date reviews). Many different chemicals are used as repelling weapons of most of the orders of millipedes (Eisner *et al.*, 1978) (Table 3.1).

3.2 MACROMOLECULES

To give an overview of the molecular data of possible use to the systematist, we shall take into account both the chemical nature and the biological significance of the molecules involved and the nature of the tools employed to study them.

Following Young (1988), we can summarize the most important techniques used to study proteins as: protein immunology; electrophoresis of proteins; and amino acid sequencing. For nucleic acids the techniques are: nucleic acid hybridization; DNA restriction fragment analysis; and nucleic acid sequencing.

There have also been developments in the techniques for analysing proteins and nucleic acids. Both protein immunology and nucleic acid hybridization provide us only with similarity estimates for macromolecules (i.e. unpolarized similarity or distance data). Protein electrophoresis and DNA restriction fragment analysis provide us with partial structural details of macromolecules without fully resolving their structures. This final step is obtained by utilizing sequencing techniques, which permit the identification of long strings of amino acids and nucleotides.

3.2.1 PROTEINS

(a) **Immunological techniques**

Broadly speaking, immunological techniques rely on the following principles. A protein (antigen) from species A is injected repeatedly into a rabbit (or other suitable animal), to elicit in it the production of antibodies against the antigen. The

Table 3.1 Toxic or repugnatory secretions of millipedes, according to Eisner *et al.* (1978). Millipedes belonging to the orders Polyxenida, Sphaerotheriida, Chordeumida, and (possibly) Glomeridesmida lack repugnatory glands

Compound	Order						
	Glomerida	Spirobolida	Spirostreptida	Julida	Callipodida	Polydesmida	Polyzoniida
Hydrogen cyanide						+	
Formic acid						+	
Acetic acid						+	
Isovaleric acid						+	
Myristic acid						+	
Stearic acid						+	
Hexadecyl acetate				+			
Δ^9-Hexadecenyl acetate				+			
Δ^9-Octadecenyl acetate				+			
trans-2-Dodecenal		+					
Benzaldehyde						+	
Benzoic acid						+	
Mandelonitrile						+	
Mandelonitrile benzoate						+	
Benzoyl cyanide						+	
Phenol						+	
o-Cresol				+			
p-Cresol					+		
Guaiacol (=2-metoxyphenol)						+	
1,4-Benzoquinone			+	+			
2-Methyl-1,4-benzoquinone		+	+	+			
2-Methyl-3-methoxy-1,4-benzoquinone		+	+	+			
2,3-Dimethoxy-1,4-benzoquinone		+		+			
5-Methyl-2,3-dimethoxy-1,4-benzoquinone		+					
2-Methyl-1,4-hydroquinone				+			
2-Methyl-3-methoxy-1,4-hydroquinone				+			
Glomerin (=1,2-dimethyl-4(3H)-quinazolinone	+						
Homoglomerin (=1-methyl-2-ethyl-4(3H)-quinazolinone	+						
Polyzonimine							+
Nitropolyzonimine							+

antibodies thus produced against the antigen from species A are tested for their differential reactivity against the same ('homologous') antigen as against corresponding ('heterologous') antigens (i.e. related proteins) from a second or third species. The (measurable) strength of the immunological reaction is used as an estimate of the (unknown) similarity between the antigenic protein from species A and the corresponding antigenic protein from other species. I am aware, however, of only one experimental demonstration of the relationship between amino acid sequence differences and immunological reactivity (Maxson and Maxson, 1986).

Immunological techniques have been mainly applied to vertebrates, with special attention being given to albumins and transferrins, as well as to the so-called microcomplement fixation. The reader is referred to Maxson and Maxson (1990) for further details and a discussion of the advantages and limits of several methods in comparative immunology.

(b) Protein electrophoresis

Starch gel electrophoresis was introduced by Smithies (1955) but only gained prominence in systematic research several years later. Electrophoresis is used mainly to detect allozymes, allelomorphs of enzymatic proteins. These allelomorphs are separated according to their differential mobilities in electric fields, because they possess different electric charges. After electrophoretic separation, the molecules are detected using specific stains. To date, stains for over 100 enzymes are available. Protein studies have been restricted to very few enzymes despite the fact that Avise (1983) suggested that at least 100 loci should be examined by this technique, in order to obtain a reliable estimation of genetic differentiation. Most studies were carried out on 20 or 30 loci.

Electrophoretic analysis of allozymes has proved particularly useful for the differentiation of closely related organisms (i.e. between populations of a species, or species of the same genus). These types of analyses have helped to detect sibling species, reproductively isolated populations, or groups of populations, apparently indistinguishable by conventional morphological characters. When two sibling species occur sympatrically (in the same locality), electrophoresis permits a clear-cut demonstration of the reproductive isolation, by revealing the existence of loci fixed for alternative alleles in the two co-occurring populations. The lack of heterozygotes for those alleles suggests the existence of two separate gene pools.

During the late 1970s and 1980s, the electrophoretic analysis of allozymes become routine analysis for many microsystematic problems. Far less developed are the systematic studies of isozymes (structurally distinct enzymes which exhibit similar catalytic functions) despite their dependence on different structural genes (Ferguson, 1988).

The genetic differences revealed by electrophoresis are not always associated with morphological differentiation. Davis and Gilmartin (1985) were able to list a dozen plant and animal genera, where at least two morphologically differentiated species do not differ, with respect to their electrophoretic patterns, more than couples of conspecific populations usually do. Sometimes, however, the reverse is true. Hewitt (1988), in studying a narrow hybrid zone between two races of the flightless grasshopper *Podisma pedestris*, found that the two forms differed in a significant proportion of their enzymatic loci, although they are very similar in karyotype and overall morphology.

(c) Amino acid sequencing

During the 1960s and early 1970s, amino acid sequencing seemed a very attractive method for the study of molecular phylogenies. At that time the first atlases of molecular sequences appeared (to be replaced eventually by computerized data banks) and the first algorithms were developed for deducing from them the first molecular trees. Dayhoff's *Atlas of Protein Sequence and Structure* (1969) and Fitch and Margoliash's (1967) paper on the construction of phylogenetic trees have both found a place in the history of molecular biology and systematics. At that time, a detailed knowledge of the structure of the genome was still out of reach and amino acid sequences of proteins were thought to provide phenotypic evidence most close to the genotype. However, this evidence did not always provide a reasonable match with the conventional classifications based on morphology. It was no surprise therefore that the first molecular trees met with enthusiasm and resistance at the same time. These contrasting responses have accompanied subsequent developments in molecular phylogenetics and systematics prompted by a precipitous shift of emphasis from proteins to nucleic acids. Only recently have we witnessed a more balanced attitude with the added possibility that old and new views on phylogeny can be reconciled.

Amino acid sequencing of molecules such as haemoglobins, myoglobins and parvalbumins has provided much interesting comparative data, still in the focus of phylogenetic studies, especially for vertebrates. Two classic studies are Goodman, Weiss and Czelusniak (1982) (vertebrates) and Miyamoto and Goodman (1986) (mammals). Especially useful, for a general discussion of molecular phylogenetics, is Goodman, Miyamoto and Czelusniak (1987).

3.2.2 NUCLEIC ACIDS

Both DNA and RNA have proved useful in molecular systematics. Many investigations have been performed on mitochondrial DNA (mtDNA) and chloroplast DNA (cpDNA), rather than on nuclear DNA (nDNA), for reasons explained later. The RNA molecules of choice have been those of ribosomal RNA (rRNA). There have been several approaches to the systematic study of nucleic acids: DNA–DNA hybridization; analysis of molecular fragments obtained by means of restriction enzymes; and sequencing of nucleotides.

(a) DNA–DNA hybridization

The hybridization technique applies to double-stranded molecules, and therefore, in principle, applies only to DNA. In this technique, the DNA of two species are mixed together and heated, so that each molecule ruptures into two single strands. By lowering the temperature, these single-stranded molecules are allowed to re-associate into conventional, double-stranded DNA. During the process, 'hybrid' double-stranded molecules form between the homologous sequences of the DNA of the two species. The hybrid molecules are carefully heated again, until they dis-associate once more into single strands. Disassociation of 'hybrid' molecules occurs at a temperature lower than that required by the original DNA molecules, because only partial bonds are established between the heterospecific strands. Single DNA strands from closely related species match each other better than those from more distant species; consequently, they require more heat to dissociate than the latter. The differences in 'melting' temperatures between conspecific and heterospecific ('hybrid') DNAs are therefore assumed to be estimates of overall similarity between the genomes of different species.

DNA–DNA hybridization has been widely discussed. The advantage it has over all other methods of analysis is that it allows the whole DNA complement of one species to be compared with that of another. Moreover, it is described as not being particularly difficult or time-consuming. There are problems, however, in cal-ibrating the method to obtain comparable results (cf. Springer and Krajewski, 1989). This technique has been extensively used by Sibley and Ahlquist (1983, 1984, 1987, 1990). Their studies on the phylogeny of birds (cf. section 7.24) and primates have been the subject of intense debate, and have attracted extremes of enthusiasm and scepticism. Recently, Sarich, Schmid and Marks (1989) evaluated the use of DNA–DNA hybridization studies as a means of evaluating the extent of divergence among the single-copy fraction of vertebrate genomes. In particular, they focused on the nature and information content of the melting profiles as a clue to understanding phylogenetic relationships. They acknowledge that DNA hybridization is a valuable and cost-effective tool for phylogenetic reconstruction. However, they point out several serious flaws in the recent work by Sibley and Ahlquist and ask for a detailed publication of primary experimental data and for a more careful evaluation of the data, in a phylogenetic perspective. Werman, Springer and Britten (1990) have given a list of the factors which may compromise the interpretation of DNA–DNA hybridization data. Many of these causes, from parallelisms to intraspecific variation, from the uneven distribution of rates of change to measurement error, are common to most types of molecular evidence, but DNA–DNA hybridization is particularly sensitive to differences in genome sizes between the two species to be matched.

(b) Restriction site analysis

This approach consists of analysing the small strings of nucleotides (restriction site fragments or RSF) obtained by treating nucleic acid molecules with the so-called restriction enzymes, which cut them at particular sites only, different for the dif-ferent enzymes. This technique (Dowling, Moritz and Palmer, 1990) has been widely applied to plant cpDNA and animal mtDNA, as well as to some nuclear sequences.

This type of analysis is evidently coarser than sequencing. Dowling, Moritz and Palmer (1990) point to the problems deriving from RSF polymorphism, when comparing recently separated species, whereas comparison of distantly related species are easily marred by problems of parallelism (homoplasy, see Chapter 2).

(c) Sequencing

Sequencing of nucleotides provides the most exhaustive way to recover information from a macromolecule. Recently, sequencing of nucleic acids has become easy at a reasonable cost, particularly since the development of techniques allowing the 'amplification' of minimum amounts of material (polymerase-chain-reaction technique or PCR). Besides the interesting applications of this technique to the study of marine bacterioplankton (cf. section 7.2), I would like to mention the work of Olson (1991), where PCR amplification of nucleotide strings, followed by sequencing, allowed the separation of the morphologically indistinguishable, co-occurring larvae of two species of holothurians (sea cucumbers) *Psolus fabricii* and *Cucumaria frondosa*. The same technique has been used by Silberman and Walsh (1992) for the specific identification of the phyllosome larvae of some spiny lobsters.

(d) Mitochondrial DNA (mtDNA)

The importance of these molecules for phylogenetic and systematic purposes (in animals) has been reviewed by Moritz, Dowling and Brown (1987). In animals, mtDNA is inherited maternally. In most species studied so far, mtDNA consist of two genes coding for ribosomal RNA (rRNA), 22 genes coding for transfer RNA (tRNA) and 13 genes coding for enzyme subunits; introns are absent. However, in *Ascaris* there are only 12 protein coding genes. When comparing mtDNA sequences of more or less closely related organisms, divergence increases steadily and rapidly with time (about 0.5–1% per million years per lineage) until some 15% divergence is attained; it then slows down dramatically. The initial phase is mainly caused by substitutions in 'dispensable' sequences, such as intergenic sequences, parts of the control region and even coding regions, but never in such a way as to cause amino acid replacements in proteins. In comparison with single copy nuclear DNA (scnDNA), the rate of mtDNA sequence evolution is not the same in all groups. Both molecules evolve at similar rates in *Drosophila* and sea urchins (Echinoidea), whereas mtDNA evolves faster than scnDNA in mammals (primates, rodents) and frogs (*Xenopus*). Moritz, Dowling and Brown (1987) point to some particularly interesting aspects of mtDNA, such as maternal inheritance and relatively rapid evolution, which explains its widespread use in the study of the dynamics of hybrid zones. mtDNA is also an excellent genetic marker of gene flow in matrilineal inheritance. mtDNA studies are not free of error. For example, back-mutations have been detected in comparisons involving conspecific individuals and the random extinction of mtDNA lineages is another potential cause of 'wrong' results. A small chance of paternal inheritance of mtDNA has been suggested.

Some of the 50–100 copies of mtDNA molecules carried by the sperm cell may penetrate the egg (which already contains 10^5–10^8 mtDNA molecules) during fertilization (Kondo *et al.*, 1990; Avise,1991; Hoeh, Blakley and Brown, 1991). An experimental study involving hybrids of two species of mussels (*Mytilus*) (Zouros *et al.*, 1992) has recently confirmed this possible paternal contribution to mtDNA inheritance.

Brown and Wright (1979) have used mtDNA to identify sexually reproducing maternal ancestors of parthenogenetic 'species'. More recently, this approach has been applied to the study of the lizard genus *Cnemidophorus*, which includes many parthenogenetic taxa with problematic affinities (e.g. Wright, Spolsky and Brown, 1983; Frost and Wright, 1988), or to identify the ancestry of the unisexual fish *Poecilia formosa* (Avise *et al.*, 1991). It has been possible to demonstrate the unique origin of each parthenogenetic taxon (because of uniform mtDNAs) as well as to demonstrate the origin of some triploid parthenogenetic forms from hybrid diploid females. mtDNA studies have also been performed in the problematic group of European green frogs (*Rana esculenta* complex), but in this case the results are much more of a puzzle (Spolsky and Uzzell, 1986).

With respect to its uses in studies of hybrid zones, mtDNA has provided some information on patterns of introgression (Avise *et al.*, 1984; Szymura, Spolsky and Uzzell, 1985; Wilson *et al.*, 1985; Carr *et al.*, 1986; Lamb and Avise, 1986; Harrison, Rand and Wheeler, 1987; Harrison, 1989). Carr *et al.* (1986) studied hybridization between two species of deer (the white-tailed deer *Odocoileus virginianus* and the mule deer *Odocoileus hemionus*), occurring locally in sympatry in west Texas. In the hybrids, Carr *et al.* found a common restriction fragment length (RFL) pattern of mtDNA, closer to that of *O. virginianus*, thus suggesting that hybridization occurred mostly between mule deer bucks and white-tailed deer does. In the context of systematics, however, the primary interest of mtDNA studies resides in its contribution to the understanding of phylogenetic relationships among closely related taxa (cf. De Salle *et al.* (1987), for Hawaiian *Drosophila*).

mtDNA data may be nucleotide sequences, or collections of restriction fragments resulting from enzymatic cleavage of mtDNAs by specific endonucleases. Data must be read very cautiously. For instance, in their study of intraspecific variation in the mtDNA of the rainbow trout (*Onchorhynchus mykiss*), Beckenbach, Thomas and Sohrabi (1990) found that intraspecific divergence estimates based on nucleotide sequences are smaller than the estimates based on RFL analysis, at variance with what happens with interspecific comparisons. The significance of restriction fragment evidence is studied more effectively as differential fragment mobility or on restriction site maps.

Restriction fragment analysis is effective only for closely related species, because comparisons are unreliable when the overall sequence divergence is greater than 10–15%, and when there is variation in the overall length of the molecule. Avise *et al.* (1987) demonstrated that mtDNA provides a powerful tool for studying relationships within species. In many species, the geographical distribution of populations and genealogical pattern of mtDNA are interestingly correlated. Avise *et al.* (1987) refer to these studies as intraspecific phylogeography and they formulate 'phylogeographic hypotheses' which can be tested by mtDNA studies. For example, 'The geographic placements of phylogenetic gaps are concordant across species', and 'Phylogenetic gaps within species are geographically concordant with boundaries between traditionally recognized zoogeographic provinces'. Interesting tests of these hypotheses have been performed on a few freshwater fish groups in North America, with encouraging results. This approach seems promising for improving our understanding of genealogical and geographical relationships between populations, both before and after speciation. Another example is given by Templeton (1987), in his studies of the interaction between cladogenesis and mode of inheritance (diploid versus. haploid) in Hawaiian *Drosophila*. He contrasts two lineages, one apparently less affected by founder events, compared with the other in which speciation is interpreted as the result of founder events. In full agreement with theoretical expectations, Templeton finds that the patterns of variation of nuclear DNA versus mtDNA is different in the two lineages; the 'haploid' mtDNA is interpreted as diverging much faster than the nuclear DNA in the cases where speciation is linked with founder effects.

There are several reports of incongruence between mtDNA and nDNA evidence. For example, the divergence between *Drosophila teissieri* and *D. yakuba*, as estimated on mtDNA evidence (either on sequencing or on RFL) was 10 times lower than the divergence estimated from DNA–DNA hybridization studies (Monnerot, Solignac and Wolstenholme, 1990).

Phylogenetic resolution of mtDNA data is better when the species to be compared are neither too recently separated nor too distantly related (Dowling, Moritz and Palmer, 1990). In the first case, it is often difficult to distinguish interspecific patterns from intraspecific polymorphism. Neigel and Avise's (1986) simulation showed that mtDNA sequences from closely related monophyletic sister taxa progressively change from apparently polyphyletic to paraphyletic to monophyletic, owing to the progressive loss of the original polymorphic variants, which at some time are replaced by unique derived forms, different in each taxon. Moreover, if multiple speciation events closely spaced in time (at least in the small populations involved) occur within a group, starting with an ancestral highly polymorphic population, it becomes very difficult to obtain the correct topology from the analysis of a single molecular sequence. According to Dowling, Moritz and Palmer (1990), this difficulty has contributed to the debate over the phylogeny of hominoids (cf. Goldman and Barton, 1992; Thorne and Wolpoff, 1992; Wilson and Cann, 1992).

Palmer (1992; see also Doyle, Lavin and Bruneau, 1992) has recently summarized the reasons for the limited interest of mtDNA analysis for plant taxonomy. mtDNA changes too slowly in nucleotides sequences, whereas large duplications are readily created and lost; in addition, plant mtdNA contains many foreign sequences (some 5–10% of the total).

(e) Chloroplast DNA

In plants, the analysis of chloroplast DNA (cpDNA) has proved very useful (Harris and Ingram, 1991; Soltis, Soltis and Milligan, 1992b).

The cpDNA of land plants is in the form of circular molecules, mainly 120–217 kilobase pairs in size (De Pamphilis and Palmer, 1990; Downie and Palmer, 1992). Several features of cpDNA explain its popularity among plant molecular systematists: small size; relatively low rate of structural and sequence evolution; lack of recombination; and uniparental inheritance. A curious pattern has been discovered in several gymnosperms, where cpDNA is inherited paternally, and mtDNA maternally (Szmidt, Alden and Hallgren, 1987; Wagner *et al.*, 1987).

Golenberg *et al.* (1990) claim to have sequenced a fossil cpDNA sequence from a Miocene *Magnolia* species. Despite the reasonable concerns expressed by Pääbo and Wilson (1991) regarding possible contamination of the sample by bacterial DNA, or possible molecular changes (depurination) induced by water, Golenberg *et al.*'s paper is a major step forward in the analysis and taxonomic evaluation of plant fossils. This study of fossil DNA has been recently followed by DeSalle *et al.*'s (1992) study of fossil rDNAs (mitochondrial 16S and nuclear 18S molecules) from the remains of a termite (*Mastotermes electrodominicus*) preserved in Oligo-Miocene amber.

A problem with the use of cpDNA evidence (which also extends to mtDNA) has been raised by Wendel and Albert (1992). They point to some putative causes of reticulate evolution in the cotton genus (*Gossypium*), thus illustrating the dangers of relying solely on maternally (uniparentally) inherited markers when reconstructing phylogeny.

(f) Ribosomal RNA (rRNA)

Ribosomal RNA (rRNA) sequences have been the favourite of many molecular systematists. rRNA is a constituent of living matter with essentially the same functions in all organisms, and can be readily isolated and sequenced. Moreover, rRNA is ideally suited for comparisons between very distantly related organisms, across conventional phyla and kingdoms, because many regions of the macromolecule are extremely conserved. That means that large segments of the sequences are virtually unchanged, whereas other regions are more divergent, and therefore more informative at relatively lower systematic levels. Also, rRNA is unlikely to have been involved in 'horizontal' (i.e. non-genealogical) transfers from one organism to a distantly related one. By comparison, the possibility of horizontal transfer of other macromolecules, of nuclear DNA in particular, is now well established. The latter phenomenon has opened up enormous possibilities for genetic engineering. The natural transfer of nuclear DNA if unnoticed may deeply bias our phylogenetic reconstructions. In a very stimulating test study, Ragan and Lee (1991) tried to 'debug' strings of molecular data by using a computer package originally developed to uncover textual contamination in medieval manuscripts.

One example of horizontal gene transfer involving eukaryotes is worth describing here (in prokaryotes horizontal gene transfers have been known for a long time). Busslinger, Rusconi and Birnstiel (1982) found that their molecular phylogenetic studies of sea urchin species failed dramatically to match phylogenetic reconstructions obtained from more conventional data. They explained their puzzling results as due to a horizontal transfer of a histone gene cluster from *Strongylocentrotus droebachiensis* to *Psammechinus miliaris*. Shortly after the publication of this paper there was, in some circles, concern over the possibility of virus-induced speciation.

Like mtDNA, rRNA has proved of relatively little use in plant evolutionary studies (Hamby and Zimmer, 1992). Of possible interest, however, is the finding that 18S and 26S rRNA point to a paraphyletic nature of gymnosperms (Hamby and Zimmer, 1992).

Returning to rRNA, it should be noted that it occurs as three, basically different types within cells. In eukaryotes, these can be described as 5S (with about 120 nucleotides), 18S (with about 1800 nucleotides) and 28S (with about 4000 nucleotides). In prokaryotes, the corresponding types have generally shorter sequences (5S, 16S, 23S). Most RNA analyses have been undertaken on the 5S fraction. Classic studies dealing with a range of organisms belonging to different phyla include Ohama *et al.* (1984), Hendriks *et al.* (1986), both on metazoans; Wolters and Erdmann (1988) on eukaryotes; and Hori and Osawa (1987) on a variety of organisms. Several studies have been published on the 18S molecules, including Field *et al.* (1988) for metazoans and Wolters and Erdmann (1988) for eukaryotes. A few papers based on partial sequences of the much more informative 28S molecules have also been published, such as those by Qu *et al.* (1988) and Baroin *et al.* (1988), on the phylogenetic relationships of some groups of protists. Many more sequences are already available for the smaller (but less reliable) 5S rRNA than for the other, larger kinds. For example, Hori and Osawa (1987) analysed 352 different 5S rRNA sequences from a wide range of organisms.

A spectacular result of rRNA phylogenetic studies has been the demonstration that king crabs (Lithodidae) are a type of carcinized hermit crab (Cunningham, Blackstone and Buss, 1992). Lithodids look very like large crabs, with a heavily calcified exoskeleton, whereas hermit crabs have a long, soft and asymmetrical abdomen depending for protection on the use of an empty gastropod shell. This unique habit, which evolved some 150 million years ago, has been given up in the lithodid lineage, where complete carcinization has occurred within a fairly short time span, without involving too much genetic change. This history has been unravelled by analysing DNA sequences for the gene encoding mitochondrial large-subunit rRNA. According to this evidence, lithodids are not only derived from hermit crab-like forms, but are even nested within the single hermit crab genus *Pagurus*, as currently understood.

When interpreting the systematic analyses based on rRNA, it is impossible to disregard the problems raised by recent discoveries concerning the 'restless' behaviour of RNA genes. Dover (1988a, pp.623–4) said: 'Nothing would seem more predisposed to generate a mess ... than the recent findings by P.H. Boer and M.W. Gray of a bizarre fragmentation and scrambling of scraps of ribosomal RNA genes (rDNA) among protein-coding and transfer RNA genes. Yet nothing of the sort happens; instead, perfectly assembled two-dimensional (and by inference three-dimensional) RNA structures, identical to the conserved rRNA 'core' structures within the large and small ribosomal subunits, emerge by intermolecular interactions between the unspliced mini-gene transcripts. ... A comparison of rRNA sequences among species has shown that they consist of modules of conserved core segments with more variable segments (also called 'expansion segments')'.

3.2.3 A CRITICAL EVALUATION OF MOLECULAR DATA

(a) Molecular clocks

Underlying most studies in molecular phylogenetics is the belief that molecular data are more reliable than morphological data because molecules are much less subject to natural selection, than morphological traits. Many molecular systematicists have explicitly embraced Kimura's neutral theory of molecular evolution (Kimura, 1983), in the light of which one could simply read the amount of difference between related molecules in two different organisms as a measure of the time elapsed since the separation of the two lineages. This is the well-known concept of the molecular clock (Zuckerkandl and Pauling, 1965).

Accordingly, molecular data have received particular attention because of their possible use in determining the chronology of evolutionary events. Until recently, chronology was based exclusively on palaeontological and palaeogeographical evidence, but very seldom supported by absolute dating using radiometric methods.

Today, several different methods are available to determine rates of molecular evolution. As excellently summarized by Wilson, Ochman and Praeger (1987), these methods rely on comparisons of the 'relative amounts of evolutionary change along molecular lineages of equal time depth in different species'.

The first method consists of comparing, for a given pair of species, the amount of sequence divergence, for different genes, compared with, for instance, the extent of divergence between the albumin genes of two mammal species to the divergence between the cytochrome c genes of the same two species. In principle, both genes began to diverge when the common ancestor of the two species gave rise to the diverging lineages to which the species under comparison belong. This method does not work well, however, if the ancestral species was polymorphic for the traits under consideration, unless we can get any evidence as to the subsequent evolution of this polymorphism in each of the derived lineages. This evidence is exceedingly difficult to obtain.

Another method for reconstructing molecular phylogenies applies to the so-called duplicated genes occurring within a given species, such as α- and β-globin genes in man and other mammals. In this case, it is possible to enquire whether the divergence between the two paralogous genes (cf. below p.61) is the same as that between the corresponding genes in other species.

A third kind of molecular comparison requires comparing genes of species under study with those of a reference, less closely related ('outgroup') species. Thus, by comparing the amount of gene divergence between species A and species C with that between B and C, it is possible to evaluate the rate uniformity of molecular evolution within the group A plus B.

Amidst the enthusiasms for the first developments in molecular phylogenetics, it has become fashionable to equate molecular evolution with radioactive decay. To what extent are we justified in assuming this 'neutral' behaviour of macromolecules is a matter of intensive research and of lively debate. The problems at stake clearly involve our general views on biological evolution, an issue far too large to discuss here for a review of related matters (see Zuckerkandl, 1987).

The neutral theory of molecular evolution (cf. Kimura, 1983) has provided a comprehensive theoretical background for this approach, but it is clear that many amino acid or nucleotide substitutions are not neutral (cf. Woese, 1987; Patterson, 1989) and in some algorithms developed for phylogenetic reconstruction some 'corrections' have been introduced to overcome this bias (Swofford and Olsen, 1990). Moreover, there is increasing evidence that even at the molecular level evolutionary matters are not simple. According to Dover (1988b), most genes are subject to some mechanism of DNA turnover, such as unequal crossing over, transposition, gene conversion, slippage or RNA mediated exchanges, thus determining the gradual change in the genetic composition of sexual populations (Dover's molecular drive). In some gene families, these molecular changes seem to be coevolutionary, involving functionally interacting components of the genome. Accordingly, it seems to be difficult to accept a clock-like behaviour as the rule for molecular changes. Rather, 'DNA can be useful as a phylogenetic tool once the variety of turnover mechanisms and their modes of behaviour are well understood for any given region of the genome.'(Dover, 1988b, p.152). Dover's view is perhaps too pessimistic, in the light of recent progress in molecular systematics, but his authoritative warning against an uncritical acceptance of the molecular clock hypothesis is certainly useful.

Some doubts regarding the behaviour of cpDNA according to a molecular clock have been raised by Jansen and Palmer (1988). They found that cpDNA sequence divergence among genera of the Asteraceae only ranges between 0.7 and 5.4%, lower than in other angiosperm groups. Accordingly, they wonder whether this should be interpreted as Asteraceae being a fairly young family, or its cpDNA having evolved at an unusually slow rate. One could also conclude from this finding that 'genera' do not have equivalent meanings in different plant families (cf. section 2.12).

There are also problems with the tempo and mode of mtDNA evolution. Heterogeneities among mammalian orders have been recorded by Hasegawa and Kishino (1989). Also, the rates of evolution of mtDNA in sharks are 7–8 times slower in comparison with mammals (Martin, Naylor and Palumbi, 1992). A detailed appreciation of these unequal rates of molecular evolution is important for the development of molecular systematics. Powell's (1991) conclusion, that neither the neutral theory of molecular evolution nor molecular clocks are required for using molecular data in phylogenetic reconstructions, is reassuring.

(b) Intraspecific variability

In most studies of molecular systematics no allowance is made for intraspecific variation of the molecular data. In a sense, this neglect amounts to embracing an old-fashioned typologism. A reaction against this attitude, however, is finally emerging.

Soltis, Soltis and Milligan (1992b) summarize examples of variation in cpDNA RFL polymorphism patterns for 59 species in 35 genera. Most impressive is the extreme intraspecific variability found within *Trifolium pratense*, where five to nine different chloroplast genotypes have been identified from single populations.

The usual analysis of RFL polymorphism is perhaps too coarse for detecting many patterns of intraspecific variability. In the *Lisianthus skinneri* complex (Gentianaceae), Sytsma and Schaal (1991) found no restriction site variation within populations or individuals, despite the site differences existing among populations.

The most convincing demonstration that phylogenetic inferences based on single specimens of each species are not safe comes from a study of mtDNA restriction site data in cyprinid fishes (Smouse *et al.*, 1991). Another example of unusually large intraspecific variability in mtDNA has been reported for the parasitic nematode *Osteragia osteragi*, with estimates of within-population mtDNA diversity 5–10 times larger than usually reported for species in other taxa (Blouin *et al.*, 1992).

If one specimen per species is not enough, one species per higher taxon (say family, or order) also provides too little evidence of phylogenetic relationships. Rosenberg, Davis and Kuncio (1992) have studied 28S rRNA sequences in seven species of truncatellid gastropods and found an unexpected variability in sequences, so high that they could not identify a preferred alignment for the sequences. In addition, *Truncatella scalaris* and *T. clathrus*, which are so similar they have been considered synonymous, differed at two positions and in a five nucleotide deletion in the latter species.

(c) Concordance between different kinds of molecular evidence

According to Soltis, Soltis and Milligan (1992b), discordance between chloroplast and nuclear DNA data is not necessarily negative or noisy. On the contrary, it can be useful to unravel intraspecific relationships or to study introgression (see also Rieseberg and Brunsfeld, 1992). A good example has been given by Dorado, Rieseberg and Arias (1992) with their analysis of cpDNA introgression in Southern California sunflowers (*Helianthus*). In Southern California, most of *H. petiolaris* individuals (i.e. with nuclear genome and morphology largely in agreement with this species) have the alien cytoplasm of *H. annuus*, as revealed by mtDNA. A more frequent exchange of chloroplast genotypes in respect to nuclear genes is also reported for oaks (*Quercus*) (Whittemore and Schaal, 1991). A similar introgression pattern was demonstrated in *Drosophila* from a mismatch of nDNA and mtDNA, by Aubert and Solignac (1990). In obligate parthenogenetic forms of the freshwater crustacean *Daphnia pulex*, the population subdivision for mtDNA genomes was found to increase with distance much faster than that for nuclear genes. The same analysis has also shown the widespread polyphyletic (multiple) origin of several parthenogenetic forms (Crease, Lynch and Spitze, 1990). In the same area, the most puzzling evidence comes from Kraus and Miyamoto (1990), that the mtDNA of a unisexual salamander of hybrid origin (genus *Ambystoma*) is unrelated to either of its two nuclear parental haplotypes.

(d) Molecules versus morphology

There has been considerable concern for assessing the comparative merits of morphological and molecular approaches to phylogeny. A few interesting papers dealing with this topic have been brought together by Patterson (1987) in a small but stimulating book with the provocative title: *Molecules versus Morphology in Evolution: Conflict or Compromise?*

It must be said at the outset, that many alleged instances of incongruence between morphological and molecular evidence are the simple effect of a flawed methodology: either because morphological and molecular data have been subjected to different types of analysis, or because only one of the data sets has been analysed adequately (Donoghue and Sanderson, 1992; Smith, 1992).

Increasing evidence is becoming available on the relative merits and limitations of both morphological and molecular approaches. Matters are very different, however, with different groups and at different taxonomic levels. Morphology is inadequate, for example, when dealing with the relationships between the major divisions (the kingdoms) of the living world, or with the systematization of prokaryotes. On other hand, it can provide exceedingly detailed evidence of relationships, as in the cases of complex chromosomal rearrangements exhibited by certain species groups, for example in *Drosophila* or *Aedes* (Coluzzi, 1982). A fine review of this topic is given by Hillis (1987). His arguments, supplemented by some additional materials, are briefly summarized in the following lines.

There are three major advantages of molecular methods: the size of the data sets potentially available for comparisons, the broad phylogenetic limits within which to extend the comparisons, and the independence of molecular traits from external influences. Nevertheless, there are also obvious advantages of morphological methods: the possibility of applying them to museum specimens and to fossil species (both increasingly closer, however, to the working range of newest tools of molecular systematics); the possibility of incorporating ontogenetic information and the usually much lower costs. Hillis identifies three major areas of conflict between the two approaches: differences in assumptions; differences in design and methods of analysis; and differences in results.

The most important differences in assumptions relate to rates of evolutionary change. Hillis observes that morphologists, at variance with many students of molecular evolution, deny that the evolutionary rates of changes are 'constant enough that they can be assumed a priori to be equal for purposes of analysis', and 'constant enough that a 'clock' model can be applied to dating lineage divergence'.

The differences in design and in methods of analysis between morphological and molecular studies concern both sample size and distribution (morphologists generally rely on larger samples and often pay adequate attention to population, or geographical, differences) and the preferred methods for the reconstruction of phylogenetic trees. By consequence, the results are also different for morphological and molecular phylogenetic reconstructions, especially when the molecular data are 'distance' data (as in DNA–DNA hybridization), instead of detailed 'sequences'. The latter are clearly amenable to conventional cladistic (or 'numerical cladistic') analysis, whereas distance data seem to escape this type of treatment. This topic is also developed by Andrews (1987).

Detailed claims of mismatch between the two types of data, however, are common, although not necessarily uninformative. Wayne *et al.* (1991), for instance, studied the Island fox (*Urocyon littoralis*) by using morphometrics, allozyme electrophoresis, mtDNA restriction-site analysis and analysis of hypervariable mini-satellite nuclear DNA to estimate variability within populations as well as distances among them. The lack of correlation among the interpopulation genetic distances estimated by the four techniques was not interpreted as pointing to the superiority of one kind of evidence over the other, but rather as the effect of differential mutation rates and of colonization history.

A very interesting comparison of molecular and morphological evidence involves the cichlid fishes of some African lakes. According to Sturmbauer and Meyer (1992), the six species of *Tropheus*, forming a lineage endemic to Lake Tanganyika, contain six times more genetic variation than the whole cichlid species flock of Lake Victoria, comprising hundreds of morphologically diverse species (cf. Chapter 4).

Other authors are more critical about morphology. Baverstock and Adams (1987), in their studies involving a wide range of Australian vertebrates, find that molecular, chromosomal and morphological evolution can occur at enormously different rates in different lineages. They agree with the widespread opinion that molecular evolution proceeds in a more clock-wise manner than morphological evolution. The latter should be regarded as very poor evidence for total genomic evolution.

In other instances, molecules and morphology are in good agreement, as in decapod crustaceans, where both kinds of evidence identify the same main lineages, for example the dendrobranchiate, the caridean, the stenopodidean and the reptant, in the same mutual relationships (Abele, 1991).

As for plants, an extensive discussion of the relationships between morphological and molecular evidence has been provided by Sytsma (1990). At the family level, Sytsma reports on the very good agreement between a morphology-based cladogram of Compositae and a cladogram for the same family based on cpDNA. On the contrary, at the species level there are many examples of mismatch, often seemingly caused by instances of dramatic increase in rate of morphological divergence. The complex case of orchids of the subtribe Oncidiinae is worth quoting, where a phylogeny based on the study of cpDNA reveals 'that strong regional evolution has produced quite similar floral morphologies in parallel among distantly related species ... For example, several sections of *Oncidium* in Brazil are phylogenetically more closely related, based on DNA, to other genera in the same region than they are to other sections of *Oncidium* with similar morphologies ... occurring in other geographical areas' (Sytsma, 1990, pp.108–9). In such a case (cf. Chase and Palmer, 1988, 1992) it is difficult not to assign the last word in phylogenetic reconstructions to molecular evidence.

More puzzling are Jansen *et al.*'s (1992) findings that in Asteraceae, cpDNA phylogeny agrees well with the non-cladistic relationships proposed by Thorne

(1983), much less so with the cladistic tree of Bremer (1987). However, composites are a traditionally difficult group and further developments are expected.

According to Doyle (1992, p.161), 'molecular systematics in plants has done much to popularize cladistic methodology, but ironically may represent an insidious new form of one-character taxonomy'. Doyle also derives a good lesson from molecular investigations in the cabbage genus (*Brassica*). Palmer *et al.* (1983) demonstrated cpDNA polymorphism in *B. napus* accessions that were homogeneous for more traditional molecular markers (isozymes and nDNA) as well as for gross morphology and cytology (karyology). Interestingly, one of the two cpDNA genome types fell completely outside the putative parental types, including *B. campestris*, to which the other cp genome could be referred. From these data, Doyle (1992) derives the conclusion that a fundamental incongruence between gene and species tree (see below) could not be suspected, had only one of the cpDNA genotypes been sampled by chance.

(e) Phylogeny of organisms or phylogeny of genes?

A further methodological caveat has been raised by Doyle (1992). Commenting on the recent developments in molecular phylogenetics, he remarks that all too often people seem to forget the actual nature of the terminal 'taxa' of a tree based on cpDNA only. These 'taxa' are chloroplast genomes, not necessarily species. This comment also applies to trees based on other types of molecular evidence.

This problem is not trivial, because of the widespread occurrence of gene duplications, leading to the origin of paralogous, rather than orthologous, molecular families. These two terms were introduced by Fitch (1970) and may be defined as follows, according to Goodman, Miyamoto and Czelusniak (1987, p.149): 'orthologous sequences ... are descendants of gene-lineages whose splittings from one another coincide with the splitting of the species-lineages within which they occur; paralogous sequences ... are descendants of gene-lineages whose splitting occurred by gene duplication prior to speciation of the species-lineages'.

The importance of distinguishing between these types of correspondence has often been stressed (e.g. Goodman, Miyamoto and Czelusniak, 1987; Patterson, 1987). As noted by Patterson, discovering the relationships between paralogous molecular sequences allows the reconstruction of gene phylogenies, rather than phylogenies of the organisms containing them. It should be pointed out that paralogous molecules commonly occur within the same organism, as products of what were originally two copies of the same gene, which underwent divergent evolution after the gene duplication event. For example, human α, β and γ (foetal) haemoglobins are paralogous molecules. Real problems in phylogeny reconstruction arise when comparing different organisms that retain alternative subsets of whole sets of paralogous molecules derived from a common (once unique) ancestral gene.

The species 4

Let us also remember that plants vary and opinions vary. One man's fish is another man's poisson. One man's moss is another man's mess.

H.A. Crum (1985, p.22)

In systematic biology, no concept has been the subject of such heated debate as the species concept. Together with systematists, students of population biology and philosophy of science have contributed substantially to the debate. The literature concerning the theoretical and operational aspects of the species problem, as well as the related field of speciation mechanisms, has grown to an enormous size. Important monographs published during the last 20 years include Grant (1971, 1981), Bocquet, Génermont and Lamotte (1976, 1977, 1980), Endler (1977), White (1978), the volumes edited by Barigozzi (1982), Vrba (1985), Willmann (1985), Iwatsuki, Raven and Bock (1986), Giddings, Kaneshiro and Anderson (1989) and Otte and Endler (1989). For a historical view, the reader is referred to Mayr (1982b), to a volume on the history of the species concept in life sciences by Atran *et al.* (1987) and to Provine's (1989) essay on the historical developments of the concepts of founder effect and genetic revolutions.

4.1 SPECIES CONCEPTS

Grant (1981) identified five different 'types' of species:

- taxonomic species (morphological species or phenetic species);
- biological species (genetic species);
- microspecies (agamospecies);
- successional species (palaeospecies or chronospecies);
- biosystematic species (ecospecies, coenospecies).

Grant's scheme covers most of the concepts and meanings of 'species', but it is not enough. For example, the successional species concept may overlap with the morphological species concept. Also, Simpson's evolutionary species concept does not fit into a class, as Grant acknowledges. In addition, some new species concepts have been developed since Grant produced his survey.

4.1.1 SPECIES AS CATEGORY VERSUS SPECIES AS TAXON

As a category, the species is commonly identified as the basic unit of the Linnaean hierarchy, at least in its modern, post-Linnaean use. To Linnaeus, as to folk taxonomists (Berlin, 1992; but see Atran (1986) for a different view), the genus concept was certainly much more important than the species concept. In so far as we can conceive it in these terms, the species is logically a class, the elements of which are individual species taxa, such as those currently known as *Homo sapiens* (man), or *Solanum tuberosum* (potato). Unfortunately, the two concepts of species as a category and species as a taxon have been confused until recently. It is to Ernst Mayr's great credit that he repeatedly pointed to the absolute necessity of discriminating between them, before approaching any discussion of the 'species-problem'. Sadly, there are still many students, both biologists and philosophers, who continue to overlook this distinction. I conclude with Mayr's (1987, p.146) definitions: 'A species taxon is an object in nature recognized and delimited by the taxonomist. The species category is the rank given to a species taxon'.

4.1.2 SPECIES AS NATURAL ENTITIES VERSUS SPECIES AS ABSTRACT CONSTRUCTS OF THE HUMAN MIND

There is no doubt that many traditional uses of the term 'species' correspond to artificial constructs of the human mind. Species as taxa are artificial by definition, in so far as we define them as groups of morphologically similar individuals, more similar among themselves than they are to individuals belonging to separate clusters of natural diversity (i.e. other species). Species taxa are also artificial constructs, in so far as we define them with reference to preserved type specimens (cf. Chapter 5), as prescribed by the International Codes of Nomenclature. Species as taxa are abstractions of our mind, so long as we believe that they are perfectly 'equivalent' to one another, without realizing what 'equivalent' might mean.

In day-to-day practice, all species taxa are commonly perceived as equivalents, because they are all named by a Linnaean binomial. Very seldom do we realize that they should only be offered a uniform taxonomic and nomenclatural status provided that they are 'biologically equivalent'.

It is universal practice to compare lists of plant and animal species collected on different islands of an archipelago or in different samples within a given type of environment. There are many difficult cases, in which the intrinsic problems are often overlooked. Let us take two examples, from ecology and biogeography respectively. Consider first the following example concerning the plant species in two grassland plots. In the first plot we might identify 45 species, compared with 50 in the second plot. However, 14 of these 50 are agamospermous 'microspecies' (cf. section 4.1.5) of dandelion (*Taraxacum*), whereas nothing of this kind occurs in the first plot. All dandelion 'microspecies' are treated taxonomically with Latin binomials. Are circumstances sufficient to consider them as contributing 14 taxa to the diversity as contributed by as many individual species, all bisexual and possibly outcrossing, of other genera?

As already noted by Rensch (1928) and Mayr (1980, p.101), a similar problem arises in zoogeography, when closely related entities with separate geographical ranges (semispecies, together forming a superspecies) are treated in the same way as more distinct species, not belonging to superspecies. In large archipelagos, or large mountain ranges, a superspecies is frequently represented by several semispecies. On the other hand, one semispecies might occur in geographically more continuous broad areas. To overcome this difficulty, Mayr (1980) proposed as a unit of comparison a new category, the zoogeographic species, the elements of which are both superspecies and distinct species that are not members of superspecies.

A few authors have taken what we could define as a nihilistic approach to the species problem. Nelson (1989b) and Lidén (1992) claim that there are no such things as 'species phenomena' or 'speciation'. They believe that no pattern of biological diversity can be distinguished, at a hypothetical species level, within the universal distribution of similarity patterns, as due to common descent, within which taxonomic rank is completely arbitrary, but for the nested pattern of subtaxa within 'higher' taxa. The same nihilistic concept is adopted by Nixon and Wheeler (1990) and Vrana and Wheeler (1992). These authors advocate individual organisms rather than any interbreeding group as the most suitable terminal entities for phylogenetic analysis.

Not very far from this is Cracraft's (1983, 1987a) species concept, recently applied (Cracraft, 1992) to a detailed revisionary work of a bird family, the Paradisaeidae (birds-of-paradise). Within this group, previous authors, working according to the biological species concept, recognized some 40–42 species, 13 of which were monotypic and 27 polytypic, with a total of about 100 subspecies. Besides describing two new species and 'synonymizing' 35 subspecies of previous authors, Cracraft has raised 65 subspecies to species rank, thus ending with a new count of 90 species in the family.

From a different theoretical background, Löve, Löve and Raymond (1957) and Löve (1964) arrived at a similar species concept, when regarding every distinct plant cytotype (or group of individuals with a given chromosome number) as a different species. This approach has been widely criticized and opposed by most botanists.

Many systematists are satisfied with a strict taxonomic concept of species, while others are not. In so far as we classify living organisms, it seems reasonable to try to identify taxonomic species corresponding to biologically meaningful units.

Efforts to define species in biological terms, primarily by recognizing them as reproductive communities, can be traced back to the work of John Ray (1627–1705) and Georges Louis Leclerc de Buffon (1707–1788). Mayr (1987) cited two very attractive old definitions: one from von Baer (1828), the species is 'the sum of individuals that are united by common descent', and the other from

Plate (1914), 'The members of a species are tied together by the fact that they recognize each other as belonging together and reproduce only with each other'.

Within population genetics, it has become fashionable to restate Buffon's concept of the species as reproductive communities in terms of gene pools, thus leading to the well-known biological species concept, slowly changed through the efforts of Dobzhansky (1951, 1970), Mayr (1957, 1963, 1969) and Ghiselin (1974, 1987). According to Dobzhansky (1970, p.354), 'a biological species is an inclusive Mendelian population; it is integrated by the bonds of sexual reproduction and parentage'. In Mayr's (1969, p.23) terms, 'the members of a species form a reproductive community. The individuals of a species of animals recognize each other as potential mates and seek each other for the purpose of reproduction ... The species, finally, is a genetic unit consisting of a large, intercommunicating gene pool ... These ... properties raise the species above the typological interpretation of a 'class of objects''.

Barton (1989, p.22) defines the biological species as 'a group of populations possessing inherited differences that prevent gene exchange with other such groups'.

Botanists (e.g. Jeanmonod, 1984) have been less interested in a biological species concept than zoologists. However, the concept has found some strong defenders among botanists (e.g. V. Grant, 1971, 1992).

Paterson (1978, 1980, 1981, 1982, 1985, 1986, 1988; also Lambert and Paterson, 1982, 1984) contrasts the traditional (Mayrian) 'isolation concept' of biological species with a 'recognition species concept', based on what he describes as Specific Mating Recognition Systems (SMRSs). SMRSs are those specific morphological and behavioural traits that allow an individual of a sexual species to differentiate between potential partners and organisms belonging to every other reproductive community. Paterson rejects Mayr's relational concept, according to which 'species are more unequivocally defined by their relation to non-conspecific populations ('isolation') than by the relation of conspecific individuals to each other' (Mayr, 1963, p.20). On the contrary, Paterson contends that each species is adequately defined, *per se*, by the unique set of recognition signals exchanged by males and females in their mutual 'recognition' behaviour.

The two species concepts (isolation concept and recognition concept) are better regarded as complementary, instead of alternative. This is Templeton's (1989) view, but he feels uneasy with both species concepts, because they do not apply to all living organisms. They fail, in particular, with those without (or with rare) sexuality, as well as with those with too frequent and extensive sexuality. Accordingly, he presents his own cohesion concept, regarding the species (Templeton, 1989, p.25) as 'the most inclusive group of organisms having the potential for genetic and/or demographic exchangeability'.

In the face of so many 'biological' species concepts, we may ask, with Mayr (1987): do biological species have reality in nature? I follow Mayr and several other students in answering: yes, despite the fact that many living organisms do not belong to species. In addition, I stress my view (see section 4.1.4), that the biological species concept is far from covering every facet of what biologists have been calling species. However, there are many species with Latin names which correspond to 'good' biological species.

The first samples of natural diversity directly investigated by Linnaeus were the living plants and animals from his homeland, Sweden. This is a limited area, within which there is seldom doubt as to the presence or absence of sterility barriers between two groups of individuals. In Linnaeus' works there is hardly a hint of the spatial or geographical dimensions of species. There is also no temporal dimension, because he dealt only with contemporary, Recent living beings. The context of Linnaean systematics is nondimensional (Mayr, 1987). This nondimensional framework is where the biological species concept works best. Therefore, it is not surprising that many Linnaean species correspond to Mayr's biological species.

The 'reality' of species has been stressed by Mayr and many others. They have identified many organizational and functional aspects of biological species that can better be appreciated by contrast with other 'kinds of species.' However, major problems arise with the biological species concept, when trying to cope with populations in spatial and temporal dimensions, as discussed later.

4.1.3 SPECIES AS CLASSES VERSUS SPECIES AS INDIVIDUALS

An evolutionary species concept was introduced by Simpson, and subsequently redefined by Wiley, Van Valen and others. Many authors have pointed out that the division into 'biological' versus 'evolutionary' species concepts is unfortunate, because both concepts (Mayr's and Simpson's) are, in a sense, biological and evolutionary, but there is no point in changing these terms. It is sufficient to note that the biological species concept envisages species as reproductive communities, whereas the evolutionary species concept is primarily ecological. Apparently, it also involves the time dimension, but there are difficulties with this point, as we will see soon.

Simpson (1961, p.153) defined the evolutionary species as 'a lineage (an ancestral–descendant sequence of populations) evolving separately from others and with its own unitary role and tendencies'. This definition has been revised slightly by Wiley (1981, p.25; cf. Wiley, 1978) in the following terms: 'An evolutionary species is a single lineage of ancestor–descendant populations which maintains its identity from other such lineages and which has its own evolutionary tendencies and historical fate'.

Evolutionary species as ecological entities have been better characterized by Van Valen (1976, p.333), who defines a species as 'a lineage or a closely-related set of lineages, which occupies an adaptive zone minimally different from that of any other lineage in its range and which evolves separately from all other lineages outside its range'.

These definitions might look attractive, but as long as they try to define species in terms of 'evolutionary roles', or 'niches', these definitions 'confound professions with organization' (Ghiselin, 1987, p.138). The evolutionary species concept regards the species taxon as a class of individuals sharing a common profession, a common role in the 'economy of nature'. In this respect, the evolutionary species concept is hardly different from that of taxonomic species: both treat species as classes. But a class cannot do what an individual can; it cannot change. Change within a class is restricted to the membership of individuals within it, whereas a species actually changes or evolves as a whole.

Therefore, in a sense, biological species are (in ontological terms) individuals, not classes. The biological individuals ascribed to a given species are not the members of a class, but parts of an integrated whole. The whole maintains its identity despite the turnover of its parts. The football team of my home town is always 'The Padua', despite the changing formations it displays on the playing field. A species can evolve, compete, speciate or become extinct. All of these phenomena are produced by integrated wholes (i.e. by individuals); classes cannot do anything at all. This view was first proposed by Ghiselin (1966a) and later developed by Ghiselin (1974, 1987, 1988b), Hull (1976, 1978), Mayr (1987) and others; see also Minelli (1988a).

In a widely cited paper entitled *A radical solution to the species problem* Ghiselin (1974, p.538) defined species as 'the most extensive units in the natural economy such that reproductive competition occurs among their parts'. This definition is adequate, but convoluted. Ghiselin subsequently (1987, pp.136–137) suggested several alternative definitions: 'Sexual populations, or reproductive communities, consist of populations and subpopulations, the largest of which are species, the smallest demes. We can, therefore, define 'species' [as a category] as the class whose members are the largest such units. And we do. Yet we can equally define 'species' by saying that species are by definition the populations that speciate. ... We have yet a third way of formulating what is fundamentally the same species concept. This is to say that species are those individuals that have to evolve independently of each other. For this to happen, it is a necessary condition that they form separate reproductive units, and a sufficient condition that they have speciated'. All of these definitions are within the domain of the biological species concept.

While agreeing that species are not classes, Mayr (1987, p.161) prefers to call them populations, instead of individuals, because the term 'population' has been traditionally used in biology, 'ever since the typological concept of natural kinds had been given up'. Moreover, 'the term conveys the impression of the multiplicity and composite nature of species'. This alternative view does not seem to have gained much favour. Indeed, there are many other traditional uses of the term population.

4.1.4 PHYLOGENETIC SPECIES CONCEPTS

I have already mentioned Nelson's (1989b) radical approach, closely related to the operational solution, to treat as 'species' (i.e. as the smallest operational unit within the cladogram) the smallest diagnosable chunks of biological diversity, as exemplified by Cracraft (1992) in his study of the birds-of-paradise.

Let me briefly review here a few additional species concepts, as discussed in the last decade under the common title of 'phylogenetic species concepts', or the like.

A much debated issue is monophyly. Does it apply to species in the same way as to supraspecific taxa? The issue is important, because species monophyly implies that a species cannot survive cladogenesis; otherwise, the hypothetical ancestral species surviving together with a new specifically differentiated branch would be paraphyletic, by definition. Paraphyletic species taxa are not a problem for Mayr (1987, pp.136–137), but things are different within a cladistic theoretical framework.

Species concepts based on monophyly were proposed by Mishler and Donoghue (1982), Cracraft (1983, 1987a), Ackery and Vane-Wright (1984), Donoghue (1985), Mishler and Brandon (1987) and McKitrick and Zink (1988). These authors argue that there is no basic difference between the species and the other taxa, some monophyletic entities simply being more inclusive than others. This is in agreement with what I have described as Nelson's nihilistic approach to the species problem.

Mishler and Brandon's (1987) phylogenetic species concept is more complex, in that it 'uses a (monistic) grouping criterion of monophyly in a cladistic sense, and a (pluralistic) ranking criterion based on those causal processes that are most important in producing and maintaining lineages in a particular case' (p.397). Such causal processes include interbreeding, selective constraints and developmental canalization. In this sense, their 'phylogenetic species concept' is wider than the traditional 'biological species concept'.

Following a similar line of reasoning, Van Valen (1988, p.53) advocated a 'polythetic species concept'. In Van Valen's terms (pp.51, 53), 'a population is a species to the extent that it has 'most' of the members of such a list of criteria as ... A single origin and final extinction, with reproductive (informational) continuity between these events ... Capacity to evolve ... A mechanism for recognizing other individuals or gametes of the same species ... Reproductive isolation from other species' and a dozen more criteria, collected from a century of literature on the species. Van Valen's suggestion looks like an extreme effort to capture a fleeting image, before it finally escapes from our hands.

Much more simply, Ridley's (1989, p.1) cladistic species concept 'defines species as the group of organisms between two speciation events, or between one speciation event and one extinction event, or (for living species) that are descended from a speciation event'. In Ridley's opinion, ecological and biological (reproductive) species concepts should be taken as sub-theories of the cladistic species concept. 'Taking the individuality of species seriously', he concludes, 'requires subordinating the biological, to the cladistic, species concept'. Things are, however, not so simple, because of the basic difficulties that arise when we try to reconcile monophyly and interbreeding, as cladistic and biological (i.e. reproductive) requirements for defining species (cf. de Queiroz and Donoghue, 1988).

According to Platnick (1977a), Willmann (1983) and Ax (1984), the concept of monophyly does not apply to the species level, but only to higher taxa. I do not see this restriction as a necessary one (see also de Queiroz and Donoghue, 1988).

De Queiroz and Donoghue (1988, 1990b) discuss at length the intrinsic difficulty of satisfying, with a single species concept, both the requirement of monophyly and that of reproductive cohesion, or interbreeding. They are not ready, however, to dispense with either, thus accepting what they would regard as a reductive species concept. They imagine (1988, p.335) 'several possible fates for the term 'species'.

One possibility is that it may become restricted to one of the classes of real biological entities, such as those resulting from interbreeding or those resulting from common descent. ... Realistically, the use of the term 'species' will be determined as much by historical and sociological factors as by logic and biological considerations. In any case, the entities deriving their existence from different natural processes are all valid objects of investigation'. I agree with them. Systematics needs increasingly sounder concepts and definitions, but we cannot quietly dispense with entire kinds of natural phenomena, as non-existent, simply because we cannot easily fit them within our comprehensive theoretical framework. For instance, Nelson's clear-cut nihilistic view of the species is coherently set within his anti-Darwinian attitude, not less than within a consequent pattern-cladistic approach to the diversity of living organisms. However, that is not enough, in my view, to embrace it as the final solution of the species problem. Nor do I have any better solution to offer.

4.1.5 DO UNIPARENTALLY REPRODUCING ORGANISMS BELONG TO SPECIES?

One of the main reason for Templeton's dissatisfaction with the biological species concept is that only sexually reproducing organisms may be part of such entities. According to White (1978), between 1000 and 10 000 animal taxa fall within the heterogeneous category of uniparentally reproducing organisms. Numbers are much higher for the plant kingdom, fungi, protists and prokaryotes.

Following Génermont (1980), we can identify five modes of uniparental reproduction in animals: asexual multiplication, self-fertilization, thelitoky, gynogenesis and hybridogenesis. In gynogenesis, there is no true fertilization, but the egg requires activation by a sperm cell. Examples are known in the planarian *Dugesia*, the freshwater oligochaete *Lumbricillus*, the fish *Poecilia formosa*, and some newts of *Ambystoma*. In hybridogenesis (a borderline case between uniparental and biparental reproduction), fertilization occurs, but the sperm cells are provided by males of a related bisexual form, whereas the hybridogenetic population proper only consists of females. The zygotes thus formed are diploid and develop into hybridogenetic females. In the latter, ovogenesis proceeds in a very peculiar way. The paternal chromosomes are discarded with the polar cells, so that the haploid egg cells only receive maternal chromosomes. The hybrid genotype complement is then restored at fertilization. This type of reproduction was first described in freshwater fishes of *Poeciliopsis* (Schultz, 1961). More recently, it has also been found in several populations of the European green frogs of the *Rana esculenta* complex (e.g. Tunner, 1970; Uzzell and Berger, 1975; Dubois, 1977; in this group, however, things seem to be more complex and geographically diverse than suggested by the earlier studies).

Three main origins of uniparental reproduction can be recognized: hybridization, polyploidy, or loss of sexual reproduction from within heterogonic cycles, where amphigony and thelytoky originally alternate. In this way, strictly thelytokous strains may develop, as in rotifers or cladocerans.

In many cases, for example the European green frogs (in Central Europe at least), it is still easy to identify the sexual form(s) involved in the parentage of a uniparentally reproducing animal. Sometimes, sexual and asexual forms are so similar that they are confused under a single species name. In other cases, the sexually reproducing species and the uniparentally reproducing strains are quite dissimilar and traditional taxonomy recognizes them as subspecies, or as distinct species. It should be borne in mind, however, that the biological species concept does not apply, by definition, to uniparental strains where hybrid fertility cannot be tested. In addition, there is often evidence of the polyphyletism of these uniparentally reproducing forms. The problem is certainly interesting but (cf. section 4.3) it is not crucial when deciding how to arrange these forms within the system of living organisms.

Mayr (1987) would place these organisms belonging to uniparentally reproducing forms in a separate type of taxon. He suggests paraspecies, but this name has already been used by Ackery and Vane-Wright (1984) for species lacking autapomorphies. Paraspecies *sensu* Mayr would probably fall within Simpson's concept of evolutionary species. They clearly represent classes, not individuals and are not 'true' biological species. In Ghiselin's (1984b, p.213) words, 'agamospecies' are like heaps of leaves that have fallen off the tree that gave rise to them'.

As so often happens with problems at the species level or below, the systematic arrangement of uniparentally reproducing forms is dealt with differently by botanists and zoologists. Facts may sometimes justify these differences, but tradition is also important. Wagner (1970, p.146) acknowledges that 'the main pillars of plant evolution are the populations of normal species – diploid, sexual and outbreeding. These basic stocks represent the diversity which has resulted from millions of years of phylogenetic change. ... Polyploids, apomicts, inbreeders, and hybrids are common today and we have no reason to believe that they did not occur even among early land plants. The fact, however, that normal species with normal life cycles still prevail today strongly indicates that such plants have always been responsible for the primary thrust of evolution. A kind of evolutionary noise is produced through the repeated origin of populations with deleteriously modified life cycles in which the advantages of diploidy, sexuality, outbreeding and species purity are counteracted or cancelled'. Some botanists give this 'evolutionary noise' detailed treatment, that some regard as fine taxonomic work, whereas others, such as Wagner, regard it as mere 'taxonomic noise'.

Consider a few examples. In the European flora, there are a few genera, where a handful of sexual species coexist together with a huge number of 'agamospecies'. In *Taraxacum* (dandelion) there are a few obligate sexual species, together with several facultative apomicts and some 2000 obligate apomicts (agamosperms). All of these forms have been named and are widely recognized as species, although the obligate apomicts are sometimes labelled as microspecies, or agamospecies. What most puzzles non-botanists is that a lot of agamospecies may grow together within a very restricted space. Sterck, Groenhart and Mooren (1983) recorded up to 19 'species' from areas of only 125 m² and 100 or more 'species' have been identified in a 1 ha area of wasteland (Richards, 1986). Assemblages of agamospecies are possibly subject to intense turnover, as might be guessed from the dynamic character of the communities where they most often occur. However, despite some previous suggestions (Janzen, 1977, 1979), it does not seem that these assemblages of different strains commonly evolve towards a prevalence of one genotype. As so often happens, matters are even more complex. For instance, there are some hints of a current trend towards re-sexualization of several apomictic lineages within *Taraxacum* (Doll, 1982).

Rubus is another genus, where a great many apomictic 'species' have been identified and named. In the Northeastern USA and adjacent areas, different authors have recognized between 24 and 381 species in this genus (Camp, 1951). In Europe, Weber (1981) observed that *Corylifolii* tend to arise sporadically from crosses with *R. caesius*, giving rise to some successful strains. Weber considers that a minimum population area of 20 km diameter is necessary before a 'species' can be acknowledged. This point of view can be regarded as operational, but what is its biological significance?

Bonik (1982) contends that conventional species concepts do not apply to diatoms. In one of the most commonly encountered groups of the genus *Nitzschia*, he finds 'roughly speaking, two different population structures definable in typological terms, i.e. a cosmopolitan type (which occurs only in common, but definite habitat types) and other rarer, endemic taxa, often confined to localities with extreme conditions ... Now, in so far as we assume ... a continuous gene flow within the whole group of lanceolate *Nitzschia*, we are confronted with a structure very close to a syngameon ... Therefore, cosmopolitan respectively endemic forms should be regarded as the result of different 'adaptive strategies', as they may result in large and ecologically heterogeneous populations.' (p.416, author's translation).

An unconventional investigation has been performed by Holman (1987), to assess whether a group of uniparentally reproducing animals consistently offers more taxonomic difficulties than a related group of bisexuals. Holman compared the taxonomy of bdelloid rotifers, with obligate parthenogenesis, with that of the monogonont rotifers, which occasionally reproduce bisexually. The degree of synonymy (number of available names for accepted genus or species taxon) is less in bdelloids than in monogononts, thus Holman contends that species in obligate parthenogens are not more difficult to recognize than in bisexual organisms. Holman also regards this finding as proof against the validity of a biological species concept. The problem remains, that Holman, together with rotifer specialists, is probably using the word 'species' in two different senses, when discussing bdelloids and monogononts. Another point is that outbreeding involves genetic exchange, hence more variability and fuzzier limits between putative taxa. Moreover, it is quite certain that bdelloid rotifers have been less extensively studied than monogononts, owing to their less attractive, uniform morphology, as well as to their preference for less frequently sampled habitats.

4.1.6 THE SPECIES IN PALAEONTOLOGY

According to Traverse (1988), the small green alga, *Botryococcus braunii*, has lived unchanged as a species, for 470 million years. Species recognized by palaeontologists are morphospecies (Hallam, 1988, p.134) hardly reconcilable with the current 'biological' species concepts.

'I have come to the conclusion that the concept of species is misleading, inappropriate and useless in professional paleontology. ... I hope to show that the species concept itself should be abandoned in professional paleontology ... because it hinders understanding of fossils' (Shaw, 1969, p.1085). These were the words of a palaeontologist who stated that a species name in palaeontology corresponds to nothing more than a class of objects sharing a certain degree of phenetic similarity. In a similar vein, although with more subdued wording, Reif (1984, p.282) stressed the view, that 'none of the proposed concepts leads to objective criteria for the diagnosis of evolutionary species, and that the fossil record is almost nowhere sufficient for the direct application of species-delimiting criteria'.

Despite these difficulties, most palaeontologists contend that recognizing and naming fossil species is justified. Several palaeontologists have also proposed various special concepts, such as chronospecies.

Willmann's (1985, 1987) species concept is close to Ridley's (1989) cladistic concept, as briefly summarized earlier. Willmann (1987, p.3) actually defines species as 'a section of an evolutionary lineage which is (objectively) delimited in time by two successive speciation events'. Accordingly, Willmann (1986, p.356) regards speciation as 'a relational term, as are 'species' ... and 'reproductive isolation''. This concept is one of the less arbitrary ways of defining species within fossils, but it is not easy to see how it works in practice.

A very interesting case documenting the problems with fossil morphospecies of sticklebacks (*Gasterosteus*) is given by Bell (1987). These primarily marine fishes have repeatedly invaded freshwater habitats, both rivers and lakes. According to Bell, invasion of lacustrine environments has been regularly accompanied by reduction of pelvic structures. On a coarser scale, pelvic-reduced sticklebacks seem to form a distinctive lineage, continuous through time. On a finer scale it is possible to appreciate that lacustrine populations are ephemeral; their characteristic phenotype goes systematically to extinction, only to be built anew in other, more recent lacustrine populations. Seemingly apomorphic traits reduce to an intricate pattern of closely packed homoplastic changes.

4.2 TAXONOMIC DIVERSITY WITHIN THE SPECIES

4.2.1 THE SO-CALLED 'BIOSYSTEMATIC CATEGORIES'

Everyone knows of cases where the boundaries between closely related species are very difficult to draw. Sometimes, two 'species' seem to be distinct in some regions where they meet, whereas they grade into each other in another part of their geographical range. Expressions, such as 'species *in statu nascendi*', are often used in the systematic literature, although seldom accompanied by a satisfactory definition or discussion. To accommodate these 'intermediate' entities, some authors have introduced non-conventional categories, to be placed just above or below what they regard as the 'true' species level. These categories are often called biosystematic categories, or taxonomic categories of 'evolutionary systematics'. Bernardi (1980) and Stace (1989) review the 'biosystematic categories' used by zoologists and botanists, respectively. For animals, they suggest the following categories.

(a) Groups of two or more 'good species', forming a monophyletic group, the mutual relationships of which offer some peculiarities deriving from the conditions in which they have speciated.

(a') Absence (or nearly so) of morphological differences between these species: sibling species (Mayr, 1942) = microspecies (Gates, 1951) = biospecies (Scheel, 1968; but the term biospecies had already been employed in a different sense by Cain (1954)).

(a") Closely related allopatric species corresponding to Mayr's (1940) semispecies = allospecies of Amadon (1966). Two or more semispecies form a superspecies *sensu* Mayr (1931), such as *Artenkreis sensu* Rensch (1928).

(a''') Genetic isolation between closely related species is not complete and irreversible: such is the case with the so-called species *in statu nascendi* (Blair, 1943) which, when grouped together, form the coenospecies (Turesson, 1922) = commiscuum (Danser, 1929) = coenogamodeme (Gilmour and Heslop-Harrison, 1954). It is interesting to note that the latter terms are all derived from the botanical literature, whereas zoologists have not felt the need to use special terms for these entities. It seems that many plant 'species', especially in the temperate floras, could be regarded as coenospecies, or parts thereof.

(b) Taxonomic levels between the species and the subspecies.

(b') Partial interfertility wherever the two forms live in sympatry: quasispecies (Schilder, 1962).

(b") Sterility barriers in some areas of sympatry, but no such barriers in other areas: vicespecies (Avinoff, 1913).

(c) Monophyletic groups of subspecies of the same species, morphologically and geographically isolated from other subspecies within the species: Formengruppe (Laubmann, 1921) = exerge (Verity, 1925) = citraspecies (Dujardin, 1965) = megasubspecies (Amadon and Short, 1976). To my knowledge, these terms have only been used for butterflies and birds.

Let me now briefly review the biosystematic categories recognized by botanists. The problem is more complex in botany than in zoology, partly because of greater flexibility in plant genetic systems, partly because the boundary between sexually reproducing populations and the widespread asexual clones is not always absolute (cf. Grant, 1958, 1981; Baker, 1959; Ehrendorfer, 1970; Briggs and Walters, 1984). Botanists (e.g. Briggs and Walters, 1984; Jeanmonod, 1984) are seldom satisfied with the biological species concept.

The main biosystematic categories recognized among sexually reproducing forms, as inferred by the results of artificial hybridization experiments, include: comparium, coenospecies, ecospecies and ecotype.

Members of two comparia cannot be crossed, because they are separated by genetic incompatibility barriers. A comparium comprises one or more coeno-species. Coenospecies belonging to the same comparium can be artificially crossed, but they produce highly or completely sterile F_1 hybrids. A coenospecies may comprise several ecospecies, which are only partially separated by sterility barriers; the F_1 hybrids are never completely sterile.

At a subspecific level, ecospecies may be subdivided into ecotypes. An ecotype was defined by Clausen, Keck and Hiesrey (1945) as formed by 'all the members of a species that are fitted to survive in a particular kind of environment within the total range of the species'. No sterility barrier exists between ecotypes of the same (eco)species.

Roughly speaking, both ecospecies and coenospecies may be equated with Mayr's biological species. However, as emphasized by Grant (1981, p.85), these and similar correlations (e.g. that between ecotype and local race) are 'imperfect enough in 'typical' groups and are more so in other groups'. This is because the biosystematic categories are defined only according to one of several genetic mechanisms effective in nature.

4.2.2 SUBSPECIES, RACES, VARIETIES

Intraspecific diversity is very conspicuous in some groups, because it affects readily observed traits, such as colour patterns, or the absolute or relative proportions of body parts. In other groups it is limited to subtle traits, particularly biochemical characters. In geographic terms, intraspecific variation is also quite diverse, sometimes continuous (clinal) sometimes less so. The latter case allows the easy identification of distinct ('racial') units.

Different opportunities to define the intraspecific differences in taxonomic terms derive from the fact that the individual populations of a given complex may represent more or less advanced stages in the speciation process. Also, our interest (more often economic than scientific) in identifying and naming forms obviously varies from group to group.

Linnaeus said that botanists were not interested in the subtler variations within species. In practice, however, Linnaeus sometimes distinguished and named varieties, in the same way as many before and after him. A more intentional, scientific approach to intraspecific variation is necessary both for understanding evolution and for abandoning the unsatisfactory typological attitudes of traditional taxonomy. For the systematist, the problems consist in identifying possibly different patterns within it and in providing a suitable nomenclature for the entities we recognize. Unfortunately, the job of naming 'varieties' had been conscientiously undertaken long before there had been any theoretical clarification of the subject. Biological nomenclature has been inundated with a huge number of names for varieties, races, morphs and aberrations. These have often been arranged into an exquisite hierarchy, to suggest both regularity and solid ontological status to these very subjective taxa. Progress in science has led to an increasing appreciation of diversity within species, as well as an understanding of the complex patterns of distribution.

Stuessy (1990, p.189) has tried to work out 'useful definitions of subspecies, variety, and form for practising plant taxonomists working with sexually reproducing angiosperm taxa', observing that 'Several criteria need to be utilized and these include morphological distinctness, geographical cohesiveness, and where known, genetic divergence, natural reproductive isolation, and degrees of fertility or sterility of natural hybrids'. Accordingly, his Table 12.1 summarizes the common use, but cannot provide clear-cut, definitive criteria and Stuessy is forced to conclude (p.193) that 'If no infraspecific classification has ever been proposed within a group, and the patterns of morpho-geographic variation leave a question as to whether subspecies or varieties should be recognized, then I favor use of subspecies as the initial category of choice. However, both categories can and should be used if judged helpful in a particular group'.

Hamilton and Reichards (1992) found that approximately 8% of the plant species discussed in 26 major journals and series published between 1987 and 1990 were subdivided into subspecific entities. Of these subdivided species, 42% were divided into subspecies, 52% into varieties, 3% into formae and the remaining 3% into infraspecific taxa of two or more levels.

For animals, the International Code of Zoological Nomenclature recognizes only subspecies, although the International Code of Botanical Nomenclature is more detailed, allowing names for subspecies, varieties, subvarieties, forms and sub-forms. Additional ranks are also provided by the International Code for the Nomenclature of Cultivated Plants (Brickell *et al.*, 1980).

The apparent clarity of the Codes, however, conceals many problems. For instance, to what extent is intraspecific diversity hierarchically arranged? What is a subspecies? How do we proceed when intraspecific variation is not hierarchical, as has frequently been recognized by many authors?

The subspecies concept was first defined within ornithological circles (Allen, 1877). The subjectivity of this category has been underlined by many students, although alternative concepts are often much worse. Böhme (1978, pp.264–265), for example, proposed to restrict the term to those cases where 'intra-specific evolutionary divergence can potentially give rise to new species'. Therefore, 'characters expressing these intra-specific divergences should be adaptive characters necessary for the evolutionary process'. The problem is, how can we forecast the future adaptive value of what is today an advantageous apomorphy developed within a population? Rather than following Böhme's proposal, I think that La Greca's (1987) plea for avoiding the use of subspecies should be supported.

4.3 HYBRIDS

Horse and donkey are 'good' biological species. In captivity, they hybridize, but mules and hinnies are sterile. Thus, no gene flow occurs between the two species. This is possibly the most often quoted example of the hybridity test for validating two species. It is still a good example, despite two instances of fertility reported recently in China in a mule and a hinny (the karyology of these animals was studied by Rong *et al.*, 1988). In many other cases, sterility barriers are not so clear-cut as those between horse and donkey. From our point of view, these cases are much more interesting. Moreover, in many cases it is not necessary to force two related species to hybridize, because they already do so in nature.

Plant hybrids are very common in nature. Many stable hybrid forms are described in floras. Knobloch (1972) summarizing the literature, was able to list nearly 3000 intergeneric crosses among a list of 23 675 intergeneric and interspecific hybrids. Things are much more intricate with artificial hybrids. For orchids alone there are over 70 000 hybrid 'grexes' derived from over 1100 natural species. A single grex may have up to 35 species in its parentage (Hunt, 1986).

There is an extensive bibliography on natural animal hybrids, aptly summarized by Remington (1968), White (1985), Hewitt (1988) and others. In a survey of natural hybridization in birds, Grant and Grant (1992) report that 9.5% of bird species are known to have bred in nature with another species and produced hybrid offspring. The percentage is lower (8.0%) for the passerine birds, but much higher for other groups (e.g. 21.5% for galliforms (game birds) and 41.6% for anseriforms (ducks and geese)).

Hewitt (1988, p.158) defines a hybrid zone as 'a cline or set of clines between two parapatric hybridizing taxa for genes and the distinguishing characters they determine. In principle, these clines may originate in two ways: primarily or secondarily'. There is a secondary hybrid zone when the hybridizing populations have been separated for some time before making new contact. An interesting feature of hybrid zones is their narrow width, but a given hybrid zone might be narrower for some genes than for others. In a hybrid zone involving *Gryllus pennsylvanicus* and *G. firmus*, patterns of variation in morphology, enzyme genotypes and mtDNA genotypes are not associated (Harrison, 1986; Harrison, Rand and Wheeler, 1987). Other well-documented cases include a hybrid zone between two races in *Chorthippus* (Orthoptera) (Barton and Hewitt, 1989) and another between two chromosomal races of the house mouse (Searle, 1991). A non-concordance among electrophoretically investigated loci was also reported for a hybrid zone between races of the newt *Ambystoma tigrinum* in central New Mexico (Jones and Collins, 1992).

Hybridization raises interesting questions as to the topology of the phylogenetic tree. Species taxa can merge, thus giving rise to anastomoses between the branches of the tree. Eigen, Winkler-Oswatisch and Dress (1988) and Maynard Smith (1989) have explored some possible geometrical methods for analysing comparative data, when we cannot confidently presuppose a branching topology. Maynard Smith wonders whether in prokaryotes there is the possibility of distant gene transfer, or extensive hybridization between distantly related 'species'. This might suggest a network as a better image of evolutionary relationships, than the conventional branching trees (Maynard Smith, 1990). De Queiroz and Donoghue (1988) have stressed the inapplicability of cladistic analysis to living organisms that are not related according to a branching hierarchical pattern.

4.4 SPECIATION

In so far as we can identify the 'species' as a core concept (or, perhaps more accurately, as a cluster of core concepts) in systematics, it is difficult to avoid a short discussion of speciation. However, that does not strictly imply a fundamental relationship between species concepts and speciation processes (Chandler and Gromko, 1989).

There are many different attitudes towards speciation. Some authors even deny the existence of speciation as a distinct natural phenomenon. That is Nelson's (1989b) opinion, clearly related to his nihilistic species concept. It is also Bock's opinion, because, in strict terms, nondimensional biological species 'cannot be born, and they are ageless as they cannot be dated from any fixed time in the past' (Bock, 1986, p.31). For other biologists, every speciation event has its individuality, being the outcome of a unique mixture of different ingredients, such as geographical context, and the number and type of genetic structures involved (Capanna, 1982). For others, there is a very limited range of fundamentally different modes of speciation, but there is no general agreement as to their number, or to the relative importance of different factors for distinguishing them. For a few researchers, all speciation events involving sexually reproducing organisms are fundamentally the same, and the differences among the individual cases concern only marginal issues.

This array of attitudes partly reflects the uneven attention paid to recognizing several different and partly independent aspects of speciation. Powell's (1982) factorial analysis distinguishes the ecological context within which the speciation process develops, the genetic changes associated with it, the causes and the effects of speciation. Other useful reviews of speciation mechanisms are provided by Bush (1975a), Endler (1977) and Barton and Charlesworth (1984).

4.4.1 RATE OF THE SPECIATION PROCESS

Two speciation modes can be distinguished when considering the rate of the process: conventional, gradual speciation occurring in the course of a long series of generations, and instantaneous speciation. In animals, instantaneous speciation might be the result of a shift to obligate thelytoky, or the sudden appearance of hybridogenesis (Mayr, 1982a). In plants, the same outcome may result from allopolyploidy: from an interspecific cross (hybridization), followed by genome duplication, a polyploid genome such as AABB arises combining the diploid genomes AA and BB of the two parental species.

In population genetic terms, Templeton (1982, p.107) similarly distinguishes between the slow speciation by divergence and the quick speciation by transilience: 'Under divergence, the isolating barriers evolve in a continuous fashion (but not necessarily slowly or uniformly over long periods of time), with some form of natural selection, either directly or indirectly, being the driving force leading to reproductive isolation. Transilience modes involve a discontinuity in which some sort of selective barrier is overcome by other evolutionary forces'.

4.4.2 GEOGRAPHICAL PATTERNS OF SPECIATION

White (e.g. 1978, 1982) and other authors have emphasized a contrast between geographical and non-geographical (particularly chromosomal) types of speciation. White's contrast is often regarded as unfortunate (e.g. Mayr, 1982a), because it confounds spatial or spatio-temporal dimensions of speciation with the types of genetic or chromosomal mechanisms involved in its causation.

Many details have to be determined before we can identify a given speciation event as allopatric, parapatric, or sympatric, because the geographic relationships between the diverging populations mostly change with time. In classifying speciation events according to their geographic pattern, we should strictly take into consideration the ranges inhabited by the diverging populations at the time they become reproductively isolated.

The geographical modes of speciation can be classified as allopatric, peripatric, parapatric and sympatric speciation.

Allopatric speciation (Mayr, 1942) occurs when reproductive isolation is reached while the diverging populations occupy distinct ranges. For example when they are isolated by geographic barriers, such as islands, or in different caves. This is the classic Darwinian scenario of speciation by geographic isolation.

Peripatric speciation (Mayr, 1982a) is speciation in peripheral founder populations (Mayr, 1954). This mode of speciation corresponds to quantum speciation *sensu* Grant (1963, 1981), but not in the sense of White (1978) and Ayala and Valentine (1979). This type of speciation has been supported by Carson's studies on Hawaiian *Drosophila* (Carson, 1971, 1975).

Many authors contend that allo- or peripatric models also apply to many speciation events within a single land mass, even in the absence of major oro- or hydrographic barriers. The continuity of a given habitat type over very large distances, as in the case of the Amazonian forest, cannot be guaranteed for long time spans. Seemingly continuous habitats are divided into small or large fragments, which act as refugia and centres of divergence. Later, as conditions change they provide a source for bursts of speciation (e.g. Haffer, 1969, 1974; Renno *et al.*, 1990). It is fair to say, however, that this model has been criticized by Connor (1986) and others.

Parapatric speciation (Smith, 1965, 1969) is the disruption of genetic flow at a given geographical point, along a cline of geographical variation within a species. According to White (1978) this is a type of *in situ* speciation, whereas Mayr (1982a) only recognizes post-speciation parapatric distribution patterns.

Sympatric speciation (Mayr, 1942, 1963) is the most hotly disputed geographical mode of speciation. It is not the same as Ford's (1964) parapatric speciation. Maynard Smith (1966) demonstrated the theoretical possibility of sympatric speciation, but his models apparently postulated many requirements unlikely to be fulfilled in nature. These theoretical requirements have been relaxed in later models (e.g. Lande and Kirkpatrick, 1988). The occurrence of sympatric speciation has been advocated for several different examples. First, to explain the simultaneous occurrence of a sometimes very large number of co-generic species (a species-flock) on the same oceanic island, or in a single lake. Species-flocks are considered to have originated from one species, possibly the first immigrant. There are many examples including the 300 or so species of *Haplochromis* (cichlid fishes) in Lake Victoria (Greenwood, 1974; Echelle and Kornfield, 1984; Barel, 1986); the 67 species of weevils of *Miocalles* on the tiny island of Rapa (Paulay, 1985); and some millipede species-flocks in Macaronesia, including some 30 species of *Cylindroiulus* on Madeira (Enghoff, 1982, 1983, 1992). The monophyletic origin of Lake Victoria cichlids has been confirmed by molecular data: as for mitochondrial DNA, their overall variation being less than within the human species, but virtually no mtDNA genotype is shared among species (Meyer *et al.*, 1990). Strict sympatric speciation is not the only possible explanation for these exceptional patterns. For example, periodic fragmentation of land (and especially forest) areas by lava flows has probably fragmented previously continuous populations on volcanic islands, such as the Hawaii or Rapa. Moreover, many of the isolated populations were probably reduced to very small numbers of individuals, thus favouring genetic changes via founder effects. Barton and Charlesworth (1984) give a critical review of this mode of speciation. Several other forms of micro-allopatry (including the subdivision of the major basin into several partially separated basins and spawning site isolation) have been advocated by Mayr (1984) to explain the intralacustrine speciation underlying *Haplochromis*-type species-flocks.

Things are different for some host-specific phytophagous insects, where the case for sympatric speciation has been proposed with even greater force. The best investigated examples are those of some fruit flies of the genus *Rhagoletis* (Diptera Tephritidae), and the small ermine moths (*Yponomeuta padellus* group) (Lepidoptera Yponomeutidae). The *Rhagoletis pomonella* group comprises three widely sympatric forms, often regarded as sibling species. *Rhagoletis pomonella* s. str. normally attacks hawthorns but, since the 1860s, it has also colonized cultivated apples in north-eastern USA. Today, three widely sympatric forms occur, one developing on hawthorn (*Crataegus* spp.), the others attacking apple (*Malus pumila*) and dogwood (*Cornus florida*). Bush (1969, 1975a,b) interprets this host shift as the new formation of host races which will eventually become host-specific species. In the absence of courtship signals, reproductive isolation is maintained by the flies mating only on the host fruits. Generally, females mate and oviposit on the fruit of the host from which they emerged.

Feder, Chilcote and Bush (1988) reported finding genetic differentiation between co-occurring hawthorn and apple populations of *Rhagoletis* at a field site in Michigan. They regard the finding as support for sympatric differentiation of 'host races'. McPheron, Smith and Berlocher (1988) report significant differences in allele frequencies between sympatric fly populations reared from apple and hawthorn trees from another field site in Illinois. Smith (1988) reports heritable differences in seasonal emergence of flies from three sympatric populations living on apple, hawthorn and dogwood in Illinois. These findings seem to overcome the difficulties raised by many against the sympatric speciation hypothesis of *R. pomonella* (Futuyma and Mayer, 1980; Jaenicke, 1981; Butlin, 1987b). What remains to be seen is whether these host-specific subpopulations will turn out to be completely isolated species. Feder and Bush (1991, p.249) regard the apple and hawthorn populations of *R. pomonella* as two partially reproductively isolated and genetically differentiated host races: a result consistent (in their opinion) with predictions of sympatric speciation models. However, the geographical pattern of allozyme variation for these flies is complex because inter-host differences are superimposed on latitudinal allele frequency clines within the races. In addition, there are pronounced allele frequency shifts among *R. pomonella* populations across three major ecological transition zones in the mid-western USA.

The case for sympatric speciation may be better established for other phytophagous insects, living on hawthorn and apple, such as the small ermine moths. Thorpe (1929) observed that populations of what was at the time regarded as one species, *Yponomeuta padellus*, reared on apple or on hawthorn, showed distinct preferences for the plant species from which they emerged. This preference was exhibited both in larval feeding and in the choice of oviposition sites. Laboratory crosses demonstrated that the two forms were able to hybridize, so Thorpe (1929, 1930) concluded that they were simply host races of the same species. However, the two species do not hybridize in nature. Allozyme studies by Menken (1980) demonstrated that the two forms were sibling species, currently known as *Y. padellus* and *Y. malinellus*. The whole complex is currently regarded as a complex of sibling species and host races, more or less clearly separated by species-specific pheromones produced by the females, partly under the control of plant secondary metabolites.

In the case of the treehoppers of the *Enchenopa binotata* group, genetic differences between sympatric subpopulations living on different hosts demonstrated the existence of fully isolated species (Wood and Guttman, 1983; Wood, 1987; Claridge, 1988).

A similar case is that of a leafhopper, *Oncopsis flavicollis*. Populations known under this name live on two species of birch (*Betula pendula* and *B. pubescens*) and have often been regarded as 'host races'. However, they differ not only in host preference, but also in karyotype and in mating calls. According to Claridge and Nixon (1986), at least three biological species are recognizable within this complex in Britain alone. The new evidence clearly suggests that sympatric speciation occurs, but the alternative hypothesis of allopatric divergence cannot be easily discounted without further evidence.

Studies by Via (1984a,b, 1986) and Futuyma and Philippi (1987) have demonstrated the genetic background behind the most important components of larval or adult behaviour of phytophagous insects with respect to their host plants. However, alleles that increase the performance of a given insect on a given host plant do not necessarily reduce the performance on other hosts (Butlin, 1987b) as required by Bush's (1975a) genetic model for sympatric speciation. Claridge (1988) regards the findings concerning leafhoppers and planthoppers as consistent with models of allopatric speciation involving sexual selection. Possibly more convincing is a sympatric scenario for speciation in parthenogenetic clones of aphids with alternative host preferences, as advocated, for instance, by Müller (1985). Further evidence, however, is badly needed.

Before concluding this short account of the geographical aspects of speciation, brief reference should be made to the recurring contention that some species have a polytopic origin (i.e. that they have more than one origin). Polyphyly is possibly less heterodox than it may seem at first sight, in so far as polytopy is advocated (Van Valen, 1988, p.57) in cases of speciation by polyploidy or hybridization. Two allotetraploid ferns (*Asplenium bradleyi* and *A. pinnatifidum*) are polymorphic at the same loci as their diploid parents: this indicates that each of them has originated more than once (Werth, Guttman and Esbaugh, 1985). The multiple origin of several polyploid plants has also been confirmed by molecular evidence. A review by Soltis, Doyle and Soltis (1992a) lists two tetraploid autopolyploids (*Heuchera micrantha* and *H. grossulariifolia*) and several allopolyploids, one hexaploid

(*Draba lactea*) and the remaining tetraploids (*Brassica napus, Glycina tabacina* race BBB₂B₂, *G. tomentella* T1 race, *Solanum tuberosum* ssp. *andigena, Tragopogon mirus, T. miscellus* and *Aegilops triuncialis*). The list is all but exhaustive, as demonstrated by the elegant studies of the recurrent formation of polyploid taxa in Nordic representatives of the genus *Draba* (Brassicaceae) (Brochmann, Soltis and Soltis, 1992a,b). These studies, mostly based on cpDNA evidence, suggest that multiple origins of polyploids are perhaps the rule rather than the exception.

4.4.3 SELECTION AND SPECIATION

In his classic book *Genetics and the Origin of Species*, Dobzhansky (1951) suggested that selection for assortative mating within each race would reinforce pre-mating isolation mechanisms, but this view has been severely criticized (e.g. Barton and Hewitt, 1981a; Nevo and Capranica, 1985). Contrary to a traditional view still defended by Otte (1989) and Coyne and Orr (1989), Butlin (1987a,b,c) contends that there is little, if any, evidence for the occurrence of reinforcement. Moreover, low fitness of hybrids is usually determined by the combined small effects of many genes, whereas theorists such as Maynard Smith (1966), Felsenstein (1981b) and Caisse and Antonovics (1978) have demonstrated that reinforcement occurs only if there is 'strong selection against the heterozygote for the fitness gene, strong assortative mating produced by the mating locus and tight linkage between them' (Hewitt, 1988, p.164).

These reinvestigations of the problem suggest that there is no special selection for isolation. In other words, speciation is the byproduct of genetic divergence of isolated populations, be it adaptive in itself, or the outcome of genetic drift, conforming to what Futuyma (1989) called a Fisherian and a Wrightian scenario. Coyne (1992) has summarized the scanty factual evidence in favour of the origin of reproductive isolation as a simple byproduct of divergent adaptation. He reports two laboratory studies and one example from nature, concerning the monkey flower *Mimulus guttatus* (Macnair and Christie, 1983; Mcnair, 1989).

Both random events and selection processes are necessary ingredients of Carson's (1971) flush-and-crash model of speciation. Carson (1982) attempts a 'reductionist' approach, by exploring 'to what extent genetic events common to any and all modes can be identified' (p.411). His model is intended to apply to all sexual, diploid species and, considered in terms of previous examples (Carson, 1971, 1973), it develops as follows: 'The crux of the argument will be that the formation of a new species from an older one occurs only following two events that must occur successively. First, the old polygenically balanced gene pool, or segments of it, suffer disorganization by forces that are essentially stochastic. This can occur in a relatively small number of generations. Following this, the concept calls for reorganization of a new balanced polygenic system over hundreds or thousands of generations under natural stabilizing selection. This 'organization theory' states that speciation requires the development of a new and major organization of polygenic balances' (Carson, 1982, p.412). Carson (1985, 1989) develops his model further, explicitly calling for the unification of speciation theory in diploid sexual plants and animals. In his view, a unified theory is possible: 'if zoologists would abandon insistence on reproductive isolation as a prime species criterion and substitute the notion of a positive fitness system under stabilizing selection'. These efforts are probably remote from the immediate interests of the working systematist. I could not ignore them here, however, as they are one of the soundest theoretical approaches to the whole subject of speciation.

4.4.4 GENETIC AND PHENOTYPIC CHANGES DURING SPECIATION

It is increasingly clear that the genetic changes occurring during speciation are the most diverse. With the widespread use of electrophoresis to measure genetic distances between populations, it has been customary to compare the mean genetic distances between local populations with those between subspecies, semispecies or full species of a given group. For example, in *Drosophila*, Ayala and Kiger (1980) gave a mean value of D (genetic distance) between local populations of 0.013, whereas the values rise to 0.163 for comparisons between subspecies, to 0.239 for semispecies and to 1.066 for species. D behaves similarly in other groups, but with some quantitative differences. The same authors gave D equal to 0.058 as the mean difference between local populations of mammals and D equal to 0.559 as the mean difference between species. For several examples it seems that a value of D between 0.2 and 1.0 might estimate the average amount of genetic differentiation associated with speciation. However, these figures do not hold in cases of rapid or 'saltational' speciation (quantum speciation *sensu* Ayala), where the average value of D is approximately 0.050 (Ayala, 1982).

The results of electrophoretic studies are possibly misleading. I do not intend to discuss the limitations of electrophoretic methods here. How many of the sampled alleles represent entire genomes? How much does one or a few population samples analysed electrophoretically for each species represent the overall variability within a species? What matters most here is not a problem of sampling or statistical technique but a conceptual problem. When we measure the genetic distance between two closely related species, we only measure how different their genomes are today. How can we distinguish between the amount of genetic divergence accumulated during speciation from the divergence derived by evolution of the two species after becoming isolated? Seen from this perspective, it is clear that the only data which can be used with some confidence are those concerning the genetic distances between populations not yet fully isolated. They can provide an indirect estimate of genetic divergence during speciation. Moreover, in so far as speciation develops under selective pressures, we cannot discount Carson's (1982, p.429) observation that 'electrophoretic similarity between species is less a measure of overall similarity than it is of the irrelevancy of this type of variation to adaptive evolution'. Perhaps electrophoretic data help us read the time elapsed since the divergence of two gene pools on the molecular clock, but they do not help in understanding how many genetic differences were involved in the speciation process.

Moreover, we seldom remember that usually polymorphism pre-dates speciation and the polymorphic condition of the ancestor is passed on to its descendants. A few studies have called this important circumstance to our attention. Hasegawa, Kishino and Yano (1987) suggested that polymorphism in mitochondrial DNA antedate speciation events in the group comprising apes and man. Figueroa, Günther and Klein (1988) have provided more compelling evidence by studying polymorphisms of the major histocompatibility complex (MHC) loci. According to these authors, some allelic differences occurring today within, say, laboratory mice or laboratory rats must have arisen before the separation of the two genera (*Mus* and *Rattus*) from a common ancestor, many million years ago. A similar study has been carried out by Lawlor *et al.* (1988) in humans and chimpanzees. Again, they hypothesized that the substantial proportion of contemporary polymorphisms for two systems (HLA-A and B) was similar in both species and already existed in the common ancestor of humans and chimpanzees. However, their arguments do not seem to be stringent, in phylogenetic terms. Further evidence has been gathered by Coyne and Kreitman (1986) and by Solignac and Monnerot (1986), while studying three closely related species of *Drosophila*: *D. simulans, D. sechellia* and *D. mauritiana*. Coyne and Kreitman demonstrated continuity of polymorphisms in the DNA region around the alcohol dehydrogenase (Adh) gene across the different species. Solignac and Monnerot obtained similar data from mtDNA studies. The persistence of polymorphisms across species implies that the founder populations were not small (Brookfield, 1986). What matters more in our context is that this may cause serious problems in phylogenetic inference (Figure 4.1).

In this context, it is worth citing Roth's (1991, p.189) remark, that 'If we wish to trace the phylogeny of population events, perhaps we should use population-level, not individual organism- level characters'. A population-level apomorphy would be the fixation of a trait x, rather than the presence of trait x, itself an apomorphy at the level of the individual organisms.

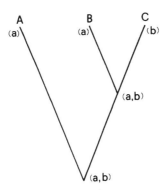

Figure 4.1 Pitfalls of phylogenetic reconstructions caused by polymorphism 'crossing' the speciation barrier. In the figure, redrawn after Brookfield (1986), a polymorphic ancestor, with two alternative states, a and b, gives rise to a monomorphic offshoot A (with a) and to a polymorphic one, which later splits into two daughter species B and C, respectively fixed for a and b. Despite the common possession of a, A and B are not cladistically closer related than are B and C.

Regarding relationships between genotypic and phenotypic changes, nobody today expects that such relationships are definite. Sibling species are known in most groups in which adequate studies have been performed. Major morphological differences which easily differentiate between species of the same genus are sometimes controlled by one or two genes only. Gottlieb (1984) quotes many examples, such as alternate versus opposite leaves (*Ulmus glabra* versus *U. montana*); lobed versus entire leaf margins (*Lactuca graminifolia* versus *L. canadensis*), presence or absence of floral spurs (*Aquilegia* species groups). Generally speaking, Gottlieb (1984, p.704) has shown that in flowering plants 'differences in structure, shape, orientation, and presence versus absence appear to be governed by one or two genes. Differences in dimensions, weight and number usually exhibit continuous variation and are influenced by numerous genes, though many of them probably act only indirectly via general effects at the whole organ or whole plant levels'.

4.4.5 CAUSES AND MECHANISMS OF SPECIATION

This is possibly the most intricate study area in speciation, not only because of the obvious difficulties arising as we proceed from description to causal analysis of natural phenomena, but also because of conceptual problems. For instance, it could be suggested that genetic changes are partly the cause and partly the effects of speciation. Powell's (1982, p.68) detailed but non-exhaustive list of causes of speciation includes selection, recombination, genetic drift, mutation, migration, a combination of all the above mechanisms and symbiosis. I shall confine my discussion to a few unusual situations which, in addition to their general interest for evolutionary biology, are also of interest to the systematist. The reader is referred to Dover (1982a,b), Dover *et al.* (1982) and Ish-Horowicz (1982) for a discussion of the speculative model of speciation by molecular drive.

(a) Chromosomal speciation

In a series of widely known works White (1968, 1978) developed non-allopatric models of chromosomal speciation. He has distinguished between two modes of invasive and stasipatric chromosomal speciation (White, 1982, 1985).

According to White (1982, p.83), there is invasive chromosomal speciation 'when a chromosomal rearrangement establishes itself in a peripheral population of a species that is then able to invade an area previously unoccupied by the species'. As examples he referred to the South American spiny rats of the genus *Proechimys* studied by Reig *et al.* (1980; Benado *et al.*, 1979) and the subterranean mole rats of *Spalax* (for a review see Nevo (1982)).

Stasipatric speciation (White, 1982, p.84; but cf. White, 1968, 1978) 'results from chromosomal rearrangements establishing themselves within the range of a species and spreading out from the point of origin'. White characterizes this mode of speciation as 'internal', giving rise to derivative species with a geographical distribution within that of the ancestral species. The classic examples are the Australian wingless grasshoppers of the *Vandiemenella (Warramaba) viatica* group. Several authors (e.g. Mayr, 1982a; Grant, 1985; Hewitt, 1988) object to White's stasipatric model. I follow Hewitt's (1988, p.164) remark that the greatest difficulty is perhaps 'the apparent immobility of the hybrid zones, whether they enclose a small bubble of a new mutant or a widely distributed race. The initial fixation of negatively herotic mutants is also a problem, although recent theoretical work suggests how this may occur in a continuous population as well as in a small one ...'.

(b) Hybrid speciation

Hybrid speciation, 'the origin of a new species directly from a natural hybrid' (Grant, 1981, p.242), has been widely recognized as a mode of speciation in plants, where many instances are known and well documented. The existence of hybrid speciation in the animal world has long been denied, but a few well-documented examples are now known (Bullini, 1985). Hybrid speciation cannot occur, whenever mechanisms exist for preventing segregation of parental characters in the F1 generation or subsequent generations. Grant (1981) identifies seven ways of stabilizing hybrid reproduction: vegetative propagation; agamospermy; permanent translocation heterozygosity; permanent odd polyploidy; amphiploidy (allopolyploidy); recombinational speciation; and segregation of new types isolated by external barriers. Vegetative propagation and agamospermy (collectively known as apomixis) bring us back to the problems of uniparentally reproducing organisms already considered in section 4.1.5.

Permanent translocation heterozygosity is a complex situation, best known in *Oenothera* (Onagraceae), although other examples occur in genera of the same family and elsewhere (e.g. *Rhoeo*, Commelinaceae). In these forms, extensive translocations of chromosome arms have occurred in most or even all the chromosomes. Consequently, most or all of the chromosomes form an entirely closed ring at meiosis. These stabilized hybrids are formally diploids and their genomes are currently described as comprised of two haploid genomes known as Renner complexes. Generally speaking, one Renner complex is transmitted through the pollen line, while the other is transmitted through the ovule line. Such unilateral transmission is assured by a system of balanced lethals. Consequently, no homozygotes can form and the plants always breed true.

In other plants, permanent 'odd' polyploids (e.g. 5n) are maintained by sexual means. One haploid genome is transmitted through the pollen and the remaining ones (4n in the pentaploids) are transmitted through the ovule. These sexually reproducing, permanent odd polyploids form a heterogamic complex, according to Grant (1981), articulated in heterogamic microspecies. One example is the *Rosa canina* complex, the microspecies of which are mostly pentaploid, although some tetraploids and hexaploids also occur.

Bullini (1985) has extensively reviewed the increasing evidence of hybrid speciation in such diverse animal groups as grasshoppers (*Warramaba virgo* complex), stick-insects (*Bacillus*), weevils (*Otiorrhynchus*), planthoppers (*Muellerianella*), freshwater snails (*Bulimus*), planarians (*Dugesia*), fishes (*Poecilia, Poeciliopsis,*

Menidia, Carassius), newts (*Ambystoma*), frogs (*Rana esculenta* complex) and lizards (*Heteronotia, Lacerta, Cnemidophorus*). For the unisexual vertebrates see Dawley and Bogart (1989). Vrijenhoek *et al.* (1989) list 51 biotypes of clonally reproducing vertebrates, distributed among 20 genera and 12 families.

It would be tempting to disregard these unisexual forms as short-living experiments with no future. Widespread parthenogenetic clones of recent origin have been demonstrated, such as the Australian gecko *Heteronotia binoei* (Moritz, 1991), but some clonal lineages in *Ambystoma* are possibly 4–5 million years old (Hedges, Bogart and Maxson, 1992; Spolsky, Phillips and Uzzell, 1992).

(c) Symbiotic speciation

This unusual mode of speciation has recently been brought to general attention, although it was first suggested by Kitzmiller and Laven in 1958, to explain the puzzling evidence obtained by Laven (1958) in his crosses between strains of the common mosquito, *Culex pipiens*. Laven had observed that frequent incompatibility, sometimes reciprocal, sometimes non-reciprocal, between strains of different geographical origins was caused by differences in the egg cytoplasm. Fifteen years later, Yen and Barr (1973) demonstrated that this incompatibility was caused by a rickettsia-like bacterium (*Wolbachia pipientis*). Powell (1982, p.71) said: 'Presumably the mechanism involved here is some kind of 'proto-immune' response. If sperm produced in a male infected with a particular strain of *Wolbachia* enter the cytoplasm of an egg infected with the same strain of *Wolbachia*, fertilization proceeds normally. If the sperm enter the cytoplasm of an uninfected egg or an egg infected with a different strain of *Wolbachia*, the egg cytoplasm responds by killing the sperm before they affect fertilization'. Incompatibility has also been demonstrated among sympatric strains (Barr, 1980).

Parasitic or mutualistic microbial symbionts have also been proposed as the direct cause of reproductive isolation in several other examples, such as: mosquitoes (genus *Anopheles*), some *Drosophila* groups (*D. paulistorum, D. pseudoobscura, D. simulans*), the weevil *Hypera postica*, the flour beetle *Tribolium confusum*, the moth *Ephestia cantella*, the embiid (lower insect) *Haploembia solieri* and the monkeys *Macaca mulatta* and *M. fascicularis*. In several instances, the microbial symbiont has not been identified. For *Haploembia* a gregarine protozoan has been suggested as the causal agent and a *Plasmodium* for the monkeys.

There are two possible mechanisms of symbiotic speciation, in addition to that identified by Powell (1982). One possibility is a selective detrimental effect of the parasite on some genotypes, with selection favouring assortative mating among those genotypes, as advocated by Wheatley (1980) and Steiner (1981). Alternatively, as suggested by Stefani (1956, 1960) for *Haploembia*, the parasite may cause widespread sterility in males of some populations, thus favouring the origin of a parthenogenetic sibling species (cf. White, 1978). Symbiotic speciation has been reviewed by Thompson (1987), who underlines the need for a theoretical refinement of these models and for more substantial empirical evidence to support them.

Resources and media 5

Nomina respondeant Methodo Systematicae; sint itaque: Nomina Classium, Ordinum, Generum, Specierum, Varietatum; Character. Classium, Ordinum, Generum, Specierum, Varietatum; Differentiis definita, nam nomina nosce oportet qui rem scire velit, confusis enim nominibus omnia confundi necesse est.

C. Linnaeus (1758, pp.7–8)

Names are needed according to the systematic method, as names for classes, orders, genera, species, varieties; according to the characters of the classes, orders, genera, species and varieties; defined by differences, because whoever wishes to know the things needs to know the names; in fact, if names are confused, everything, by necessity, is confused.

5.1 HUMAN RESOURCES

According to Wilson (1988b, p.14), some 4000 systematists are currently working, full or part time, in North America. A larger figure (8000–10 000) is given by Gaston and May (1992), following an estimate provided in 1985 by the US National Science Foundation. By comparison, Gaston and May say 588 taxonomists are working in Australia today. I estimate that there are 20 000–30 000 systematists working all over the world today. It is difficult to obtain a more accurate figure, because of the different people who can be considered worthy of the title of systematist. The figures given include both professionals and amateurs.

In the present state of systematic research, these human resources are too scanty. The biggest problem is one of organization and division of labour. There is an enormous bias towards a few taxonomic groups. A large percentage of taxonomists are amateurs who generally do not collect and study little known invertebrates, but prefer showy insects, such as butterflies, dragonflies and beetles. This bias towards 'beautiful' organisms is also felt within the better-studied insect orders, with the preference given to tiger-beetles or stag-beetles compared with other families of Coleoptera. Strong biases towards particular groups are also widespread among professional systematists, although their preferences are not of an aesthetic nature.

The geographical distribution of systematists is also uneven. This problem is partly overcome by the existence of a few major collections, where many groups can be studied on a worldwide basis. However, no museum collection can supplement the wealth of information obtained by specialists collecting or working in the field, particularly when their work continues in the same region for years.

Kevan (1973, p.1213) estimated that 'the satisfactory initial identification of undescribed insects alone would require, at a conservative estimate, 75 000 man-years to complete'. I agree with his subsequent affirmation that larger human resources could be made available, but I think that Kevan has largely under-estimated the efforts required to achieve his descriptive goal. Kevan says that such a descriptive task is not undertaken, because most of the working time of practising taxonomists is spent on routine identification of specimens belonging to known species. I believe that matters are more complicated. It may be true that routine identification of specimens does not lead to the discovery of new species, but it is also true that new species are recognized through continuous scrutiny of large series, where specimens belonging to both named and unnamed species are mixed. Preliminary sampling of large unidentified collections may lead the specialist to concentrate efforts on the more promising items, but a careful scrutiny of long series, where known species represent a substantial percentage of the whole, is difficult to avoid. Moreover, the search for new species is obviously subject to the law of decreasing returns: the more new species we recognize in a collection (or in a given fauna, or flora), the more time-consuming it becomes to find new ones. Therefore, the human resources needed to recognize, describe and study adequately all existing species are far larger than those estimated by Kevan. Perhaps, better estimates can be obtained by drawing a curve, for a given taxonomic group, to estimate the number of existing species in terms of the rate of description of new species. The reliability of these analyses, however, is not great (cf. Chapter 6).

In many groups, the description of new species is increasingly dependent on refined but not necessarily clear-cut species concepts. Consequently, in some groups in which the species descriptions were thought to be complete, new species are described to accommodate a shift in ideas (e.g. Cracraft's (1992) study of birds-of-paradise). The work awaiting our efforts is of enormous proportions. It requires serious discussion of the research priorities, not only for the relationship of systematics to other sciences, but also within the disciplines of systematics. I believe that a substantial improvement could be obtained by a more concerted, co-operative effort. The species diversity on earth is decreasing at a greater pace than can be matched by our current (or even foreseeable) descriptive efforts. The most author-itative students in the field of conservation and management of natural resources estimate that by 1999 a substantial percentage of living species (9–50%, according to the estimates; cf. Lugo, 1988) will completely disappear as a consequence of habitat destruction (for a comprehensive series of essays on this topic see Wilson (1988a)). How many will become extinct before being recognized and described?

Jacobs (1969) estimated that Augustin Pyramus de Candolle (1778–1841) was able to describe 300 species a year, whereas George Bentham's (1800–1884) upper limit was 1000–2000 species a year. Conditions are different today. According to Jacobs, it is now virtually impossible to study more than 100 plant species satisfactorily in a year. This figure is high for groups in which the researcher has to deal with extensive synonymy. The work is easier in those groups where the materials are concentrated in a few collections. However, a thorough record requires examination of collections distributed among a number of institutions. A good deal of time is thus spent in travelling to them and arranging to study them. To demonstrate that *Allophylus* (Sapindaceae) reasonably includes only one species, Leenhouts (1967) had to evaluate 225 different descriptions and type specimens.

Another important point was raised by Jacobs (1969); the importance of working on projects that can be finished in a reasonable time. For example, A.P. De Candolle began publishing the *Regni Vegetabilis Systema Naturale* which proved too large to be completed. Later he revised his project to the form of the *Prodromus Systematis Naturalis Regni Vegetabilis* that was eventually completed in 1873, many years after his death. Similarly, the *Flora Indica* project of Hooker and Bentham proved too large to be completed. However, Hooker later finished the smaller revised *Flora of British India* in seven volumes (1872–1897).

Some writers are prodigious, for example R. Ciferri (1897–1964) the author of more than 1000 mycological publications, M.J. Berkeley (1803–1889) who named about 6000 new species of fungi, and W. Nylander (1822–1899) who described some 5000 new species of lichens. No-one described more new taxa than the entomologist Edward Meyerick (1854–1938), who is credited with describing more than 15 000 species. However, he 'refused to use a microscope until his later years, and he refused to acknowledge that genital characters were worthwhile. He based much of his classification on the variation as seen through a hand lens' (Hodges, 1976).

5.2 INSTITUTIONS

The major natural history museums have long been centres for systematics and they continue to perform their leading roles in systematic science. A well preserved and well studied collection still serves a central role as the single most important resource of systematics. Miller (1985) reviews the many roles of natural history museum collections in biological research.

There are few comprehensive accounts of the historical developments of natural history museums and their collections (for a semi-popular account see Mannucci and Minelli (1987)). However, there are many interesting issues linked with the histories of museums including the history of zoological, botanical and palaeontological explorations and the development of techniques for preserving and studying specimens. Museum collections are also an important reference system of the diversity of living organisms; to be used as material for new evolutionary ideas.

It is difficult to get an overall picture of natural history museum collections, especially within zoology. Botanical and palaeontological collections are much better documented than their zoological counterparts. Most important herbaria are included in *Index Herbariorum* or *Taxon*, the official journal of the International Association of Plant Taxonomy (IAPT), which regularly updates references to those institutions. For fossils, there is Cleevely's (1983) book on *World Palaeontological Collections*. Information about zoological collections is scattered. It is often difficult to know whether an old collection of centipedes or leeches exists or not and, if it is still extant, where it is preserved, and how well it can be accessed. Useful information can be gained from books such as Dance's (1986) account of shell collections and Horn *et al.*'s (1990) treatment of insect collections. An important role in developing a co-ordinated account of museum resources has been undertaken in recent years by the Washington-based Association of Systematic Collections. Two publications issued by that Association are worth citing, Knutson and Murphy's (1988) *Systematics: Relevance, Resources, Services, and Management: A Bibliography*, and the proceedings of the symposium on the *Foundations for a National Biological Survey*, edited by Kim and Knutson (1986).

It would be interesting to know the overall size of museum collections, but no relevant figures are available. However, Arnett and Samuelson (1986) give a checklist of world collections of insects and spiders, although the coverage of private collections is incomplete. From Arnett and Samuelson's detailed account, it is possible to estimate that 500 million insect specimens are stored worldwide in public collections. Accordingly, as an estimate of the number of natural history specimens preserved in world collections, I would suggest 2000 million specimens. For this calculation, specimen does not necessarily correspond to an individual organism. It is possible that trillions of individual diatoms are preserved in the collections of the Academy of Natural History in Philadelphia, but I would regard the slides prepared for microscopic examination as the unit specimens, rather than the individual diatom cells. Thus, the diatom collection in Philadelphia (the largest of its kind in the world) contains some 400 000 specimens.

Sadly, ironic or even contemptuous comments about human and material resources employed in preserving, improving and studying natural history collections are common, among lay people and in the academic world. Indifferent or overtly hostile attitudes towards these treasures (and people working on them) are one of the major problems. However, museum staff and taxonomists are accustomed to dealing with these problems. Recently, however, matters seem to have become worse, because difficulties have been raised within the community of practising taxonomists. From the columns of *Nature*, Clifford, Rogers and Dettman (1990, p.602) suggested 'rationalizing' taxonomy and dispensing with all but a few 'truly valuable specimens [to be] retained and preserved into the future'. The best of the world's museums and taxonomists answered these arguments in the strongest terms (Gyllenhaal *et al.*, 1990; West *et al.*, 1990). However, one feels discomforted when confronted with such a disregard for the enormous treasures of information stored in natural history collections and for the danger into which they may fall in the wrong hands.

Special attention is generally given to type specimens. These represent the permanent voucher material corresponding to published names of new taxa. Their study is imperative, when compiling new treatments. However, there are often signs of dissatisfaction with the current practice of linking names of species (or other taxa, e.g. subspecies) to one type specimen. There is a widespread view that this practice reinforces typologism, an attitude we should have rejected from systematics since an evolutionary and populational paradigm has been established in biology. However, this criticism can be rejected by observing that type specimens have a special status from a nomenclatural, rather than a systematic, point of view. A type specimen is a link between the original description of a taxon and a real object, so that the observations of the authors can always be checked.

Papp (1988) has also criticized the unending worship of types from a strictly curatorial point of view, and suggested revisions to the Codes of Nomenclature. More technical objections are also sometimes raised against the conventional types. Brinkhurst (1980, p.508), for example, when discussing problems and trends in the systematics of aquatic oligochaetes, tackled 'the thorny issue of the single holotype when a species may be described from whole animals (placement of pores) whole mounts (setae, penis sheath) dissections (complex convoluted male duct, proportion of parts) and sections (precise insertion of a number of prostate glands and other details)'. However, the importance of type specimens has also been stressed (e.g. Fauchald, 1990).

Old museum collections are becoming increasingly useful for molecular studies, for example for studying mitochondrial DNA (Diamond, 1990). In addition, molecular studies are less destructive than many conventional morphological investigations. The amplification of single DNA molecules by the polymerase chain reaction (PCR) technique allows molecular studies to be performed on very small amounts of tissue, such as the root of a single feather shaft from a stuffed bird (Ellgren, 1991).

Additional or alternative sources of systematic information are the basis of what we should call living collections. For animals, such collections are virtually non-existent. However, living collections of vascular plants in botanic gardens have always played an important role in providing information which cannot be obtained from herbarium specimens. However, living plants have always been regarded as secondary sources of taxonomic information and the International Code of Botanical Nomenclature clearly states that taxa must be 'typified' by herbarium specimens, not by living ones.

For bacteria, protists, some fungi and algae systematic investigations are undertaken on clones cultured in standardized conditions. The International Code for the Nomenclature of Bacteria prescribes that the type material for a bacterial taxon be a culture of a well-defined strain. This practice is followed increasingly by students of unicellular and few-celled eukaryotes. Matters are rapidly progressing and new international agreements, both on experimental procedures and on systematic and nomenclatural matters, are expected. The last edition of the International Code of Botanical Nomenclature (Greuter et al., 1988) recommends (Recommendation 9A.1) that: 'whenever practicable a living culture should be prepared from the holotype material of the name of a newly described taxon of fungi and algae and deposited in a reputable culture collection. (Such action does not obviate the requirement for a holotype specimen under art. 9.5)'. Students of these organisms may profit from the directories of the American Type Culture Collection, the Cambridge Collection of Algae and Protozoa and others.

A very unusual case of description of a new species from living material only is that of *Laniarius liberatus*. This African bird, belonging to the shrike family (Laniidae), was described from a single specimen from Somalia. This putative 'type specimen' was released in the field, after a short time in captivity, only leaving a DNA sample, in the hands of its authors, as a proof of its specific distinctness. This species, described by Smith *et al.* (1991), is therefore without a conventional type specimen. This procedure, apparently suggested by over-zealous conservation concerns, has been rightly rejected by Peterson and Lanyon (1992); but see Hughes (1992) for a more favourable comment. In the meantime, Reynolds and Taylor (1991) have suggested that the existing rules of the International Code of Botanical Nomenclature (ICBN) already apply to the species to be described on DNA samples as the only available type material. These are possibly the first signs of a trend that will develop in the future. As for my opinion, I welcome molecular evidence, not only for organisms poor in morphology, but I cannot subscribe to a reductive approach to systematics, where living beings are simply equated with molecular fingerprints.

Taxonomists of many plant and animal groups belong to specialized associations, or working groups. Links are also provided by official institutions where many systematists work, or by wide-ranging associations such as the Systematics Association, the International Association of Plant Taxonomy (IAPT), the Society of Systematic Biologists (SSB) (until 1991, the Society of Systematic Zoology) and the American Society of Plant Taxonomists (ASPT). There are many specialized associations with a national base; some of these (especially those specializing in birds or insects or plants) are very old, such as the British Ornithologists' Union, or the Société Entomologique de France. Others are much younger, such as the national societies of nematologists founded in many countries, ranging from India and Pakistan to Japan and Italy. All these associations perform useful tasks, both in spreading knowledge and co-operation within their countries and in publishing their journals. However, the best co-operation is found in international working groups, which regularly exchange information through informal newsletters. Members of these groups also meet periodically to discuss ideas.

5.3 LITERATURE

This section is written particularly for beginners, but I think that some of the comparative figures below might also encourage reflection from more experienced colleagues.

The conventional distinction between primary sources, secondary sources, abstracting and indexing services is one way of looking at the literature. However, other classifications are important for our purposes.

The bulk of the primary literature is given in papers published in scientific journals. It is difficult to estimate their number because there are no comprehensive lists and the limits between scientific and technical journals and other kinds of media are uncertain. Moreover, a huge number of new journals appear every year, while other titles disappear. The lifetime of a journal may be quite short. Kronick (1985) estimates that the total number of scientific and technical journals published in the world is between 30 000 and 100 000, although the higher figure is obviously for a very broad concept of 'journal'. Nudds and Palmer (1990) give a similar estimate of 50 000. Of these titles, several thousand may be of occasional interest for biological systematics, although most pertinent information appears in a few hundred journals. Specialists find most new information in a few journals. Hawksworth, Sutton and Ainsworth (1983) estimate that the mycological and lichenological literature already consists of over 300 000 books and papers, whereas the annual output is approximately 5000 titles, scattered through some 35 000 publications. For the life sciences as a whole, Dextre Clark (1988) says that *Biological Abstracts* was scanning 9000 journals in 1986, whereas CABI scans 150 00 titles from serial publications. The *CAB Abstracts* database grows by about 130 000 titles per annum (Dextre Clarke, 1988, p.307). Approximately 600 000 papers are indexed annually by *Biological Abstracts* scanned from 9000 sources including journals, books and conference proceedings from over 100 countries (Kelly, 1991). The *Zoological Record* indexed 70 446 references in 1989 (Kelly, 1991). Even the largest and more efficient indexing and abstracting services have important gaps in their literature coverage, for example *Biological Abstracts* only covers 50% of the current botany literature (Kelly, 1991).

Secondary literature sources of interest to systematists span from the review papers published in monographs series such as *Evolutionary Biology*, to species catalogues or checklists that often incorporate previously unpublished information. They also share many aspects with the abstracting and indexing services.

Systematics is very dependent on abstracting and indexing services. For example, a zoologist could not undertake systematic work without the *Zoological Record*. Kronick (1985) estimates that 1000 different indexing and abstracting services are currently at work, in the fields of science and technology. Many of them employ sophisticated information technology for storage, retrieval and dissemination of information to users. Several services are now offered on-line or on compact discs and this trend is rapidly developing. In my experience, many abstracting and indexing services are not very useful to the working systematist, for many reasons. The most important reason is the selective coverage of the abstracting services. They seldom cover adequately literature that does not appear in journals, although many primary sources appear in book form, or as occasional issues. Most services exclude coverage of irregularly published journals, especially from museums. Many journal sources are also deliberately discarded as 'minor', according to the most diverse standards, for instance language, or because their contents only occasionally match the fields of the indexing services.

However, the indexing and abstracting services are not responsible for the excessive scatter of published systematic information. Before criticizing the output of abstracting systems we should examine how systematists contribute to their potential input. I fully acknowledge the universal right to a free choice of media for making known the results of our studies. However, because there is one scientific community, there is also a universal responsibility for the circulation of knowledge. Authors and editors, as well as publishers, are equally involved in this affair. The exponential increase in the size of the literature forces us to acknowledge its priority over most other problems of our science.

There are several different types of inventories of the diversity of living organisms:

- comprehensive works, where all taxa belonging to the group are not only listed, but also discussed in some detail;
- checklists or catalogues, where taxa are listed in systematic order, down to genus, species or subspecies level;
- nomenclators, where taxa are alphabetically listed, without any attempt to arrange them critically into a system.

There is no comprehensive and up-to-date list covering all animal taxa (Recent and fossil) at the species level. However, Sherborn's *Index Animalium* (1902, 1922–1933; cf. also Poche's (1938) supplement) covers the period from 1758 to 1850. Genera are more comprehensively covered in Neave's *Nomenclator Zoologicus* (1939–1940, 1950), as well as in Schulze, Kükenthal and Heider's *Nomenclator Animalium Generum et Subgenerum* (1926–1954). Neave's work covers the period from 1758 to 1945 (published and forthcoming supplements will soon bring it to 1980), whereas Schulze, Kükenthal and Heider's nomenclator extends only to 1926. For those years not covered by these works, the only comprehensive reference work is the yearly *Zoological Record*, which started in 1864. None of these publications is complete. Levine (1962) estimated that only about 76% of the living protozoan species described between 1864 and 1958 are listed in the *Zoological Record*, but Raup (1976) suggests that for non-protozoan groups the coverage is probably much better. Comprehensive works exist for several animal groups. Many taxa have been covered by monographs in series such as *Das Tierreich*, or *Genera Insectorum*. However, only a few volumes of these series have been published in recent decades. Although most of them are outdated monographs, they have yet to be superseded by better ones.

Major monographs or checklists covering large animal groups are listed in Appendix 1.

Several groups of extinct animals have been covered by issues of the *Fossilium Catalogus* but, like the major insect catalogues, this work also mainly consists of old and outdated volumes. A nearly complete list of the genera of extinct vertebrates has been published by Carroll (1987). Many invertebrate groups are adequately covered, at genus level, by volumes of the large, and still incomplete, Moore's *Treatise on Invertebrate Palaeontology*.

A major resource, for botanists, is Farr, Leussink and Stafleu's *Index Nominum Genericorum (Plantarum)*. There are also several other generic lists, such as Mabberley (1987) for vascular plants, Grolle (1983) for liverworts, Hawksworth, Sutton and Ainsworth (1983) for fungi, and Round, Crawford and Mann (1990) for diatoms. Generic and specific names are covered by *Index Kewensis* for seed plants (see also *The Kew Record of Taxonomic Literature*, for a current updating of the *Index*), and by Wijk, Margadant and Florschutz's *Index Muscorum* (1959–1969) (periodically updated by supplements published in *Taxon*), by Lamb's (1963) *Index Nominum Lichenum* and by Stace (1989) and Hawksworth, Sutton and Ainsworth (1983). There are also some outstanding bibliographical resources such as Stafleu and Cowan's (1976–1988) monumental *Taxonomic Literature: A Selective Guide to Botanical Publications and Collections, with Dates, Commentaries and Types*, in seven volumes (Supplement I, by Stafleu and Mennega, was published in 1992) and Greene and Harrington's (1988, 1989) *Conspectus of Bryological Taxonomic Literature*. In addition to general works, for most groups there are important monographs covering one of the major zoogeographical or phytogeographical regions. In this way, a few groups are more or less comprehensively covered, although not by one single work.

Until 1988, approximately 943 878 botanical names had been recorded by *Index Kewensis*, including 42 434 generic and 861 421 specific names; only 33 099 infraspecific names were recorded because this taxonomic rank was not covered in the *Index* prior to 1971 (Anonymous, 1989).

Major projects to develop worldwide databases for plant species names are quite advanced. International effort is being devoted to the Species Plantarum Project (SPP) and to the Global Plant Species Information System (GPSIS), as documented in the recent issues of *Taxon* and elsewhere (cf. Hawksworth, 1991a).

On a smaller geographical scale, there are the regional or national faunas, systematic treatments of single taxonomic groups within more or less restricted limits. Despite their limited scope and their consequently biased perspective, the single volumes of the major faunas are often the best, or even the only guide to the study of several groups. This type of publication is much less widespread in zoology than in botany. Most tropical areas are covered by more or less advanced floras, whereas faunas are still mainly a European enterprise. Major projects have now started in non-European countries, such as Canada, Israel, China, New Zealand and Australia.

For the prokaryotes, the most comprehensive work is Balows *et al.* (1992).

All these works are very important and their authors (and publishers) should be praised for them. However, most of these works should be used very carefully. In many cases, they are, of necessity, the product of a thorough and time-consuming bibliographical survey, not the result of a much more useful, but also much more difficult, process of monographic revision. We have already mentioned how difficult such work might be. However, we should never refrain from insisting on the importance of revisionary studies, a necessary requirement to develop a more organized systematic literature. When refusing to tackle major revisions, taxonomists, as so colourfully stated by Jacobs (1969), are 'nibbling at the crumbs of their task, without being able to digest the substantial chunks'. Davis and Heywood (1963) say 'Monographic work goes on painfully slowly. This is most unfortunate for two reasons: firstly, because floras could be written much better if more monographs were available; secondly, because systematics cannot make real progress without monographic research'. In the same context, it is also useful to remember Watson's (1971) comments about the taxonomic literature. Nearly two decades since they were written, these comments still retain their full interest. 'Taxonomic literature is quantitatively an embarrassment, but it is generally very incomplete: when studying a large genus at a world scale, or trying to review the systematic arrangement of a large family, adequate data of a quite elementary kind will be either more or less inaccessible or simply unavailable. ... Even information on universally popular floral attributes can prove remarkably difficult to obtain.' Needless to say, the same difficulties are also commonly experienced by zoologists.

However, rather than lamenting this state of affairs, I believe that an estimate of the current level of revisionary work has seldom, if ever, been tried. For major groups, data are not easily available, but in the case of vascular plants some calculations are readily made. My source is Mabberley (1987), who provides references to generic reviews, whenever available, genus by genus. Of the 14 162 genera of vascular plants listed by Mabberley, the existence of a worldwide monographic study is claimed for 1596 (11.27%). There are approximately 44 700 (i.e. 17.88%) species within the monographed genera. On the whole, this picture may not look bad, but it should be specified that many of these reviews are old and possibly unsatisfactory in view of current knowledge (Figure 5.1). Moreover, a large number of genera of primary economic or scientific interest have never been comprehensively revised. In spite of the efforts over the last years, things are not much better than in 1971, when Watson lamented that 'some of the most important genera have not been reviewed, let alone, comprehensively studied, since de Candolle's *Prodromus* and the most comprehensive accounts of larger families mostly date back to the same period' (Watson, 1971, p.131). Matters are even worse in many other groups of living beings.

5.4 NOMENCLATURE

A few useful accounts of the origin and developments of the current systems of biological nomenclature have been published (e.g. Heppell, 1981; Ride and Younès, 1986; La Vergata, 1987). Historical information has also been included in some editions of the International Codes of Nomenclature.

Biological nomenclature is primarily regulated by three codes (see Jeffrey (1989) for a readable summary of their contents).

- The *International Code of Zoological Nomenclature* (ICZN), now in its third edition (International Commission on Zoological Nomenclature, 1985);
- The *International Code of Botanical Nomenclature* (ICBN), currently in the edition adopted at the XIV International Congress of Botany at Berlin (1987) and published by Greuter *et al.* (1988);
- The *International Code of Nomenclature of Bacteria* (Sneath, 1992).

For viruses, previously covered by the Bacterial Code, a definitive Code of Nomenclature has yet to be written. A series of Rules was agreed upon by the International Committee on Taxonomy of Viruses (Matthews, 1981), but they have not generally been accepted by plant virologists (e.g. Kingsbury, 1986). A new set of proposals, recently edited by Franki *et al.* (1991), will probably meet with more success.

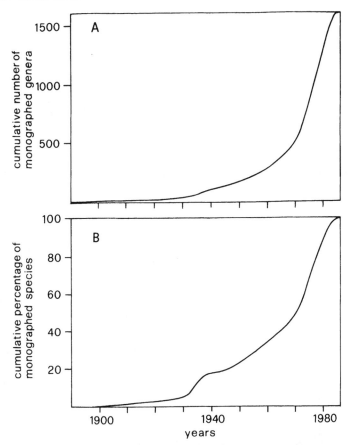

Figure 5.1 Genera of vascular plants covered by a comprehensive monograph, according to the data in Mabberley's checklist (1987). (A) illustrates the distribution of these generic monographs according to publication year, whereas (B) shows the corresponding numbers of species covered by these monographs. In many cases, the last (or only) monograph for a given genus is already several decades old. Still worse, most plant genera have never been monographed, in this century at least.

The primary aim of these codes, as stated in the Introduction (p.xiii) of ICZN (1985): 'is to provide the maximum universality and continuity in the scientific names ... compatible with the freedom of scientists to classify animals according to taxonomic judgements'. A few lines below we read that: 'Zoological nomenclature is the system of scientific names applied to taxonomic units (taxa; singular, taxon) of animals known to occur in nature, whether extant or extinct, including names based on fossils that are substitutions (replacements, impressions, moulds and casts) for the actual remains of animals (ichnotaxa), and names proposed before 1931 based on the work of extant animals. ... Excluded from the provisions of the Code are names proposed ... for hybrids as such ...' (ICZN, 1985, art.5). In this context, it is interesting to quote Walker (1986), who proposes both formal and informal names for the parthenogenetic (uniparental) lizards of the genus *Cnemidophorus* (family Teiidae), the hybrid origins of which have been firmly established in recent years (cf. Chapter 3).

With fewer words, the ICBN (Principle I), after stating that 'botanical nomenclature is independent of zoological nomenclature', declares that 'the Code equally applies to names of taxonomic groups treated as plants whether or not these groups were originally so treated'. A footnote further specifies that 'for the purposes of this Code, the word 'plants' does not include bacteria'.

From these first statements, it could be wrongly deduced that the botanical and zoological codes (ICBN and ICZN) are independent, though equivalent. One major area of overlap and conflict is given by several groups of protists, which because they do not fall into any reasonable modern concepts of animals or plants (cf. Chapter 7) were never consistently treated as animals (protozoans) or as plants (unicellular algae, and the like), even in conventional terms. On the other hand, it would not be easy to merge ICBN and ICZN to form a universal Code of Biological Nomenclature. The main problems have been illustrated by Jeffrey (1986).

Some standardization of nomenclature in flagellate groups currently covered by both ICZN and ICBN has been advocated by Taylor *et al.* (1987), who propose amendments to both Codes. The nomenclature of higher taxa of protists has been discussed by Rothschild and Heywood (1987, 1988) and Heywood and Rothschild (1987). They suggest that to avoid confusion caused by a mixed use of botanical and zoological nomenclature for these organisms, the higher protistan groups recognized as genuine monophyletic taxa should be named with the suffix protists. They hope that 'nomenclatural prejudice is eliminated, and by keeping the system informal and thus circumventing the Linnaean system, our system remains flexible in accommodating new data yet should retain a great degree of stability...' (Rothschild and Heywood, 1988, p.197). Accordingly, they have proposed the following names: chloroprotists, euglenoprotists, chlorarachnioprotists, rhodoprotists, cryptoprotists, chrysoprotists, xanthoprotists, haptoprotists, bacillarioprotists, raphidoprotists, phaeoprotists, dinoprotists, eustigmatoprotists and glaucoprotists. However, it is very difficult to comprehend 'protists' as a natural group of organisms. A positive aspect of Rothschild and Heywood's proposal is the informal character of their names (without capital initial letter). The same cannot be said of Margulis and Sagan's (1985) concurring proposal, where 17 of the 42 protist phyla they recognized were given similar, Latinized names, such as Chloroprotista and Mycetoprotista. More conventional terms are used in the more recent *Handbook of Protoctista* (Margulis *et al.*, 1990), where the overall systematic arrangement seems to have been mostly shaped by Lynn Margulis.

Another tentative proposal has been advanced by Corliss (1983), when suggesting the creation of independent Codes of Nomenclature for the different eukaryote kingdoms. This proposal may be worthy of attention, but are we sure that our current partitioning of living organisms into kingdoms (cf. Chapter 7) will survive long enough to justify the creation of new Codes? In a later paper, Corliss (1984) tackled the code problem again and proposed four possible pragmatic solutions:

(1) the 'ecumenical' approach by agreement on one, unified Code;
(2) modification of one Code, or both, in order to overcome disagreement by solving the individual difficult cases;
(3) arbitrary allocation of individual protist phyla to the Zoological or Botanical Codes;
(4) development of separate Codes for the individual kingdoms.

I agree with Corliss (1984, p.98; see also Corliss, 1990) that 'something has to be done', despite the 'grave problems in attempting to successfully bring about any of the above solutions, in light of the many controversial issues and cries of distress that would immediately arise among the specialist taxonomists inevitably involved and affected'. Greuter, McNeill and Nicolson (1989) perceived as small and vanishing the number of taxa the nomenclatural status of which continues to trouble the boundary between the ICBN and ICZN. Patterson and Larsen (1991) are of the contrary opinion and claim that virtually any protist taxon is potentially cross-kingdom and thus subject to two Codes. Protists are therefore involved in puzzling questions of homonymy, or in other nomenclatural problems, such as the genus *Pseudoperanema*, which has two different type species, one (*P. hyalinum*) under the ICBN, the other (*P. trichophorum*) under the ICZN.

The prospect of unifying the different systems of nomenclature currently in use has been studied in depth (Ride and Younès, 1986): unfortunately, several differences between the codes are not easy to overcome (Jeffrey, 1986; Ride, 1988).

Another point worth discussing briefly is the alleged independence of nomenclature and taxonomy. Close scrutiny reveals that such independence is illusory. Phylogenies, said Corner (1981, p.82) 'are overwrought and obscured by the artificial code of nomenclature which can ascribe with learned names ranks which do not exist except in [our] fantasy'. This critique may sound excessive, but all taxonomic categories officially recognized by the Codes impose themselves on taxonomy as sound reference points that people do not question easily.

Insofar as the Codes provide adequate rules for assigning species names to every animal or plant, there seems to be no place for arguing that apomicts or other living organisms may not be organized into species. Some problems of close dependence of the nomenclatural rules on our systematic concepts are easier to understand; for example, the question of whether we should retain the Linnaean hierarchy.

In some circles, there is increasing dissatisfaction with the Codes. Some feel that the rules are too rigid and their application is often cumbersome and contrary to the stated purpose of guaranteeing the stability of biological nomenclature. Some of these objections are justified. Strict adherence to the principle of priority is often the reason for extremely long searches through the literature. That is often of dubious use both to the person who undertakes them and to readers of the papers. Moreover, in well collected and studied groups such as birds, butterflies or flowering plants, the search for older names is seldom definitive. Who can be sure that no minor paper has been overlooked? The worst effect of blind application to the principle of priority is a degree of nomenclatural instability. Worse still, the species most likely to be affected by it are often the most common and well known, the scientific names of which are most frequently used outside systematic circles. In this context, Oldroyd (1966) wrote that the provisions of the Zoological Code are most useful when applied to species cited rarely. Nearly all instances of common and oft-quoted species must be ruled on by special deliberations of the International Commission on Zoological Nomenclature, allowing for the rejection of newly discovered old names, to the advantage of universal recent names.

The results of the activity of the International Commission on Zoological Nomenclature are published in the *Bulletin of Zoological Nomenclature*. A comprehensive index to the names so far considered by the Commission has been published (Melville and Smith, 1987). There are 9900 entries; the largest list is for decapod crustaceans, but the Commission has also deliberated on a large number of cases concerning molluscs, insects (particularly Hemiptera, Lepidoptera and Coleoptera) and vertebrates. Official lists of conserved and rejected plant names are also published as appendices at the end of the ICBN.

Nomenclatural stability is probably threatened in a more pervasive way by progress in systematics. This apparent paradox is a major concern of some (Anonymous, 1986; Crisp and Fogg, 1988). The problem does not reside in the progress of systematic knowledge alone, but in the circumstance that one part of the scientific name of a species (the generic name) is likely to change according to taxonomic judgement. Generic names may be changed to acknowledge that a species has been misplaced within genus A, when it clearly relates to a species currently in genus B. Changes in generic names may also be the result of a revision which broadens or narrows the generic concept within a group, with many new combinations of names being produced as a consequence. I do not think that an author would dispute a change of generic assignment for a species owing to misplacement, although that implies nomenclatural changes. However, in the case of a revision, authors such as Crisp and Fogg (1988) suggest that the current binomial expression should be 'frozen' and new taxonomic opinions expressed by the use of subgeneric names, so as not to affect the stability of the original binomial. Newman (1989), however, has strongly reacted to this proposal, protesting (p.24) 'that taxonomic recognition of the relationships between species and species groups, at and above the generic level, is necessary if evolutionary pathways and patterns are to be recognized. Continuing adjustments to such a system, usually enacted when knowledge has increased to the point where relationships are no longer being accurately portrayed, are inevitable inconveniences'.

These comments lead me to discuss the broader perspective of developments in systematic research and its nomenclatural outcome. Are we sure that an increasingly refined phylogenetic system is what the 'consumers' of taxonomic information and nomenclature want? I agree with those who acknowledge an increasing separation between the two demands, those of comparative biology and phylogenetic systematics, and those of the consumers of systematic products (cf. Haskell and Morgan, 1988). People working in agriculture, medicine, veterinary science, biotechnology, applied stratigraphy, as well as in nature conservation and education, need sound reference frameworks for their operations, together with some degree of nomenclatural stability. According to Hawksworth and Bisby (1988, pp.19–20), the only answer to this demand is 'for the systematists themselves to work towards consensus systems which can be commended for general use. It is essential to distinguish these 'generalist systems' intended for consumption by non-specialists, and the 'specialist systems' essentially proposed for discussion amongst specialists and which are the lifeblood of systematic debates'. Hawksworth and Bisby refer to the particularly unsettled situation in ascomycetes. In 1982 a journal (*Systema Ascomycetum*) was founded to develop a generalist system for the over 6000 genera identified so far (cf. section 7.4). Another attempt to develop a consensus system for general use is currently being tried for legumes, where the International Legume Database and Information Service (ILDIS) Project is developing and improving a species checklist for general use by non-systematists (Bisby, 1988). The problem, however, is to find a suitable balance between taxonomic freedom and the users' needs (Humphries, 1991; Minelli, 1991a; and papers in Hawksworth, 1991a).

These efforts are worthy of discussion, but it is important not to offer a 'consensus system' as an undisputable pillar of scientific knowledge. Moreover, it is increasingly clear that our efforts must be directed equally seriously to another aspect of the question. It is one thing to improve the management of information already published, but it is another to arrange for more rational production and distribution of new systematic information.

Both aspects of the problem have apparently been resolved by bacteriologists. They fixed a point in time (1 January 1976), as the starting point for a new treatment of systematic literature pertaining to their group. For generic and specific names proposed before that date, an international committee was appointed to screen the whole literature, so as to provide an official list of accepted names, eventually published in 1980 (Skerman, McGowan and Sneath, 1980; Hill, Skerman and Sneath, 1984). After the publication of this list, no further searches for old names or a use of old systematic literature on bacteria may be regarded as meaningful. The list, which recognizes 2300 names (1800 of them specific) from within 30 000 published names, provides the initial pool of names currently available to the systematic bacteriologist. This is already a remarkable achievement, but the 'revolution' in systematic bacteriology could not be complete without further provisions for handling the forthcoming literature. We have already seen that the greatest problem for systematists is being aware of all systematic literature published in a field. In zoology and botany, there is no answer to this problem. The Codes prescribe that one should take into consideration as equally valid all published descriptions and names. (I will avoid discussing what 'published' actually means, in our technological age.) The radical solution adopted by bacteriologists consists of officially recognizing privileged media only. All new names of bacteria published between 1976 (the starting point of the new nomenclature) and 1980 (publication of the *Approved Lists of Bacterial Names*) are retained as valid only if they are published in the *International Journal of Systematic Bacteriology*. After 1980, students are free to describe new taxa elsewhere, but the proposed names only receive nomenclatural validation when a corresponding 'announcement' is published in the *International Journal of System-*

atic Bacteriology, under the column title 'Validation of the publication of new names and new combinations previously effectively published outside the IJSB' (Sneath, 1986). From this, it is very easy for bacteriologists to be aware of the full range of literature of nomenclatural importance.

Why do we not apply the same procedures of registration of names to eukaryotes? There are more problems with animals and plants than those encountered by bacteriologists. Preliminary steps in that direction have already been taken. A special committee has been established (Hawksworth, 1988, 1989) charged with working on official proposals for registration of the names of organisms covered by the ICBN. For genera the task is not too difficult with approximately 36 500 names in use, out of 79 000 available. However, the problems are much greater for species, with approximately 400 000 names in use, out of some 1 700 000 available in the literature. For the animal kingdom, matters are much worse, because of the greater number of taxa known, and a lack of comprehensive, updated works, even at the generic level. At the species level the problems are large, even for the higher plants, owing to the lack of updated monographs for most genera.

Despite the enormous difficulties associated with these projects, some steps should be taken towards registration. However, it should be made clear to everybody that those provisions should only affect nomenclature without setting any limit to the free development of systematic knowledge. In this context, it is necessary to reiterate the distinction between system and classification. It is to the latter, I think, that what follows applies.

For animal names, the International Commission on Zoological Nomenclature has begun working in that direction (International Commission on Zoological Nomenclature, 1990; Savage, 1990). The new edition of the Code is expected to incorporate major changes in policy, to provide for registration of names (or works). Clear guidelines to be followed in establishing lists of names in current use and in providing some acceptable and workable form of registration have been traced by Ride (1991), who has aptly revived the old but still up-to-date Dwight's (1909) dictum, that 'it is not justice for the dead zoologist we seek so much as justice for the living'.

A different concern for nomenclature arises when acknowledging that the outcome of phylogenetic systematics is a system of hierarchical, nested monophyla, as discussed in previous chapters. The conventional (Linnaean) arrangement is inadequate for this system, with serious consequences for nomenclature (Minelli, 1991a). The worst problem perhaps derives from the critical attitude towards the genus category, thus shaking the traditional link between generic assignment of a species and binomial species name. The problems arising at suprageneric level should be regarded as comparatively minor, although worthy of a rational revision before encumbering the nomenclature with new, possibly unnecessary names. New provisions are to be developed. I cannot follow Bretfeld (1986) who, after rejecting the Linnaean hierarchy and nomenclature, feels it is necessary to name all monophyla anew, outside a formally defined nomenclatural system. In many cases, he creates new names by a slight modification of existing family names (e.g. monophylum Sminthuridida pro family Sminthurididae).

In a more abstract vein, de Queiroz and Gauthier (1990; also de Queiroz, 1992) have suggested a 'phylogenetic definition of taxon names'. They suggest departing from the traditional character-based definitions, replacing them with definitions based on cladistic relationships only. In principle, three kinds of phylogenetic definitions of taxon names are possible. For each of them, I quote the example offered by de Queiroz and Gauthier (1990, p.310), who distinguish: (a) a node-based definition, as in defining Lepidosauria as 'the most recent common ancestor of *Sphenodon* and squamates and all of its descendants'; (b) a stem-based definition, as when defining Lepidosauromorpha as '*Sphenodon* and squamates [Lepidosauria] and all saurians sharing a more recent ancestor with them than with crocodiles and birds'; (c) an apomorphy-based definition, as in defining Tetrapoda as 'the first vertebrate to possess digits ... and all of its descendants'. In my view, the last definition seems more attractive than the others, because it does not imply the presence of a given character in all members of the taxon, thus allowing for its loss in some of the descendants of the specified ancestor. Besides the conceptual questions at issue, de Queiroz and Gauthier suggest that the nomenclatural advantages of such a definition of taxon names would be that a given name would not be associated with the first use of that name at a particular rank, but simply on its first use in association with a specified ancestor, irrespective of possible shifting in rank, possibly caused by developments in phylogenetic knowledge of the group.

The Codes prescribe (details differ between ICZN and ICBN) that the scientific name of a species should be accompanied by reference to the name of the scientist who first recognized the species under the current specific name (ICZN) or epithet (ICBN). In a future, hypothetical world of standardized nomenclature, shall we continue this practice? Twenty years ago, Jacobs (1969) wrote the following, perhaps prophetic lines: 'With progress of critical revision, citation of author's name, at any rate in the phanerogams, will lose whatever function it had, except, of course, the function of satisfying those taxonomists who cling to the sacred cow of individualism which fed them all their life'. Citation of authors' names serves a useful reference function, at least in the cases of homonymy. However, Jacobs' words should warn us about putting excessive weight onto our own contributions in the development of a body of scientific knowledge shaped by the co-operative efforts of many workers.

PART TWO
The State of the Art

We already have a 'finished' system of all known animal species, although it is not presented anywhere in its entirety but is scattered through countless individual papers. This is a hierarchic system, but not all of its parts express the phylogenetic kinship of species. For some parts of the system there are no studies that make any attempt to express phylogenetic kinship. For other parts of the system there are several 'constructions', some of which actually express the phylogenetic kinship of the species ...

W. Hennig (1966, p.235)

The inventory of natural diversity

6

Hardly any aspect of life is more characteristic than its almost unlimited diversity. No two individuals in the sexually reproducing populations are the same, no two higher taxa, nor any associations, and so ad infinitum. Wherever we look, we find uniqueness, and uniqueness spells diversity.

E. Mayr (1982a, p.133)

6.1 HOW MANY SPECIES DO WE KNOW?

How many species of living organisms are known and have been named to date? How many species are still awaiting discovery? The systematist is often asked such questions, especially in these years of growing sensibility towards the tremendous decrease of biological diversity experienced on earth. Obtaining an estimate of existing diversity is a necessary step towards understanding the amount and value of the natural wealth we are currently destroying. It is not by chance that these questions repeatedly emerge from the pages of the most comprehensive review on biodiversity and its loss crisis (Wilson, 1988a). That such detailed estimates of biodiversity are gaining increasing attention because of their impact on conservation issues is probably a major opportunity for taxonomy to obtain more adequate consideration than that enjoyed until recently. Taxonomy contributes to conservation at the most elementary level; it is easy (but important) to recall that no species can be protected unless identified and named: bad taxonomy may kill! (Daugherty *et al.*, 1990; May, 1990; Heywood, 1991). But systematics can also contribute in a more sophisticated and professional way to the adoption of more adequate conservation measures, as exemplified by several recent approaches to measuring biodiversity in terms of cladistic relatedness of the species involved (e.g. Altschul and Lipman, 1990; Williams, Humphries and Vane-Wright, 1991).

Wilson (1988a) offered a summary of taxonomic diversity similar to Table 6.1, and estimated the total number of described species at 1 392 485. Another table has recently been published by May (1988). A simple listing of numbers, however, would be seriously misleading, because of unavoidable oversimplifications, smoothing and intrinsic defaults. Therefore, before describing Table 6.1, I will briefly discuss the quality of the data.

Table 6.1 Estimated number of described species of Recent living beings

Taxon	Species	References
'Monera' (Prokaryotes)	4670	Wilson (1988b)
Karyoblastea	1	Corliss (1984)
Amoebozoa	<5000	Corliss (1984)[a]
Acrasia	6	Corliss (1984)
Eumycetozoa	>700	Corliss (1984)
Plasmodiophorea	35–40	Corliss (1984)
Granuloreticulosa	8000	Corliss (1984)[b]
Xenophyophora	36	Corliss (1984)
Hyphochytridiomycota	25	Corliss (1984)
Oomycota	800	Corliss (1984)[c]
Chytridiomycota	900	Corliss (1984)
Chlorophyta	3200	Corliss (1984)
Prasinophyta	270	Corliss (1984)[d]
Conjugatophyta	5000	Corliss (1984)[e]
Charophyta	100	Corliss (1984)[f]
Glaucophyta	12	Corliss (1984)
Euglenophyta	>1000	Corliss (1984)
Kinetoplastidea	600	Corliss (1984)
Pseudociliata	4	Corliss (1984)
Rhodophyta	<4500	Corliss (1984)[g]
Cryptophyta	200	Corliss (1984)
Choanoflagellata	140	Corliss (1984)
Chrysophyta	>600	Corliss (1984)[h]
Haptophyta	450	Corliss (1984)[i]
Bacillariophyta	10 000	Corliss (1984)[j]
Xanthophyta	650	Corliss (1984)
Eustigmatophyta	10–12	Corliss (1984)
Phaeophyta	1600	Corliss (1984)
Proteromonadea	>25	Corliss (1984)
Bicosoecidea	35–40	Corliss (1984)
Heterochloridea	15	Corliss (1984)

[a] 'Nearly 5000 species described, but perhaps half (mainly testaceous forms) should be considered doubtful until redescription; some fossil known.' (Corliss, 1984, p.104).
[b] 'More than 37 500 species described, with perhaps 80% fossilized forms (many of extinct lines); some require restudy. New 'foram' species description continue to flood the specialized literature.' (Corliss, 1984, p.104).
[c] 'More than 800 species described, but many should be considered of doubtful validity until further studied.' (Corliss, 1984, p.105).
[d] 'Some 350 species, with approx. 85 described as fossils.' (Corliss, 1984, p.106).
[e] 'There may be nearly 5000 valid species (>80% desmids), with an additional 5000 or even more desmids with taxonomic validity uncertain until restudied; fossils rare.' (Corliss, 1984, p.106).
[f] 'Some 400 species described, with approx. 300 of them as fossils.' (Corliss, 1984, p.106).
[g] 'Some 5000 included species ..., with approx. 750 described as fossils.' (Corliss, 1984, p.108).
[h] 'About 850 species described, with >200 as fossils.' (Corliss, 1984, p.109).
[i] In addition, some 1100 fossil species have been described (Corliss, 1984, p.109).
[j] 'Recorded species said to number between 50 000 and 100 000; conservative estimate of valid species is 10 000 living and 15 000 fossil forms.' (Corliss, 1984, p.109).

Table 6.1 (contd)

Taxon	Species	References
Raphidophyceae	27	Corliss (1984)
Labyrinthulea	8	Corliss (1984)
Thraustochytriacea	28	Corliss (1984)
Metamonadea	200–300	Corliss (1984)[k]
Parabasalia	350	Corliss (1984)[l]
Opalinata	>200	Corliss (1984)[m]
Heliozoa	>100	Corliss (1984)[n]
Taxopoda	1	Corliss (1984)
Acantharia	>200	Corliss (1984)[o]
Polycystina	2000?	Corliss (1984)[p]
Phaeodaria	>650	Corliss (1984)[q]
Peridinea	2000	Corliss (1984)[r]
Syndinea	40–50	Corliss (1984)
Ebriidea	3	Corliss (1984)[s]
Ellobiophyceae	<10	Corliss (1984)
Ciliophora	8000	Corliss (1984)
Sporozoa	5000	Corliss (1984)
Perkinsida	1	Corliss (1984)
Microsporidia	800	Corliss (1984)
Haplosporidia	25	Corliss (1984)
Myxosporidia	>1200	Corliss (1984)
Actinomyxidea	22	Corliss (1984)
Marteiliidea	4	Corliss (1984)
Paramyxidea	1	Corliss (1984)
Zygomycotina	765	Hawksworth, Sutton and Ainsworth (1983)
Ascomycotina	28 650	Hawksworth, Sutton and Ainsworth (1983)
Basidiomycotina	16 000	Hawksworth, Sutton and Ainsworth (1983)
Deuteromycotina	17 000	Hawksworth, Sutton and Ainsworth (1983)
Musci	15 800	Miller (1982)
Hepaticae	6890	Miller (1982)

[k] 'Vickerman (in Parker, 1982) suggests approx. 200 as valid; perhaps as many as 300 total ... described, although many needing redescription.' (Corliss, 1984, p.111).
[l] 'About 350 valid species comprise the two major included groups (trichomonads and hypermastigotes), although larger number has been described.' (Corliss, 1984, p.112).
[m] 'Some 400 of which perhaps only 50% valid and acceptable.' (Corliss, 1984, p.112).
[n] 'Approx. 175 species described (perhaps only 100 acceptable to stringent specialist).' (Corliss, 1984, p.112).
[o] 'Some 475 species described, with perhaps only 200 valid, and no fossil forms known.' (Corliss, 1984, p.113).
[p] 'Some 9750 species described, 70% as fossils, but possibly only 4800–5000 acceptable as valid. Polycystine paleontology now 'in explosive phase' (E.G. Merinfeld, personal communication), so numbers of fossil species will very likely continue to rise rapidly.' (Corliss, 1984, p.113).
[q] 'Some 1100 species described, very rarely from fossilized material, but perhaps only 650 acceptable as valid.' (Corliss, 1984, p.113).
[r] Another 2000 species known as fossils (Corliss, 1984).
[s] In addition, some 75–80 fossil species (Corliss, 1984).

Table 6.1 (*contd*)

Taxon	Species	References
Anthocerotae	270	Miller (1982)
Psilophytophyta	4	Mabberley (1987)
Lycopodiophyta	1225	Mabberley (1987)*
Equisetophyta	15	R.L. Hauke (personal communication)
Filicophyta	8400	Mabberley (1987)
Pinophyta	742	Mabberley (1987)
Magnoliophyta	240 000	Mabberley (1987)
Placozoa	2	Grell (1982)
Porifera	6000	M. Sarà (personal communication)
Cnidaria	15 000	[t]
Ctenophora	80	Werner (1984)
'Turbellaria'	2988	Crezée (1982)*
Trematoda	8000	S.C. Schell (personal communication)
Cestoda	3850	G.D. Schmidt (personal communication)
Gnathostomulida	100	W. Sterrer (personal communication)
Mesozoa	50	Kilian (1984)
Nemertea	950	R. Gibson (personal communication)
Gastrotricha	450	M. Balsamo and P. Tongiorgi (personal communication)
Rotifera	2000	B. Pejler (personal communication).[u]
Kinorhyncha	74	Higgins (1982)*
Nematoda	20 000	Platt and Warwick (1983)[v]
Nematomorpha	230	Swanson (1982)*
Acanthocephala	750	Hartwich (1984)
Priapulida	9	Hartwich (1984)
Mollusca	130 000	F. Giusti (personal communication)[w]
'Polychaeta'	12 000	K. Fauchald (personal communication)[x]
Hirudinea	600	Author's estimate
Oligochaeta	>6000	Sims and Gerard (1985)[y]
Pogonophora	>100	Author's estimate
Echiura	140	Edmonds (1982)
Sipuncula	320	Rice (1982)
Merostomata	4	Levi (1982a)
Scorpiones	1200	Francke (1982)
Uropygi	85	Levi (1982b)
Schizomida	80	Levi (1982b)
Amblypygi	70	Levi (1982b)

[t] 7700 species according to Werner (1984); 5000 in Hydrozoa only according to F. Boero, (personal communication).
[u] Platt and Warwick (1983) speak of 'nominal species'. An estimate of 12 000 species is given by Maggenti (1982).
[v] R. Wallace (personal communication) gives an estimate of 1800 species; R. Shiel (personal communication) gives, with question marks, 350 species for Bdelloidea and 2500 for Monogononta; C. Ricci (personal communication) 363 for Bdelloidea.
[w] Fewer than 50 000 species according to Boss (1982).
[x] Approximately 8000 species according to George and Hartmann-Schröder (1985).
[y] 1250 species of aquatic oligochaetes according to R. Brinkhurst (personal communication).

Table 6.1 (contd)

Taxon	Species	References
Palpigradi	60	Levi (1982b)
Araneae	35 000	Levi (1982b)
Ricinulei	33	Levi (1982b)[*]
Pseudoscorpiones	2300	G. Gardini (personal communication)
Solifugae	900	Muma (1982)
Opiliones	4000	J. Martens (personal communication)
Acari	30 000	D.E. Johnston (1982)
Pycnogonida	1000	Hedgpeth (1982)
Cephalocarida	9	Bowman (1982b)
Branchiopoda	1000	Belk (1982); F. Margaritora (personal communication).[*][z]
Ostracoda	10 000	[a]
Mystacocarida	10	Bowman (1982a)
Copepoda	13 000	Marcotte (1983); Z. Kabata (personal communication).[*][b]
Branchiura	125	Fryer (1982)
Cirripedia	1220	Newman (1987); Foster and Buckeridge (1987)
Malacostraca	30 000	[c]
Protura	270	Palissa (1989)
Collembola	3500	Palissa (1989)
Diplura	500	Palissa (1989)
Archaeognatha	280	Kugler (1982)
Zygaentoma	>300	Kugler (1982)
Ephemeroptera	2100	Edmunds (1982)
Odonata	4875	S. Brookes (personal communication); D.A.L. Davies (personal communication)
Blattaria	3684	L.M. Roth (1982)
Isoptera	2000	Brown (1982)[*]
Mantodea	1900	M. La Greca (personal communication)
Notoptera (Grylloblatt.)	24	T. Yamasaki (personal communication)
Orthoptera	20 000	Kevan (1982)

[z] 640 species of Cladocera according to F. Margaritora (personal communication).
[a] Very rough estimate. Some 55 000 species have been described to date (K.G. McKenzie, personal communication, most of them as fossils.
[b] 1500 species of Calanoida, 2000 of Cyclopoida and 4500 of Harpacticoida, according to Marcotte (1983); 5000 species of parasitic Copepoda according to Z. Kabata (personal communication).
[c] The figure is a rough estimate. Malacostraca are here intended in the traditional sense, including Leptostraca (< 20 species; Bousfield, 1982), Hoplocarida (Stomatopoda) (350 species; Manning, 1982), Syncarida (145 species; Schminke, 1982), Thermosbaenacea (16 species; H.P. Wagner, personal communication), Spelaeogriphacea (1 species; Bousfield, 1982); Mysidacea (780 species; G.L. Pesce, personal communication), Cumacea (1000 species; Watling, 1982), Amphipoda (8600 species; E.L. Bousfield, personal communication), Isopoda (10 000 species; R. Argano, personal communication), Tanaidacea (800 species; Holdich and Jones, 1983), Euphausiacea (85 species; Laubitz, 1982), Amphionidacea (1 species; Abele and Felgenhauer, 1982), Decapoda (10 000 species; Abele and Felgenhauer, 1982).
*Brown, W.L. Jr (1982) Isoptera, in Synopsis and classification of living organisims (ed. S.P. Parker), McGraw-Hill, New York, pp. 349–51.

Table 6.1 (*contd*)

Taxon	Species	References
Phasmatoptera	2500	Kevan (1982)
Dermaptera	1840	S. Sakai (personal communication)
Embiidina	250	E.S. Ross (personal communication)
Plecoptera	2000	P. Zwick (personal communication)
Zoraptera	>20	Brown (1982a)
Psocoptera	3000	C.N. Smithers (personal communication)
Phthiraptera	12 000	Eichler (1989)
Thysanoptera	4910	R. zur Strassen (personal communication)
Hem. Coleorrhyncha	20	Dolling (1991)
Hem. Heteroptera	32 500	Dolling (1991)[* d']
Hem. Homoptera	49 500	Dolling (1991)[* d']
Megaloptera	200	Aspöck, Aspöck and Hölzel (1980)
Rhaphidioptera	168	Aspöck (1986)
Planipennia	4000	Aspöck, Aspöck and Hölzel (1980)
Coleoptera	340 277	Lawrence (1982)[* e']
Strepsiptera	400	Brown (1982c)
Mecoptera	470	Brown (1982b)
Siphonaptera	1740	Smit (1982)
Diptera	125 000	Bickel (1982)[* e' f']
Trichoptera	7000	Wiggins (1982)
Lepidoptera	136 278	Munroe (1982)[* e' g']
Hymenoptera	143 000	Brown (1982d)[* h']
Chilopoda	2500	Author's estimate
Symphyla	160	Scheller (1982a)
Diplopoda	10 000	Hoffman (1982)
Pauropoda	500	Scheller (1982b)
Pentastomida	60	Self (1982)[*]
Onychophora	70	Peck (1982)
Tardigrada	531	Ramazzotti and Maucci (1983)
Phoronida	10	Emig (1982)
Bryozoa (Ectoprocta)	5000	J.-L. d'Hondt (personal communication)
Entoprocta	150	Nielsen (1982)
Brachiopoda	335	Foster (1982)
Chaetognatha	110	Salvini-Plawen (1986)
Echinodermata	6700	Fell (1982); Pawson (1982); Gale (1987)[*]

[d'] See Dolling (1991) for species numbers for the individual families.
[e'] Species numbers by families are given in Tables 6.2–6.4.
[f'] In introducing the order, Bickel (1982) estimates the known species as numbering between 100 000 and 150 000. However, the sum of the species numbers given by him for the individual families gives a total very close to 100 000.
[g'] 'The insect order Lepidoptera totals perhaps 200 000 described species' (Holloway, 1987, p.326).
[h'] In introducing the order, Brown (1982d) gives a figure of approximately 130 000 species. However, the sum of the species number given by him for the individual families gives a total of some 143 000 species.

Table 6.1 (contd)

Taxon	Species	References
Hemichordata	<100	Benito (1982)[*]
Tunicata	3000	Goodbody (1982); C. Monniot (personal communication)[*][i]
Cephalochordata	23	Azariah (1982)
Myxiniformes	32	Nelson (1984)
Petromyzontiformes	41	Nelson (1984)
Chondrichthyes	793	Nelson (1984)[j]
Dipnoi	6	Nelson (1984)
Crossopterygii	1	Nelson (1984)
Brachiopterygii	11	Nelson (1984)
Actinopterygii	20 839	Nelson (1984)
Amphibia	4014	Frost (1985)
'Reptilia'	5954	Duellman (1982)[*][k]
Aves	9091	Howard and Moore (1980)[l]
Mammalia	4216	Corbet and Hill (1986)[*]

[*] Figures calculated by adding the separate data given for individual subtaxa (mostly families, as for Turbellaria, Coleoptera, Lepidoptera etc.).
[i] Ascidiacea: 2500–3000 species (C. Monniot, personal communication).
[j] More precisely, McEachran (1982) gives a figure of 797–843.
[k] Duellman (1982) gives the following partial figures: Testudines, 222 species; Rhynchocephalia, 1; Sauria, 3407 (Gekkota, 769; Iguania, 1016; Scincomorpha, 1509; Anguinomorpha, 113); Serpentes, 2267; Amphisbaenia, 135; Crocodylia, 22. G. Underwood (personal communication) gives a figure of some 2500 species for Serpentes.
[l] 9016 species according to Morony, Bock and Farrand (1975); 9021 according to Bock (1982a); 9881, 'allospecies' included, according to Sibley and Monroe (1990).

The first problem is the choice of a unit of measure. The most usual estimate of biological diversity is the number of species. However, it seems (cf. Chapter 4) that a major portion of biological diversity does not belong in 'biological' species, although there are species names available for those organisms too. In my opinion, to add such named 'agamospecies' to the biological species would be the same as adding apples and cherries. Is it justifiable to include within the same column of Table 6.1 the 9091 'species' of birds and the 363 'species' of parthenogenetic bdelloid rotifers? Is it justifiable to add the few dozen 'sexual species' of *Rubus* and *Taraxacum* with the thousands of 'microspecies' in the same genera?

Exacting precise species numbers within small taxa, numbering tens or hundreds of species each, is certainly useful if considered *per se*, but it is completely immaterial in the face of the enormous errors accompanying other estimates. Thus, I avoid adding up the figures given in Table 6.1. Instead, I simply offer an estimate of 1 800 000 as the total number of living 'species' recognized and named at this time. A similar estimate (1 820 000) has recently been published by Stork (1988, p.322).

Table 6.1 provides some detail as to how these numbers are distributed among individual phyla and the classes and orders of some of the larger phyla. Of some interest also is the distribution of species numbers, by family, within a few large orders, such as butterflies (Table 6.2), flies (Table 6.3) and beetles (Table 6.4).

Table 6.2 The families of Lepidoptera arranged in decreasing size order.
Nomenclature and figures according to Munroe (1982)

Species	Genera	Family
25 000	4000	Noctuidae
20 000	1500	Pyralidae
20 000	1500	Geometridae
6000	100	Psychidae
4000	500	Gelechiidae
4000	500	Tortricidae
4000	400	Oecophoridae (including Metachandidae)
4000	200	Nymphalidae (including Heliconiidae, Apaturidae)
3000	500	Hesperiidae
3000	400	Satyridae
3000	300	Tineidae
3000	300	Lycaenidae
2500	200	Lymantriidae (= Liparidae, Orgyiidae, Dasychiridae, Ocneriidae)
2000	650	Notodontidae
2000	300	Lithosiidae
2000	200	Ctenuchidae
2000	70	Pieridae
1500	150	Lasiocampidae
1200	30	Cosmopterigidae (including Diplosaridae, Hyposmocomidae, Scaeosophidae, Chrysopelliidae, Walshiidae)
1000	100	Riodinidae
1000	100	Saturniidae
1000	65	Gracillariidae
1000	50	Sesiidae
1000	36	Yponomeutidae
850	190	Sphingidae
800	200	Limacodidae
700	30	Stenomathididae
600	100	Thyrididae
600	40	Xylorictidae
600	25	Epiplemidae
500	80	Hepialidae
500	50	Coleophoridae
500	50	Pterophoridae
500	45	Ithomiidae
500	40	Dioptidae
500	30	Cochylidae
500	25	Papilionidae
400	60	Drepanidae
400	50	Eupterotidae
400	40	Choreutidae
400	12	Nepticulidae
350	70	Heliodinidae
340	27	Incurvariidae
300	100	Agaristidae

Table 6.2 (contd)

Species	Genera	Family
300	50	Zygaenidae
300	30	Pericopidae
300	12	Lyonetiidae
250	17	Apatelodidae
250	15	Nolidae
250	3	Glyphipterigidae
240	1	Ethmiidae
225	6	Acraeidae
220	20	Megalopygidae
200	50	Thyatiridae
200	30	Elachistidae
200	26	Mimallonidae (= Lacosomidae = Perophoridae)
200	25	Thyretidae
200	25	Carposinidae
200	8	Danaidae
150	30	Symmocidae
150	4	Castniidae
125	4	Batrachetridae
150	3	Immidae
120	23	Thaumatopoeidae
100	35	Scythrididae
100	16	Metarbelidae
100	16	Amathusiidae
100	15	Bombycidae
100	15	Hypsidae
100	11	Brachodidae
100	10	Blastobasidae
100	9	Heliozelidae
100	8	Micropterigidae
100	8	Callidulidae
100	2	Argyresthiidae
100	1	Acrolophidae
100	1	Phyllocnistidae
80	11	Brassolidae
80	1	Momphidae
75	8	Anthelidae
71	6	Epermeniidae
70	3	Acrolepiidae
65	1	Tischeriidae
60	14	Dalceridae
60	3	Eriocottidae (including Deuterotineidae, Compsoctenidae)
50	15	Uraniidae
50	13	Blastodacnidae
50	5	Alucitidae
50	5	Megathimiidae
50	1	Opostegidae
40	6	Copromorphidae

Table 6.2 (*contd*)

Species	Genera	Family
40	5	Sematuridae
34	3	Oxytenidae
30	1	Morphidae
23	1	Ochsenheimeriidae
20	7	Amphitheridae
20	7	Brahmaeidae
20	6	Epipyropidae
20	5	Eriocraniidae
20	3	Chrysopolomidae
20	2	Douglasiidae
20	2	Hyblaeidae
15	3	Lemoniidae
10	6	Tineodidae
10	4	Dudgeoneidae
10	4	Pterothysanidae
10	4	Cercophanidae
10	2	Heterogynidae
10	2	Libytheidae
10	2	Oxychirotidae
10	1	Pterolonchidae
10	1	Epicopeiidae
9	2	Ratardidae
7	3	Neopseustidae
7	2	Prototheoridae
6	3	Pseudarbelidae
6	3	Arrhenophanidae
6	1	Mnesarchaeidae
5	2	Axiidae
5	1	Acanthopteroctetidae
5	1	Schreckensteiniidae
5	1	Cyclotormidae
4	1	Agonoxenidae
3	3	Palaeosetidae
3	2	Cocytiidae
3	1	Lophocoronidae
3	1	Somabrachyidae
2	1	Agathiphagidae
1	1	Neotheoridae
1	1	Anomosetidae
1	1	Coelopoetidae
1	1	Apoprogonidae
1	1	Endromidae
1	1	Carthaeidae

Table 6.3 The families of Diptera, arranged in decreasing size order. Nomenclature and size figures according to Bickel (1982)

Species	Genera	Family
13 000	300	Tipulidae
Several thousands	150	Calliphoridae
6000	1000	Tachinidae
6000	150	Dolichopodidae
5000	400	Asilidae
5000	150	Syrphidae
5000	120	Chironomidae
4000	600	Cecidomyiidae
4000	200	Bombyliidae
4000	70	Tephritidae
3000	130	Tabanidae
3000	100	Empididae
3000	100	Muscidae
3000	34	Culicidae
2700	220	Phoridae
2000	150	Mycetophilidae
1800	30	Agromyzidae
1500	40	Drosophilidae
1400	80	Stratiomyiidae
1400	60	Sarcophagidae
1200	130	Lauxaniidae
1200	70	Ceratopogonidae
1100	60	Simuliidae
1000	250	Chloropidae
1000	70	Ephydridae
1000	50	Platystomatidae
1000	50	Anthomyiidae
800	45	Conopidae
700	56	Sphaeroceridae
700	15	Bibionidae
500	60	Sciomyzidae
500	50	Sciaridae
500	45	Psychodidae
500	10	Therevidae
500	8	Lonchaeidae
?	50	Scatophagidae
450	45	Acroceridae
420	40	Micropezidae
400	60	Heleomyzidae
400	50	Otitidae
400	30	Pipunculidae
330	50	Pyrgotidae
330	19	Hippoboscidae
300	42	Mydidae
300	19	Milichiidae
300	15	Rhagionidae

Table 6.3 (*contd*)

Species	Genera	Family
250	16	Scenopinidae
250	15	Nemestrinidae
240	21	Sepsidae
200	22	Clusiidae
200	21	Blepharoceridae
200	15	Scatopsidae
200	10	Nycteribiidae
200	5	Psilidae
160	31	Richardiidae
160	27	Streblidae
150	13	Diopsidae
150	3	Dixiidae
110	6	Trichoceridae
110	4	Apioceridae
100	20	Platypezidae
100	18	Chamaemyiidae
100	10	Asteiidae
100	10	Xylophagidae
100	7	Athericidae
90	10	Celyphidae
85	23	Rhinophoridae
80	5	Thaumaleidae
80	4	Anisopodidae
75	9	Chaoboridae
70	20	Neriidae
70	6	Cuterebridae
67	23	Piophilidae
65	20	Oestridae
60	10	Canaceidae
60	3	Ptychopteridae
50	8	Pallopteridae
50	8	Tethinidae
50	7	Anthomyzidae
50	4	Opomyzidae
45	7	Gasterophilidae
40	8	Odiniidae
40	4	Carnidae
40	2	Curtonotidae
37	11	Tanyderidae
35	4	Chyromyiidae
34	1	Pelecorhynchidae
34	1	Lonchopteridae
30	9	Rhopalomeridae
30	1	Strongylophthalmyiidae

Table 6.3 (*contd*)

Species	Genera	Family
25	3	Pantophthalmidae
23	1	Fergusoninidae
22	2	Tanypezidae
22	1	Glossinidae
20	7	Coelopidae
20	3	Diastatidae
20	2	Teratomyzidae
20	1	Cryptochaetidae
15	7	Pseudopomyzidae
15	5	Periscelidae
14	4	Helcomyzidae
13	2	Dryomyzidae
12	4	Hyperoscelididae
12	3	Rhinotoridae
11	4	Megamerinidae
10	4	Aulacigastridae
9	1	Nothybidae
8	1	Syringogastridae
8	1	Camillidae
7	1	Deuterophlebiidae
5	1	Perissommatidae
4	4	Pachyneuridae
4	4	Nymphomyiidae
4	2	Axymyiidae
3	3	Tachiniscidae
3	2	Cypselostomatidae
3	1	Neurochaetidae
3	1	Braulidae
2	2	Sciadoceridae
2	1	Acartophthalmidae
1	1	Ironomyiidae
1	1	Mystacinobiidae
1	1	Mormotomyiidae

As for fossils, an estimate of the number of 'species' described so far is more difficult to obtain than that for living organisms. A lack of updated and comprehensive monographs, as well as problems in the application of suitable species concepts are the major causes of this difficulty. A comprehensive figure of about 300 000 fossil species of plants and animals was given by Stearn (1981), but Raup (1987) reduced it to 250 000. I suspect that the actual figures are slightly larger, although not by very much. Raup (1976) estimated the grand total of fossil 'invertebrate' species (protozoans included) described between 1760 and 1970 as something less than 200 000 species. However, some 3700 new species of fossil 'invertebrates' were annually described between 1961 and 1970, according to Raup

Table 6.4 The families of Coleoptera arranged in decreasing size order. Nomenclature and data according to Lawrence (1982)

Species	Genera	Family
50 000	4500	Curculionidae (including Cossonidae, Platypodidae, Rhynchophoridae, Scolytidae)
35 000	4000	Cerambycidae (including Disteniidae, Hypocephalidae, Parandridae, Spondylidae)
35 000	2500	Chrysomelidae (including Cassididae, Cryptocephalidae, Sagridae)
30 000	1500	Carabidae (including Brachynidae, Cicindelidae, Omophronidae, Paussidae, Pseudomorphidae, Trachypachidae)
30 000	1500	Staphylinidae (including Brathinidae, Scaphidiidae)
25 000	2000	Scarabaeidae (including Cetoniidae, Glaphyridae, Hybosoridae, etc.)
18 000	1700	Tenebrionidae (including Alleculidae, Cossyphodidae, Lagriidae, Nilionidae, Rhysopaussidae, Tentyriidae)
15 000	400	Buprestidae (including Schizopodidae)
9000	400	Elateridae (including Dicronychidae, Lissomidae, Plastoceridae)
5000	650	Pselaphidae (including Clavigeridae)
5000	200	Melyridae (including Malachiidae, Dasytidae, Rhadalidae, Prionoceridae)
5000	135	Cantharidae (including Chauliognathidae)
4500	500	Coccinellidae (including Cerasommatidiidae, Epilachnidae)
4000	150	Cleridae (including Corynetidae)
3500	150	Lycidae
3000	200	Histeridae (including Niponiidae)
3000	160	Nitidulidae (including Smicripidae, Cybocephalidae)
3000	120	Dytiscidae
3000	120	Meloidae (including Tetraonychidae)
3000	100	Anthicidae (including Cononotidae, Pedilidae)
2600	325	Anthribidae (including Platystomidae, Urodontidae, Bruchelidae)
2500	30	Erotylidae (including Dacnidae)
2300	325	Brentidae (= Brenthidae)
2200	26	Apionidae (including Cyladidae)
2100	100	Attelabidae (including Apoderidae, Pterocolidae, Rhynchitidae)
2000	250	Leiodidae (= Anisotomidae, including Camiaridae, Catopidae, Colonidae, Cholevidae, Leptodiridae)
2000	125	Hydrophilidae (including Helophoridae, Hydrochidae, Sphaeridiidae, Spercheidae)
2000	100	Lampyridae
2000	75	Scydmaenidae
1600	140	Anobiidae
1500	60	Bruchidae (= Lariidae, Mylabridae)
1300	180	Colydiidae (including Adimeridae, Bothrideridae, Monoeridae)

Table 6.4 (contd)

Species	Genera	Family
1300	120	Endomychidae (including Mycetaeidae, Merophysiidae)
1200	190	Eucnemidae (= Melasidae, Phylloceridae)
1200	100	Lucanidae
1200	100	Mordellidae
1200	75	Cucujidae (including Laemophloeidae, Passandridae, Silvanidae, Scalididae)
1100	35	Euglenidae (= Aderidae, Hylophilidae, Xylophilidae)
1000	100	Oedemeridae
900	80	Languriidae
850	45	Dermestidae (including Thorictidae, Thylodriidae)
700	90	Elminthidae (= Elmidae, Helminthidae)
700	90	Bostrichidae (including Psoidae, Lyctidae, Endecatomidae)
700	11	Gyrinidae
650	55	Cerylonidae (including Aculagnathidae, Anommatidae, Dolosidae, Euxestidae, Murmidiidae)
600	60	Trogossitidae (including Peltidae, Ostomidae, Lophocateridae, Temnochilidae)
600	55	Phalacridae (including Phaenocephalidae)
600	45	Geotrupidae
600	30	Scirtidae (including Cyphonidae, Helodidae)
550	40	Ciidae (= Cisidae)
500	27	Passalidae
500	25	Lathridiidae
450	80	Melandryidae (= Serropalpidae)
450	40	Ptinidae (including Gnostidae, Ectrephidae)
430	67	Ptiliidae (including Cephaloplectidae, Limulodidae)
400	35	Corylophidae (= Orthoperidae)
400	35	Rhipiphoridae
400	30	Scraptiidae (including Anaspididae)
400	18	Discolomidae (= Notiophygidae, including Aphanocephalidae)
400	15	Hydraenidae (= Limnebiidae)
350	42	Salpingidae (including Aegialitidae, Dacoderidae, Elacatidae, Eurystethidae, Inopeplidae, Othniidae, Tretothoracidae)
300	35	Ptilodactylidae
300	30	Cryptophagidae (including Hypocopridae, Catopochrotidae)
300	28	Byrrhidae (including Syncalyptidae)
300	15	Heteroceridae
300	5	Trogidae
300	2	Chelonariidae
250	20	Rhizophagidae (including Monotomidae)
225	12	Monommidae
200	35	Phengodidae

Table 6.4 (*contd*)

Species	Genera	Family
200	30	Limnichidae
200	20	Cerathocantidae (= Acanthoceridae)
200	18	Mycetophagidae
200	12	Dryopidae (including Chiloeidae)
200	6	Biphyllidae (= Diphyllidae)
200	4	Haliplidae
190	4	Throscidae (= Trixagidae)
175	14	Silphidae
175	12	Noteridae (including Phreatodytidae)
170	11	Cebrionidae
160	30	Mycteridae (including Hemipeplidae)
160	3	Aglycyderidae (= Proterhinidae)
150	18	Rhysodidae
150	13	Belidae
150	8	Callirhipidae
125	26	Zopheridae (including Merycidae)
100	20	Psephenidae
100	10	Pyrochroidae
100	1	Helotidae
80	15	Dascillidae (including Karumiidae)
80	6	Drilidae
60	8	Agyrtidae
60	8	Artematopidae (= Eurypogonidae)
50	16	Pythidae
50	6	Lymexylidae (including Atractoceridae)
50	5	Clambidae (including Calyptomeridae)
50	5	Rhipiceridae (including Sandalidae)
50	1	Nosodendridae
50	1	Sphaerosomatidae
43	3	Micropeplidae
40	10	Nemonychidae (= Rhinomaceridae)
35	6	Sphindidae (including Asphidiphoridae)
35	2	Propalticidae
35	1	Pleocomidae
30	8	Oxycorynidae (including Allocorynidae)
30	5	Eucinetidae
25	7	Byturidae
25	6	Torridincolidae
25	6	Tetratomidae
25	1	Georyssidae
22	5	Archeocrypticidae
21	4	Cupedidae
20	7	Cephaloidae (including Nematophidae, Stenotrachelidae)
20	2	Prostomidae
19	4	Derodontidae
18	1	Sphaeriidae

Table 6.4 (*contd*)

Species	Genera	Family
15	2	Trictenotomidae
15	1	Lutrochidae
13	3	Hydroscaphidae
12	2	Eulichadidae
11	1	Dasyceridae
10	6	Omethidae
10	2	Synchroidae
10	1	Homalisidae
8	6	Phloeostichidae
8	4	Perimylopidae
8	2	Pterogeniidae
7	6	Jacobsoniidae
7	4	Leptinidae (including Platypsyllidae)
7	1	Cerophytidae
7	1	Cneoglossidae
6	4	Chalcodryidae
6	2	Telegeusidae
5	3	Ommatidae
5	2	Boridae
5	1	Amphizoidae
4	4	Boganiidae
4	4	Cavognathidae
4	1	Hygrobiidae (= Pelobiidae)
4	1	Synteliidae
4	1	Phycosecidae
3	1	Sphaeritidae
3	1	Diphyllostomatidae
3	1	Perothopidae
3	1	Brachypsectridae
3	1	Protocucujidae
2	2	Chaetostomatidae
2	1	Lepiceridae (= Cyathoceridae)
1	1	Micromalthidae
1	1	Phloiophilidae
1	1	Achantocnemidae
1	1	Lamingtoniidae
1	1	Ithyceridae

(1976), as based on the *Zoological Record*. This high rate of species description is certainly increasing, despite some variations in comparison to Raup's compilation. For example, the number of new species of fossil 'animals' (protozoans included) described in 1983 was 5572, as determined by Minelli from the *Zoological Record*. In the same year, some 11 132 new species of living animals were described.

Obtaining this type of information about the rate of species description for the whole animal 'kingdom' is a time-consuming job. However, for the genera it is much easier. As far as I can extrapolate from genera to species, no consistent trend in the rate of description of new animal taxa emerges (Figure 6.1).

6.2 CONTINUING DISCOVERY

Progress varies from group to group. However, the inventory is still incomplete, even in groups such as birds, mammals and butterflies, which are comparatively well known in comparison to, say, nematodes or mites.

Relatively recent discoveries of previously unknown organisms include many spectacular examples, e.g. the okapi (*Okapia johnstoni*) (1901), the giant forest hog (*Hylochoerus meinertzhageni*) (1904), the Komodo dragon (*Varanus komodoensis*) (1912), the Congo peacock (*Afropavo congensis*) (1936), the living coelacanth

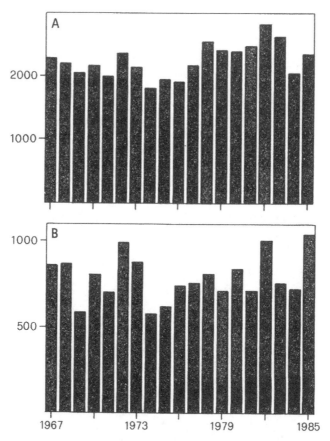

Figure 6.1 Number of new genera of animals (including 'Protozoa') (A) and number of new genera of insects (B) described in each of the years 1967 to 1985, according to the *Zoological Record*.

(*Latimeria chalumnae*) (1939) and the dawn redwood tree (*Metasequoia glyptostroboides*) (1940). New additions have even been made at the phylum level, with a group of tiny marine invertebrates, the Loricifera (Kristensen, 1983).

As for birds, the developments between 1930 and 1978 have been documented by Mayr (1980). He summarized the changes at species level concerning those groups originally included in the first volume of Peters (1931). Species listed in that volume represent about 9% of all known bird species. Of the 853 species recognized by Peters, 101 have since been reduced to subspecies or synonyms, whereas 34 taxa then listed as subspecies or synonyms have changed rank to 'good species'. Moreover, 13 valid new species have been described in the interval between 1930 and 1978, giving a final figure of 799 species instead of 853. On a worldwide basis, 9 new genera and 134 new species of birds have been described in the 50 years between 1934 and 1984 (Diamond, 1985).

Of the 1048 new genera of living mammals, listed by Corbet and Hill (1986), 134 have been discovered and named since 1900 (Diamond, 1985). The most recent genus of mammals described for Europe is the rodent *Dinaromys*. It was established by Kretzoi (1955) for a species living on the mountains of Yugoslavia (*D. bogdanovi* (Martino)) first described in 1922. Within the USA, the most recently described mammal genus is a bat, *Idionycteris phyllotis* (Allen). This species was first described in 1916 by Allen but transferred into a separate new genus by Anthony in 1923. Less well-known faunas have even contributed two mammals described as worthy of placement into new families within the last half century. These are the rodent family Seleviniidae, with *Selevinia betpakdalaensis* Belosludov and Bashanov from Kazakhstan, described in 1938 and the bat family Craseonycteridae, comprising the bumblebee-bat (*Craseonycteris thonglongyai* Hill) from Thailand, described in 1974. This tiny bat weighing only 2 g and just 3 cm in length deserves mentioning as the smallest known warm-blooded animal in the world.

Mammals described in the last few years include primates (night monkeys (*Aotus*) and squirrel monkeys (*Saimiri*)), an antarctic species of killer whale (*Orcinus glacialis*) (Berzin and Vladimirov, 1983), a musk deer from China (*Moschus fuscus*) (Li, 1981) and a new species of gazelle from Yemen (*Gazella bilkis*) (Groves and Lay, 1985). Most recently, the discovery of a new species of Malagasy prosimian, *Hapalemur aureus* (Anonymous, 1988), raised general interest, even in the popular press. The last big addition is probably *Mesoplodon peruvianus*, a new beaked whale described by Reyes *et al.* in 1991.

Unexpected additions to the list of living species come from discoveries of living specimens of species previously known only from fossil remains and believed to be extinct. Two well-known cases include the mountain pygmy possum (*Burramys parvus*), described from 20 000 year old fossils (Broom, 1896) and rediscovered, as living animal species, in 1966 (cf. Ride, 1970) and the Chaco peccary (*Catagonus wagneri*), described in 1930 from subfossil remains but rediscovered as a living animal less than 20 years ago (Wetzel *et al.*, 1975).

The inventory of vascular plants is largely incomplete. Most new additions are expected from South America: 'the most remarkable concentration of species in the world is that found in the three northern Andean countries of Colombia,

Ecuador and Peru, which together are home to about 45 000 plant species, or about a sixth of the world total, in an area just over a quarter of the size of Europe, with its approximately 12 000 plant species' (Raven, 1988, p.549). The large efforts as contributed by major projects, such as *Flora Neotropica,* have recently revealed the incompleteness of our knowledge of plants. Between 1972 and 1984 58 new species and three new subspecies of Chrysobalanaceae were described (an 18.2% increase) (Prance and Campbell, 1988, p.538). According to recent studies (cf. Prance and Campbell, 1988, p.537) only 10 000 of the estimated 18 000 plant species thought to occur in Bolivia have so far been collected. A sizeable portion of those still undiscovered will prove to be new to science.

Some groups are especially worthy of attention. Orchids, for instance, are 'amazingly poorly known taxonomically', according to Gentry and Dodson (1987). They contrast the 12 or so orchid taxonomists with the 200 or so systematists specializing on the other very large family, Compositae.

6.3 HOW MANY SPECIES ARE STILL TO BE DISCOVERED?

Estimates of the numbers of plant and animal species existing on earth have been put forward since the time of Linnaeus. Mayr (1982b) credits Linnaeus with the following estimates: 10 000 existing plants, in addition to the 6000 or so he described in *Species Plantarum* (1753) and 10 000 animals, in addition to the 4374 listed in the tenth edition (1758) of *Systema Naturae.* These estimates were soon overthrown by the flood of discoveries and descriptions witnessed by Linnaeus in his later years. In the year of Linnaeus' death, Zimmermann (1778) ventured an estimate of the number of existing species as 150 000 plants and 7 000 000 animals. For plants, Zimmermann's large estimate has been surpassed by subsequent discoveries, whereas his estimate for animals is still larger (by a factor of five) than the actual inventory. It is even larger than most estimates of the total diversity of animals proposed until very recently. What estimates can we propose today, with a factual knowledge enormously larger than that available to Linnaeus or Zimmermann?

The number of birds cannot reasonably be expected to grow by more than a few dozen truly 'new' species; in birds and in other well-known groups, major changes in species numbers are only to be expected because of the application of different species concepts (cf. Sibley and Monroe, 1990; Cracraft, 1992). For other groups (e.g. protozoans, nematodes or mites) the number of species awaiting discovery is confidently estimated to exceed those already named and described. Johnston (1982), for instance, estimated that between 500 000 and 1 000 000 species of mites exist, in addition to the 30 000 already described. These estimates may be reasonable but they are seldom founded on explicit arguments.

Much less hasty than the usual estimates of global species diversity were the calculations made by Steyskal (1965, 1973, 1976), White (1975) and others. These authors have investigated, for several taxa, the trend in species description over the years, i.e. the temporal changes in the rate of species description for a given taxon. Graphic examples are provided here for mammals (for the class as a whole, as well as for a few major orders; Figure 6.2) and for the subterranean harpacticoid copepods (Figure 6.3). From this type of curve, Steyskal and others felt confident in extrapolating an asymptotic value. However, there are problems with such curves. Species description is not a continuous and uniform process. For most groups, there

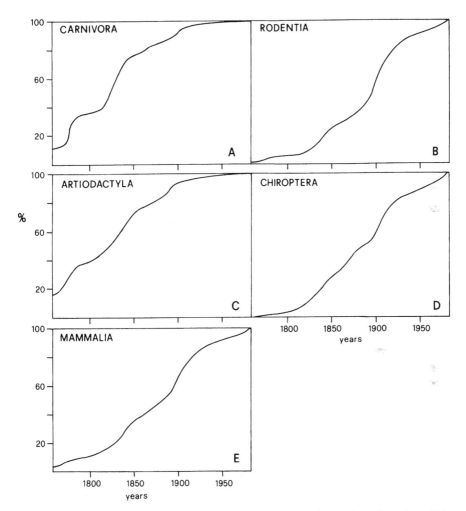

Figure 6.2 Temporal development of our knowledge of species diversity within living mammals. The curves describe the cumulative number of species known in a given year, expressed as a percent of the number of species known in 1981 (as listed in Honacki, Kinman and Koeppl, 1982). For carnivores (A) and artiodactyls (C), there was a good start with Linnaeus (*Systema Naturae*, 1758); for these groups, most species were described by the end of the nineteenth century. On the contrary, very few species of rodents (B) and bats (D) were known to Linnaeus and a large percentage of the species belonging to these last orders has been described in this century. Moreover, the very large size of these two orders heavily influences the overall trend of species description for the whole class (E).

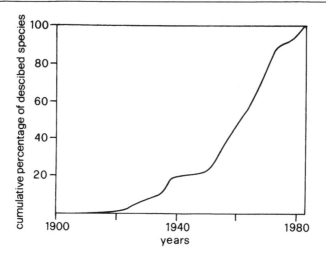

Figure 6.3 Subterranean harpacticoid copepods were virtually unknown at the beginning of this century, but Rouch (1986) lists as many as 441 species-level taxa as described up to 1983. Two thirds of these species were first described between 1952 and 1973.

are sizeable bursts of intensified description when an active specialist comes on the scene, or as a consequence of exceptional collecting efforts in previously less explored areas. Therefore, the method of estimating biological diversity from the study of trend curves has been criticized by several authors, such as Frank and Curtis (1979).

A better alternative was proposed by Erwin (1982). Erwin accurately sampled the beetles living in the canopy of a few trees in Costa Rica, then used the number of species thus collected as a starting point for his calculations. How many arthropods live in a tropical forest? Erwin's answer to this question was based on the calculation

$$n = \frac{(a \times b \times c)}{(d \times e)}$$

where a is the number of beetle species living on a given species of tree, b is the probability that a beetle species lives exclusively on that tree species, c is the number of different tree species growing in the tropics, d is the probability for an arthropod species of the forest canopy to be a beetle, and e is the probability for a forest arthropod species to belong to the animal communities of the canopy. Accordingly, Erwin (1982) estimated that some 30 000 000 arthropod species live in the tropical forests of the world; many more than previously guessed. More recently, Erwin (1988) has raised his estimate to 50 000 000 or more. Stork (1988) reconsidered Erwin's calculations and applied the same criteria to his data from Bornean forests. Stork's numbers are also impressive: 'More than 2800 species were represented in the 24 000 arthropods collected from the 10 Bornean trees. ... 1455 Chalcidoidea [small hymenopterans] are represented by 739 species, 437 of which were singletons with only eight species having more than 10 individuals – the commonest species having 19 individuals' (Stork, 1988, p.325). In Stork's opinion, the number of arthropod species living in the tropical forest could be estimated between 10 000 000 and 80 000 000. All living organisms, other than tropical terrestrial arthropods, are not covered by these impressive calculations. However, it is very unlikely that they could significantly modify this new perception of the size of species diversity on the earth.

Another analysis of the possible number of living species has been performed by May (1988). He has reviewed several factors affecting diversity, such as the structure of food webs, the relative abundance of species and the number of species and individuals in different classes of body size, to conclude with estimates in the same order of magnitude as Erwin and Stork.

A different estimate of the number of insect species worldwide has been proposed by Hodkinson and Casson (1991), based on their accurate survey of the Hemiptera of a topographically diverse area in the tropical rain forest of Sulawesi. In this investigation, collecting was carried out over a 1-year period and many different collecting methods were used. Hodkinson and Casson used the number of described and undescribed species thus collected to predict the total number of extant species of Hemiptera, and of insects at large. Their global estimate of the number of insects is 1.84–2.57 million species only. In the meantime, Thomas (1990) re-examined Erwin's and Stork's results, producing an estimate of 'only' 6–9 million species of insects in tropical forests. According to new evidence and additional ecological arguments, May (1991) has recalculated insect diversity to be of the order of 5–8 million species.

For fungi, Hawksworth (1991b) offers an estimate of 1.5 million species, mostly based on comparisons of better and lesser known mycofloras.

A further, poorly known environment, where many undescribed species still await discovery, is the deep sea floor (Chapter 8). Extrapolating from local samplings and regional comparisons, Grassle and Maciolek (1992) have suggested a possible global diversity of 10 million animal species on the sea floor, but May (1992) disagrees with their calculations and suggests a more conservative guess of some 500 000 species only.

To sum up, an accurate estimate of global species diversity is still far away. However, we begin to understand what type of probes we must use and what types of calculations we need to perform, in order to improve the accuracy of current estimates, the balance of which is possibly of the order of 10 million species worldwide.

Towards the system 7

Revision of the primary classification of organisms is a taxonomic operation, governed by taxonomic principles; it differs from revision of a family or genus only in wider scope. Because of the human need of an arrangement of organisms which will express as fully as possible existing knowledge and opinion concerning them, all groups are always subject to revision.

H.F. Copeland (1938, p.383)

Mayr (1982b, 1988) has stressed the contrast between macrotaxonomy, the science of classifying, and microtaxonomy, the science of species. I believe that this distinction rests largely on tradition, rather than on sound theoretical reasons. However, it is true that, until recently, people interested in macrotaxonomy and people interested in microtaxonomy have been different. This dichotomy has been unfortunate, as little exchange of ideas has occurred from one field to the other. Things are changing (Nelson, 1989a), however, in so far as this distinction mirrors the prevailing state of affairs, it justifies a separate treatment of what systematists are doing at the two levels of investigation.

7.1 KINGDOMS AND PHYLA

During the 1980s and 1990s, one of the most restless areas of systematics has been the arrangement of higher taxa. Ultrastructural and molecular studies have caused and are still causing a continual revision of concepts and this part of the system will probably remain in a state of change for many years to come.

The first signs of uneasiness with the traditional division of living organisms into the two kingdoms of animals and plants, can be traced back to the mid-nineteenth century scientific literature. A separate kingdom, Protoctista, was proposed by Hogg (1860) to accommodate fungi, sponges, lower algae, bacteria and protozoa. Haeckel (1866) gave the name Protista to virtually the same group of organisms, but later (1894) redefined this kingdom to include only unicellular organisms. Despite this early interest in higher classification, no substantial improvements occurred until the 1930s. Novak (1930) divided living organisms into Akaryonta and Karyonta, which correspond to the later divisions of Prokaryota and Eukaryota.

The need to separate bacteria and unicellular eukaryotes from both plants and animals and from each other was most clearly expressed by Copeland (1938). He recognized four kingdoms, one for prokaryotes (Monera), one for green algae and land plants (Plantae), one for metazoans (Animalia) and one for the remaining eukaryotes (flagellates (excl. Volvocales), rhizopods, sporozoans, ciliates, diatoms, red and brown algae, and fungi (Protista)).

There was virtually no progress for 20 years until Whittaker (1959, 1969) increased the number of kingdoms from four to five. He recognized the Monera, Protista, Plantae, Fungi and Animalia. Whittaker's work marked the beginning of a spasmodic investigation of the relationships between the major groups of living organisms. His five-kingdom system soon gained wide popularity, in part through the papers and books of Lynn Margulis (e.g. Margulis and Schwartz, 1982). However, it soon became clear that the five-kingdom system required further improvement.

Leedale (1974) disregarded unicellularity as a criterion of systematic affinity and distributed Whittaker's Protista among the remaining eukaryote kingdoms. In the same paper he also proposed an alternative system with 19 kingdoms (Monera and 18 eukaryote kingdoms), but the interrelationships of all these higher taxa were left unresolved.

Jeffrey (1971) expressed the view that the kingdom does not represent the rank best suited for the 'highest' taxon in the taxonomic hierarchy. He recognized three superkingdoms: Acytota (viruses, already separated as Aphanobionta by Novak, 1930); Procytota (prokaryotes, with two kingdoms: Bacteriobiota and Cyanobiota); and Eucytota (i.e. eukaryotes, with five kingdoms: Rhodobiota, Chromobiota, Zoobiota, Mycobiota, Chlorobiota). This system included some interesting modifications, especially the very isolated position of Rhodophyta within eukaryotes. Jeffrey's classification was not altogether new because Dougherty, Gordon and Allen (1957) had already characterized the red algae as Mesoprotista, and regarded them as intermediate between Monera and the true Eukaryota. Also, Christensen (1962) had contrasted red algae (as Aconta) with the remaining eukaryotes (Contophora).

Several authors, such as Edwards (1976), Cavalier-Smith (1978, 1981), Whittaker and Margulis (1978) and Möhn (1984) (cf. Appendix 2) have worked to develop these schemes, although only a few contributions can be mentioned here. At the kingdom level, Whittaker and Margulis (1978) essentially reproduced the already fashionable five-kingdom scheme, but accommodated these kingdoms within two superkingdoms, Prokaryota and Eukaryota. These schemes, however, were fairly arbitrary since they still contained many artificial taxa which had been recognized as such for a long time. For instance, in Whittaker and Margulis (1978), the classification of the superphyla of the kingdom Fungi included both Deuteromycota (the so-called Fungi Imperfecti) and Mycophycophyta (lichens) (see section 7.4).

In the meantime, a true revolution has affected the classification of prokaryotes. Significant progress was made with Gibbons and Murray's (1978) new classification of the bacteria. These developments were rapidly obscured by the discovery by

Woese and Fox (1977) and Fox, Pechman and Woese (1977) that prokaryotes comprised two distantly related groups, later named Eubacteria and Archaebacteria. The relationships of these two taxa are no closer to each other than either of them with eukaryotes. This discovery was based on the comparative study of 16s rRNA. The antiquity of the split between Eubacteria and Archaebacteria soon became accepted. Woese's Archaebacteria comprise a group of prokaryotes much smaller than the Eubacteria, but very diverse regarding their living conditions and metabolic requirements.

An alternative view of the archaebacteria has been developed by Lake *et al.* (1984). They considered that in Sulfolobales and Thermoproteales the structure of the ribosome is different from that of the other archaebacteria and that they are possibly closer to eukaryotes than to the other archaebacteria, although remaining at an 'archaebacterial' grade of organization. For Sulfolobales and Thermoproteales Lake *et al.* (1984) introduced a new kingdom, Eocyta. This concept was later developed by Lake (1988, 1989), following his studies on rRNA sequences, and supported by the additional molecular evidence provided by Rivera and Lake (1992). According to Lake, the deepest bifurcation within the phylogenetic tree of the living organisms separates 'a proto-eukaryotic group (karyotes) and an essentially bacterial one (parkaryotes)'. In his 1988 tree, the first group includes three typical (archaebacterial) eocytes, *Thermoproteus tenax, Sulfolobolus solfataricus* and *Desulphurococcus mobilis*, together with typical (and diverse) eukaryotes, such as man, brine shrimp, yeast and corn. In the second group fall the Eubacteria together with (archaebacterial) halobacteria and methanogens. Lake's phylogenetic speculations suggest that 'the last common ancestor of extant life, and the early ancestors of eukaryotes, probably lacked nuclei, metabolized sulphur and lived at near-boiling temperatures'. According to Searcy (1987), the thermophilic Archaebacteria have developed some eukaryotic features, such as histones, a cytoskeleton, calcium modulation and acetogenic metabolism, in response to the problems posed by their harsh environment. Furthermore, he regards both amoeboid shape and internal cytoskeleton of a few Archaebacteria as a preadaptation for the subsequent evolution of endocytosis and amoeboid movement, as typical of eukaryotic cells. The molecular phylogeny of major archaebacterial groups has been recently re-evaluated by Sidow and Bowman (1991).

Archaea are not confined to extreme environments, however. De Long (1992) has demonstrated that these microorganisms are also widespread in oxygenated coastal marine waters. The ensuing new appreciation of their ecology is likely to suggest a revised evaluation of Archaea and of their possible interest in reconstructing the first steps of life on our planet.

As a synthesis of the research developed on these matters during the last 15 years, Woese, Kandler and Wheelis (1990) have formally proposed to divide living organisms into three domains, Archaea (grouping together the Euryarchaeota, i.e. methanogens and relatives, and the Crenarchaeota, i.e. the thermophilic archaebacteria), Bacteria and Eucarya. This proposal immediately caused Mayr (1990) to suggest an alternative arrangement:

Domain Prokaryota (Monera)
    Subdomain Eubacteria
    Subdomain Archaebacteria
        Kingdom Crenarchaeota
        Kingdom Euryarchaeota
Domain Eukaryota
    Subdomain Protista
    Subdomain Metabionta
        Kingdom Metaphyta
        Kingdom Fungi
        Kingdom Metazoa

The contrast between these two schemes is much deeper than a cursory glance suggests. What are contrasted here are not two different sets of data, or two different phylogenetic analyses. The ideas behind them reflect two opposing views of the meaning of a biological classification, two opposite ways of relating a classification to phylogeny. Woese, Kandler and Wheelis (1990, 1991) have tried to suggest a phylogenetic (cladistic) classification, whereas Mayr (1990, 1991) is deliberately suggesting an 'evolutionary' one, attempting to incorporate additional information beyond that which can be expressed by a cladogram.

In the following pages I have ignored viruses. 'Viruses are probably more closely related to their hosts than to each other. They may have originated as nucleic acids that escaped from cells and began replicating on their own. ... Thus, the polio and flu viruses are probably more closely related to people, and the tobacco mosaic virus (TMV) to tobacco, than polio and TMV are to each other' (Margulis and Schwartz, 1982, p.14). Dramatic instances of virus exchange between man and other primates notwithstanding, there seems to be no better choice, for systematics, than that suggested in the preceding lines.

7.2 'PROKARYOTES'

Two recent unconventional studies point to the huge diversity of prokaryotes still awaiting discovery. Giovannoni *et al.* (1990) have studied the diversity of 16S rRNA genes from samples of Sargasso Sea bacterioplankton. Their analysis, completely bypassing the usual procedures of bacteriology (establishing cultures and so on), relies on the straightforward techniques of molecular cloning and sequencing. In this way, they have found a novel microbial group (they call it the SAR11 cluster) as well as evidence that this bacterioplankton may comprise many independent lineages derived from very distant common ancestors. Other studies of bacterial diversity performed by the same method include those of Britschgi and Giovannoni (1991); Schmidt, DeLong and Pace (1991) and Ward, Weller and Bateson (1990). This last study has demonstrated the huge diversity of still uncultured bacteria from a hot spring community. Along the same lines, Fuhrman, McCallum and Davis (1992) have studied marine planktonic archaebacteria.

Blue-green algae, or Cyanobacteria, have been the objects of far-reaching systematic controversies. A comprehensive revision of herbarium material of blue-green algae was completed in the 1970s by Drouet and co-workers (Drouet, 1978, 1981). Drouet and others strongly criticized the traditional taxonomic treatment of the group, largely based on the structure of the cell sheaths, because these structures are formed by materials which as soon as they are secreted from cells, either disperse, or consolidate in a form which is dependent on the surrounding conditions. They reduced some 2000 named species to 62. This conclusion was strongly criticized by Stanier, Stam, Holleman and others (see Friedmann, 1982), who had begun to gather completely new kinds of descriptive evidence. Instead of relying on dried herbarium specimens, they worked exclusively with pure cultures, as currently practised by bacteriologists. In fact, the blue-green algae are a group of photosynthetic bacteria. In the meantime, another new classification of Cyanobacteria had been proposed by Rippka *et al.* (1979).

If blue-green algae are prokaryotes, a place has to be found for them within the system of bacteria. The first proposal was made by Gibbons and Murray (1978), who placed them as the order Cyanobacteriales within Gracilicutes. At the same time, Stanier *et al.* (1978) proposed to consider their nomenclature under the rules of the International Code of Nomenclature of Bacteria. One important consequence of this inclusion was that the types should be pure cultures representing clones instead of herbarium specimens. To date, the contrasting views of botanists and microbiologists have not been resolved. The Nomenclature is still covered by both the Bacterial and the Botanical Codes. Those who regard blue-green algae as plants must continue to typify species with herbarium specimens, as for flowering plants, ferns and fungi. Those who regard them as bacteria typify them with living cultures, as for (typical) bacteria.

Before leaving prokaryotes it is necessary to say a few words about *Prochloron* and *Prochlorothrix*. *Prochloron didemni* is a photosynthetic prokaryote living as an exosymbiont on didemnid ascidians. First described by Lewin and Withers in 1975, *Prochloron* has often been regarded as a living representative of a very old branch of photosynthesizing prokaryotes, an assemblage thought to have given rise to the chloroplast of the green algae, higher plants and euglenoids. What is certain, the chloroplasts of these photosynthesizing forms contain the same pigments as *Prochloron*, chlorophylls *a* and *b*, but no phycobilins. Alternatively, conventional cyanobacteria are better candidates for the origin of chloroplasts of rhodophytes which, like them, lack chlorophyll *b* but contain chlorophyll *a* and phycobilins. Research on *Prochloron* has therefore been intensive, so far as allowed by the obvious difficulties of culturing it, owing to its symbiotic life-style. Matters have developed, however, with the subsequent discovery (Burger-Wiersma *et al.*, 1986) of *Prochlorothrix hollandica*, a free-living filamentous form, reasonably close to *Prochloron*. According to Turner *et al.* (1989), the evidence derived from 16S rRNA places *Prochlorothrix* within cyanobacteria. And the same conclusion results from analysis of the green chloroplasts. Within this cyanobacterial assemblage, however, *Prochlorothrix* does not appear to be specifically close to the green chloroplasts. On the other hand, such a relationship is supported by *psbA* genes, which encode for the photosystem II thylakoid protein D1 (Morden and Golden, 1989a; but cf. Morden and Golden, 1989b). In the tree derived from this latter molecular evidence, *Prochlorothrix* plus green plants form a clade within a (paraphyletic) group of cyanobacteria. Evidently, matters are still unsettled (Penny, 1989). Recent studies seem to disprove any close relationship of Prochlorales to green plant chloroplasts (Palenik and Haselkorn, 1992; Urbach, Robertson and Chisholm, 1992).

7.3 THE MAJOR GROUPS OF EUKARYOTES

Are eukaryotes truly monophyletic? A few authors, such as Walter (1987, pp.27–8), have expressed some doubts, but the problem is not an easy one to answer, or even formulate, in the light of the well-known theory of the symbiotic origin of the eukaryote cell. What do we really mean when we speak of eukaryote monophyly? Can we consider monophyly of the so-called host cells? That is, those cells contributing a eukaryotic nucleus to a symbiotic chimaera? Can we also consider the uniqueness of the symbiotic relation *per se* as monophyletic? That is the host cell, the chloroplast precursor, the mitochondrial precursor, and the symbiotic bond between them? Recent versions of the symbiotic theory (Cavalier-Smith, 1987) suggest that eukaryotes may be monophyletic. According to Cavalier-Smith, the cell that was invaded by previously independent prokaryotes (that became mitochondria and chloroplasts) was a non-amoeboid, biciliated protozoan rather than an amoeboid prokaryote. Seen another way, the transition from prokaryote to eukaryote preceded the formation of modern eukaryote cells, with all their included organelles. Mitochondria and chloroplasts then originated once only, from prokaryote symbionts. The incorporation of purple nonsulphur photosynthetic bacteria (the prospective mitochondria) and cyanobacteria (the prospective chloroplasts) occurred simultaneously in the same host cells, about 960 million years ago, some time after the origin of the first phagotrophic eukaryote. Cavalier-Smith's reconstruction is perhaps too simple to be true. It is attractive, however, because it seems to reassure us about the monophyly of Eukaryota, at least in the restricted sense of the monophyly of the host cell. As for the chloroplasts, there are still some doubts, as we saw earlier.

On these matters, we can confidently expect some major news in the near future. Sprague (1991) has recently summarized the current knowledge about genetic exchanges between kingdoms, including the experimental demonstration of 'conjugation' between bacteria and yeasts. Genes of quite different origins are probably present even within the genome of a single organelle. According to Markowicz and Loiseaux-de Goër (1991) and Assali *et al.* (1991), the rhodophyte and chromophyte plastid genome would derive from at least two different ancestors, a cyanobacterial and a β-proteobacterial one, because of combined horizontal and vertical gene transmission.

An improvement to the systematization of eukaryotes, in respect to Whittaker's previous arrangement, was provided by Cavalier-Smith (1981), who recognized nine kingdoms:

Eufungi: non-flagellated fungi
Ciliofungi: flagellated fungi
Animalia: choanoflagellates and metazoans, including sponges
Biliphyta: Rhodophyta and Glaucophyta
Viridiplantae: Chlorophyta, Bryophyta and Tracheophyta (vascular plants)
Euglenozoa: Euglenoidina and Kinetoplastida
Protozoa, including Dinophyta
Cryptophyta
Chromophyta

A different distribution was proposed by Möhn (1984, cf. Appendix 2), with a primary division within eukaryotes of Aconta and Contophora. Aconta are recognized as eukaryotes primarily lacking both flagella and centrioles and possessing plastidia similar to the cyanobacteria, and pigments intermediate between those of the cyanobacteria and the green plants. Aconta include Rhodophyta, the red algae, and Cyanidiophyta, which comprises one species, the unicellular *Cyanidium caldarium*, some 4 μm in diameter. *Cyanidium caldarium* possesses one chromatophore with chlorophyll *a*, allophycocyanin and C-phycocyanin (the pigments of blue-green algae) and lives in hot, acid springs (45–60°C, pH 2–3).

Contophora possess flagella and centrioles, although either may be secondarily lost, as in several major lines. Within Contophora, Möhn (1984) identified 11 major groups: Chlorobionta, Flagelloopalinida, Euglenophytobionta, Eumycota, Dinophytobionta, Cryptophytobionta, Colponemata, Chloromonadophytobionta, Chromophytobionta, Cormobionta and Animalia. Möhn's system was devised mainly using phylogenetic criteria, but these were abruptly discarded when he elevated Cormobionta (plants) and Animalia both to the rank of 'Oberreich', taking them from within Chlorobionta and Chromophytobionta, respectively. Accordingly, Möhn's Chlorobionta and Chromophytobionta are both paraphyletic.

Similar results were obtained by Lipscomb (1985) in her cladistic analysis of the major groups of eukaryotes. Lipscomb identified 10 major clades. The basal clade is represented by the Rhodophyceae, which is the sister-group of all remaining eukaryotes. The remaining taxa, in turn, split into two clades. The first includes three major groups, Gameophyceae, Prasinomonadida together with Phytomonadida and a more heterogeneous assemblage including Cryptomonadida, Chloromonadida, Prymnesiida, Phaeophyceae, Chrysomonadida, Xanthomonadida and Eustigmatomonadida. The other major clade is split into two branches. One leads to 'fungi' (Ascomycetes, Zygomycetes), 'euglenoids' (Trypanosomatina, Bodonina, Euglenida, Stephanopogonida), and a 'protozoan' assemblage including Dinoflagellida, Ciliophora, Schizophrynida and Opalinida. The final branch includes three subgroups: Choanoflagellida together with animals, the 'aquatic fungi' (Oomycetes and Chytridiomycetes) and the Hypermastigida. Some eukaryote groups are not explicitly included in Lipscomb's analysis, but it does not seem too difficult to find a place for the higher plants near the Phytomonadida or Gameophyceae.

These arrangements, however, are not definitive. Consider the findings of Sogin *et al.* (1989) on the rRNA of the well-known 'protist' *Giardia lamblia*. The evolutionary distances they estimated from sequence comparisons between the 16S-like rRNAs of *Giardia* and other eukaryotes are larger than similar estimates of evolutionary diversity between archaebacteria and eubacteria. The *G. lamblia* 16S-like rRNA seems to have retained many features that may have been present in the common ancestor of eukaryotes and prokaryotes. Accordingly, Sogin *et al.* (1989) raise some doubts as to the monophyletic nature of eukaryotes.

In the meantime, the Woese group (Vossbrinck *et al.*) demonstrated that another 'protist' group (Microsporidia) is also very divergent from the typical eukaryotes. A higher taxon, Archezoa, has thus been used (cf. Cavalier-Smith, 1989) for Microsporidia plus Metamonada, the latter group including *Giardia* and its relatives. Archezoa are thus the sister-group to Metakaryota (all the remaining eukaryotes). What is particularly puzzling is that the divergence between Archezoa and Metakaryota seems to be, in molecular terms, larger than that existing between Archaebacteria and Eubacteria (Cavalier-Smith, 1989). A few amoeboid eukaryotic unicellulars, the so-called archaeamoebae (*Mastigamoeba, Pelomyxa*), should perhaps be referred to the same assemblage (Cavalier-Smith, 1991).

7.4 FUNGI

Soon after the early claims to recognize Fungi as an independent kingdom, serious doubts were raised as to its monophyletic status. Flagellate fungi were soon recognized (Cavalier-Smith, 1981) as worthy of separate status from the main fungal assemblage, but this first distinction of Ciliofungi and Eufungi proved unsatisfactory.

In the meantime, Klein and Cronquist (1967), Kubai (1978) and Bold, Alexopoulos and Delevoryas (1980) proposed a range of taxa as the possible sister-groups to fungi, including heterosiphonalean algae (comparable to the recent *Vaucheria*, Xanthophyta). Barr (1983, p.70) believes that only the aquatic Oomycetes and Hyphochytridomycetes have their origins in heterokont 'algae', whereas the origin of all the other terrestrial fungi remains unknown.

The systematization of fungi is elucidated by Tehler's (1988) cladistic analysis. Tehler recognized two synapomorphies, molecular weight of 25S (instead of 28S) for rRNA and chitinous cell walls, to diagnose a revised Eumycota which comprises the following traditional groups: Hyphochytriomycetes, Chytriomycetes, Zygomycetes, Ascomycetes and Basidiomycetes. The monophyly of Eumycota seems to be supported by different kinds of molecular data, including the 5S rRNA sequence (Walker, 1985, p.274). However, there are many divergent opinions. At the opposite end of the spectrum, Eriksson (1981, p.182) suggested that the Eumycota is a polyphyletic group and there are at least five sister-groups of five different algal taxa. In particular, he and other authors favour the hypothesis that Ascomycetes and Basidiomycetes are more closely related to red algae. The monophyly of a group comprising Ascomycetes plus Basidiomycetes is fairly well accepted.

In Tehler's (1988) analysis, the outgroup of Eumycota is Oomycetes. Eumycota and Oomycetes share several characters such as the presence of hyphae, heterotrophism and similarities in the ultrastructure of the flagellar apparatus.

However, the Oomycetes are still retained within Eumycota by many recent authors. Within Eumycota, Tehler (1988) characterizes several traditional taxa as monophyletic including Amastigomycota, Dicaryomycotina, Ascomycetes, Protobasidiomycetes, Basidiomycetes, Hymenomycetidae and Homobasidiomycetales. At the same time, he discards many other traditional taxa as paraphyletic, such as Mastigomycotina, Hemiascomycetes, Ustomycetes, Holobasidiomycetes, Heterobasidiomycetes and Phragmobasidiomycetes. Ascomycetes is retained, but only by excluding Taphrinaceae and the yeast fungi. These two groups are united together as the Protobasidiomycetes and are regarded by Tehler as close relatives to the Basidiomycetes rather than the Ascomycetes.

In the meantime, Ascomycetes systematics is being studied in detail with attention being paid to previously neglected characters, such as different types of development, and to major shifts in grouping such as the incorporation of the lichens.

It has been known for a long time that the fungal partner of the lichen symbiosis is more specific than the alga. In other words, every lichen 'species' contains a different fungus, but only a small number of algal species actually represent their photosynthetic partners. Therefore, in terms of systematization it becomes obvious that lichens should be accommodated within the fungi. With some 525 genera and 13 500 species (Hawksworth, Sutton and Ainsworth, 1983), lichens represent about 20% of all described fungi. The vast majority of lichen-forming species are ascomycetes and about 20 genera are currently recognized as including both lichenized and non-lichenized species.

Lichens (and the eukaryotic cell), however, are probably not the only evolutionary chimaeras. Molecular evidence suggests that some unicellular 'protists', such as the cryptomonads, are also the product of a symbiosis of two different unicellular eukaryotes (Douglas *et al.*, 1991; Penny and O'Kelly, 1991).

The suprageneric arrangement of Ascomycetes is under much discussion. As mentioned in Chapter 5, an effort to develop a 'consensus system' of fungi was started in 1982 by O. Eriksson, with his journal *Systema Ascomycetum*, now jointly edited by him and by D.L. Hawksworth. They publish a periodically updated *Outline of Ascomycetes* providing a revised classification of the group, by incorporating newly published information (e.g. Eriksson and Hawksworth, 1986). This effort is certainly worth mentioning, but it should be said that some mycologists regard it as more authoritarian than authoritative (cf. Reynolds' (1989) objections and Eriksson's (1989) reply). I believe that consensus is dangerous in science, but can sometimes be useful in practice. Accordingly, consensus approaches cannot shape systematics, but they can help to obtain a classification for applied purposes. Both aims (science and applied use of scientific knowledge) are worthy of effort, provided that their non-equivalence is always borne in mind.

Within Basidiomycetes, there is some agreement at the family level, but ordinal and supraordinal classification is very unstable. The status of the smuts (Ustilaginales) and the rusts (Uredinales) is particularly unstable. A closer relationship between the two taxa, often accepted under a taxon Hemibasidiomycetes, is currently rejected as unnatural.

A perpetual problem in fungal systematics is the Fungi Imperfecti. This is a 'dustbin' taxon comprising forms known only in asexual, conidial condition (the so-called anamorphs, as compared with the sexual ascus- or basidium-producing forms known as teleomorphs). These asexual forms are currently classified, formally, as the Deuteromycotina, with orders, families, genera and species. It is considered that most Fungi Imperfecti are anamorphs of ascomycetes. However, as long as their teleomorphs (sexual stages) are unknown, it is virtually impossible to identify true affinities at a more detailed taxonomic level. The problem is huge because some 1680 'genera' and 17 000 'species' have been described and many of them are economically important. The knowledge varies from group to group. For some deuteromycetes the teleomorph is understood but for others it is possible that all parts of the life cycle are known but the connections between anamorphs and teleomorphs have yet to be demonstrated. According to the International Code of Botanical Nomenclature, the use of separate names for the different states of these pleomorphic fungi is allowed. However, it is prescribed that the teleomorph name be used to refer collectively to the species in all states when they become known.

What appears to be the worst difficulty is that there is an increasing awareness that many imperfect fungi have lost their teleomorphic stage and replaced true sexuality with parasexuality. Therefore, some anamorphs will never be associated with corresponding teleomorphs. As far as is possible, one should try to associate them with groups from which they are probably derived. This procedure has already been applied to some anamorphs without teleomorphs, that clearly belong to Erysiphales, Uredinales and other groups (Hawksworth, Sutton and Ainsworth, 1983).

7.5 'PROTISTS'

Today, it is impossible to recognize a taxon Protozoa (or Protista, or Protoctista) without acknowledging that it is paraphyletic or even polyphyletic. Both animals and plants had their origin from within this heterogeneous assemblage of mostly unicellular organisms, perhaps from two different groups not so distantly related to each other. Furthermore, it is sometimes difficult to say what is a flagellated 'fungus' and what is a flagellated 'protist'; the so-called 'gut fungi', for instance, were regarded as 'zooflagellates' until 1975, to be then recognized as chytridiomycetes (Heath, 1988). We still have insufficient knowledge of these organisms to obtain a satisfactory phylogenetic arrangement of all major protist taxa. At present, it is perhaps safer to identify a number of major monophyletic units and to investigate their mutual relationships without attempting to formalize the resulting hierarchy. To document the state of affairs in protozoology, it is useful to quote a paper by Wolf and Markiw (1984), where actinosporeans and myxosporeans are demonstrated, for the first time, to be the alternating life forms of a single organism, rather than two separate classes.

Corliss (1984) has no difficulty in recognizing 45 protozoan phyla (Appendix 3), but he clearly confesses that only the Ciliophora (the ciliates) is clearly or demonstrably monophyletic. An alternative classification of Protists, suggested by Margulis (1990), is summarized in Appendix 4.

Electron microscopy has provided the most powerful tool for obtaining fresh information about Protozoa. One of the most prominent discoveries in this field has been the recognition of six new organelles (conoid, micronemes, micropore, polar ring, rhoptries and subpellicular tubules) in the 'Sporozoa'. This discovery led Levine (1970) to define a new group Apicomplexa, now currently regarded as a phylum. Ultrastructural work is still very active and provides many new suggestions and grounds for ever revised classifications, as documented by the handbook of Margulis *et al.* (1990).

It is from molecular biology, however, that the most stimulating new insights come, regarding the affinities between 'protozoan' groups and between them and the other eukaryotes. For example, an analysis of partial sequences of 28S rRNA from species classified in 12 different protozoan phyla reveals that Euglenozoa emerged very early from the common eukaryote stem, whereas the two idiosyncratic groups, ciliates and dinoflagellates, emerged late, thus suggesting (Qu *et al.*, 1988) that some of their characteristics, often regarded as primitive (e.g. the kind of mitosis in dinoflagellates), are derived. Despite the morphological evidence, close affinities for dinoflagellates and ciliates are suggested by molecular evidence (e.g. Gunderson *et al.*, 1987, who compared small-subunit rRNA sequences) and already indicated in Möhn's (1984) and Lipscomb's (1985) systems. Another interesting and perhaps unexpected result is the close relationship (although already suggested by Bovee (1971) with much insight) between euglenoids and trypanosomids, a grouping strongly supported by molecular evidence (e.g. Gunderson *et al.*, 1987; Hori and Osawa, 1987).

Less expected is Wolters' (1991) finding that molecular evidence would place the Apicomplexa or Sporozoa within the dinoflagellate-ciliate clade. This is confirmed by Gajadhar *et al.* (1991), after extensive analysis of rRNA sequences. They identify this clade (with ciliates as the sister-group to dinoflagellates plus apicomplexans) as one of the five major clades within eukaryotes, the remaining being, in their opinion, animals, plants, fungi and Cavalier-Smith's (1981) Chromista, i.e. oomycetes, brown algae, diatoms and a few other 'algal' groups (see also Aritzia, Andersen and Sogin, 1991; Sogin, 1991). A cladistic analysis of the Chromista has been recently provided by Williams (1991).

Molecular approaches have also been employed within the lower groups, for example, within kinetoplastids (Lake *et al.*, 1988) or within ciliates (Lynn and Sogin, 1988), where the results differ from those based on ultrastructural evidence.

To illustrate the state of flux characterizing protozoan systematics, for ciliates no less than five different classifications have been proposed during the last decade (Corliss, 1979, 1986c; Jankowski, 1980; de Puytorac *et al.*, 1984; de Puytorac, Grain and Mignot, 1987; Small and Lynn, 1981, 1985). A paper by Lynn and Small (1988) tries to identify the consensus of some taxa, as a background for future developments.

7.6 METAZOANS

The monophyly of Metazoa has been put in doubt by some recent authors (Anderson, 1981; Walter, 1987; Bergström, 1991), but this view is a minority opinion. For positive statements on metazoan monophyly, see Salvini-Plawen (1978), Nielsen (1985), Ghiselin (1988a), Ax (1989a), Lake (1990) and Schram (1991): their arguments cover morphology, molecular biology and palaeontology.

However, our appreciation of metazoan phylogeny could change, if future investigations substantiate the molecular phylogenetic trees elaborated by Field *et al.* (1988) from the sequences (more than 1000 nucleotides for each species they looked at) of 18S rRNA. In their study, 22 classes in 10 phyla are represented.

Cnidaria seems to be the sister taxon to a protozoan ancestor different from the sister-group of the remaining Metazoa. The available data do not allow resolution of the branching order of cnidarians, plants, fungi and ciliates. Field *et al.* (1988) remark that cnidarians have always been regarded as animals, partly because they have muscles and nervous tissue, but these two features can be traced back to molecular systems already present in different protozoan groups (e.g. actin–myosin systems as a basis for motility). The same workers, however, have raised some doubt as to the reliability of their tree, because of problems in resolving the branching order of several important clades (Field *et al.*, 1990).

The hypothesis of a separate origin for diploblastic and triploblastic metazoans from different unicellular ancestors has also been raised, on the evidence provided by sequencing 28S rRNA, by Christen *et al.* (1991a,b). However, their trees also suffer from lack of resolution power as for several important, deep branchings.

Many papers have been published on metazoan phylogeny and classification. It is not possible to summarize them adequately in a few lines. I will cite the proceedings of three international symposia (House, 1979; Anonymous, 1981; Conway Morris *et al.*, 1985) and Möhn's (1984) huge, though incomplete, treatise (see Appendix 5 for his new arrangement of metazoans).

A book-length discussion of metazoan phylogeny is given by Willmer (1990): an essay of interest, for its strong objections to the phylogenetic value of nearly all types of morphological, palaeontological and molecular evidence accrued until now, when assessing the affinities between phyla or other major groups. Her criticisms, however, are not followed by reasonable alternative proposals.

Nielsen and Nørrevang's (1985) approach to metazoan phylogeny is also worth a mention. They depict an adaptive scenario of metazoan evolution, further developed by Nielsen (1985) to suggest a new arrangement of animal phyla. In his scheme (Appendix 6), choanoflagellates are united with Metazoa to form the kingdom Animalia. Metazoa (Porifera plus Placozoa plus Gastraeozoa or Eumetazoa) are characterized as possessing collagen, septate/tight junctions and spermatozoa, whereas Gastraeozoa are indicated as possessing basal lamina, nerve cells and synapses. Nielsen's classification is certainly attractive insofar as it incorporates ultrastructural data and arguments from functional comparative morphology. However, some idiosyncratic innovations, such as the unusual placement of Ctenophora, or his concept of Articulata (with Annelida s.l., Echiura, Gnathostomulida, Onychophora, Arthropoda s.l., Mollusca and Sipuncula) require further investigations.

Many well-known terms have been traditionally used to design major supraphyletic groupings within Metazoa, such as Bilateria, Protostomia, Deuterostomia, Coelomata, Spiralia, Pseudocoelomata (Aschelminthes, Nemathelminthes),

Archicoelomata, Tentaculata, and others. Many of these terms are currently employed, but very often as informal, vernacular names, so as to avoid enumeration of phyla under discussion.

As for coelomates, there are at least two major open problems (Barnes, 1985). First, how many times has the coelom evolved? Second, are all acoelomate groups relatively primitive, with respect to coelomates? In answer to the first question, I am inclined towards accepting a polyphyletic origin for the coelomate body plan, but the evidence is inconclusive. In answer to the second question, the postulated origin of acoelomate groups (Platyhelminthes, Nemertea, Gnathostomulida) from coelomates could be conceived as the result of progenesis from larval and juvenile stages of coelomates. This was suggested by Rieger (1985) on ultrastructural evidence, but I think that we must be very cautious before giving too much credit to hypotheses requiring very extensive regression. This point is also acknowledged by Nielsen (1985).

7.7 PLACOZOANS

The phylum Placozoa was established to accommodate the small multicellular organism *Trichoplax adhaerens* (Schulze, 1883), which has only been adequately described in the last two decades. *Treptoplax reptans* (Monticelli, 1896) has also been tentatively ascribed to the same group, but its true identity still remains doubtful. *Trichoplax* is a free-living marine animal, with a flattened body composed of two epithelial layers (dorsal and ventral), with fluid and a network of fibre cells between. Owing to its simple morphological structure, and to the small number of cell types recognizable in it, *Trichoplax* has become a favourite for phylogenetic speculation. The body plan of *Trichoplax* has been regarded (e.g. Grell, 1971, 1974, 1982), as a critical step towards multicellularity, in a scenario different from the traditional Haeckelian one. Grell describes *Trichoplax* as diploblastic and homologizes the dorsal epithelium with the ectoderm and the ventral epithelium with the entoderm of higher animals. Viewed in this way, *Trichoplax* is the hypothetical flattened *Placula* suggested by Bütschli (1884) as the ancestor of metazoans.

For Ax (1989a), the Placozoa are the sister-group to Eumetazoa, together forming the sister-group to the sponges. This arrangement seems to be supported by Haszprunar, Rieger and Schuchert (1991).

7.8 SPONGES

Sponges are currently regarded as true Animalia (Metazoa), mainly as the sister-group of all the remaining metazoans (Eumetazoa or Histozoa). Möhn (1984) divided sponges into three phyla (Calcara, Demospongea and Hexactinellidea, the last two together forming the infrakingdom Silicosponga, as opposed to Calcarosponga). However, these phyla merely correspond to the usual classes, simply raised to a higher rank. A slightly different arrangement was proposed by Reiswig and Mackie (1983), who contrasted the siliceous sponges as the subkingdom Symplasma, with all other sponges. Bergquist (1985) raised this subkingdom to the rank of an independent kingdom, on the grounds of syncytial (instead of cellular) organization of these sponges. To date, this proposal does not seem to have gained much support.

An interesting field of enquiry is the problem of several recent sponges (about 15 species) possessing an unusual solid calcareous skeleton. This skeleton bears a striking resemblance to those of several coral groups. According to Vacelet (1985), these sponges are the survivors of previously more widespread reef builders known in palaeontology as stromatoporoids, tabulate 'corals' and sphinctozoans. These were generally classified within Cnidaria and believed to be extinct. A separate class of sponges was erected for them, but Vacelet (1985) rejected this arrangement as unjustifiable and incorporated these forms partly within Calcarea and partly within Demospongea.

7.9 CNIDARIANS

The monophyly of Cnidaria does not seem to be in dispute. Major improvements to the system of this taxon are based on an improved knowledge of life cycles. The life cycle of the Cubomedusae was unknown until Werner (1973, 1975) studied the Japanese species, *Tripedalia cystophora*. He recognized a major difference with the conventional Scyphozoa, where the polyp gives rise to the medusa by strobilation (a form of asexual reproduction); in the Cubomedusae it directly metamorphoses into a medusa. This developmental trait strongly reinforces the isolated position of Cubomedusae within the phylum. Accordingly, Werner has proposed a new class for them, Cubozoa, now universally accepted. Möhn (1984) also placed the sessile, polyp-like Stauromedusae within Cubozoa, but this placement is less convincing.

As for Hydrozoa, knowledge of life cycles is rapidly improving the system, which was traditionally biased by the separate classifications of polyps (hydroids) and medusae. New studies (Boero and Bouillon, 1987) also suggest an alternative interpretation of the metagenetic cycle, currently understood as alternating between two generations, that of the sexual medusa and that of the asexual polyp. It is now suggested that the polyp is a prolonged larval stage, whereas the medusa represents the only true adult in the cycle. Accordingly, Boero and Bouillon propose to retain as subclasses the traditional medusa taxa (Anthomedusae, Leptomedusae and Limnomedusae) and to use these names for both adults (medusae) and larvae (hydroids). Therefore, the names Athecatae, Thecatae and Limnopolypae, already used for hydroids, are banished from the system.

The hydroids are of little phylogenetic importance, because similar hydroids often produce different medusae. Very similar polyps, all of them traditionally classified within the Campanulinidae, give rise to very different medusae, which, according to Bouillon, should be distributed among 16 families at least (Boero and Bouillon, 1987).

By-the-wind-sailors (*Velella* and relatives) are traditionally misplaced taxa. They have been removed from the Siphonophora and are now regarded as members of the separate order Chondrophora, or ranked with the Anthomedusae. It is interesting to note that 36 species in nine genera recognized by Haeckel (1888) a century ago mostly represent simple growth stages of just three species (Kirkpatrick and Pugh, 1984).

7.10 CTENOPHORANS

The phylogenetic relationships of Ctenophora, itself a reasonably well-founded monophyletic group, are still poorly understood. However, most authors place them somewhere in the proximity of cnidarians. Alternative placements, such as Nielsen's (1985; cf. Appendix 5), appear to be tentative and insufficiently supported. The question of a possible occurrence of mesoderm in Ctenophora is generally rejected (Siewing, 1977; Ortolani, 1989), which reinforces their placement near Cnidaria. A contrary opinion is given by Schram (1991), who regards the ctenophores as triblastic and as more closely related to the other metazoans than to cnidarians. I would question, however, the ability to homologize germ layers throughout the whole animal kingdom.

Ctenophora is one of the smallest phyla, but its internal arrangement is still provisional. Attempts to arrange the higher taxa cladistically (Harbison, 1985) are not conclusive. Many species described as belonging to Cydippida are probably larval stages of species belonging to other orders (Harbison and Madin, 1982). Moreover, there are newer discoveries of ctenophorans very different from those known to date. Madin and Harbison (1978) described *Thalassocalyce inconstans*, an extraordinary medusoid form which they accommodated within the new order Thalassocalycida. Two exceptionally well preserved fossil ctenophorans (*Palaeoctenophora brasseli* and *Archaeocydippida hunsrueckiana*) have been discovered (Stanley and Stürmer, 1983, 1987) in Devonian rocks. Both strongly resemble modern cydippids.

7.11 PLATYHELMINTHS

A major revision of the system of platyhelminths was published by Ehlers (1985). In addition to the traditional evidence and to the phylogenetic analysis previously developed by Ax (e.g. Ax, 1961, 1963), Ehlers has elaborated the results of his extensive ultrastructural researches, especially on the tegumentary system. The data are analysed cladistically and from the resulting system (Appendix 7) conventional taxa such as Turbellaria, Archoophora and Cestodaria have disappeared as paraphyletic. Although some of the monophyletic groups established in this work are based on a small number of apomorphies, Ehlers' effort should nevertheless be regarded as sound and courageous. One final feature of Ehlers's system deserves attention, although it is not new. That is the separation of Monogenea from the remaining Trematoda in the old classifications (Aspidobothrii plus Digenea). Instead, Monogenea are united with Cestoda to form a seemingly well-founded taxon Cercomeromorphae. Further improvements to the system of these parasitic groups can be found in Brooks (1982, 1989a,b) and Brooks *et al.* (1989).

At present, it seems that the system of Platyhelminthes has been improved at higher levels so much more than other phyla. However, systematization of lower taxa still needs more work. Particularly opaque areas are the acoels, the terrestrial triclads, the triclads of the Baikal Sea and the polyclads (Crezée, 1982). In addition, there are two groups of epibiont flatworms of very uncertain placement within the phylum, temnocephalids and udonellids. The most difficult case, however, is *Xenoturbella bocki*, a marine 'worm' described by Westblad in 1949 and often regarded as the most primitive of flatworms. Very different affinities have been suggested by some authors, for example Reisinger (1960), who regarded it as the neotenic larva of some hemichordate or echinoderm. In the latest review on the subject, Haszprunar, Rieger and Schuchert (1991) report that even recent ultrastructural research has failed to offer a clue regarding whether to place *Xenoturbella* safely with acoelomorph flatworms or with enteropneusts.

Unfortunately, flatworm taxonomy does not attract many workers, largely because of the technical difficulties of their study, which involves histological sections, whole-mount stains and the examination of living specimens for routine identification.

7.12 GNATHOSTOMULIDS

First described by Ax in 1956, Gnathostomulida are a small group of marine 'worms' of uncertain affinities. According to Ax (1985a), Gnathostomulida and Platyhelminthes are sister-groups, together forming the Platyhelminthomorpha. These are regarded, in turn, as the sister-group to all remaining bilaterian metazoans (the Eubilateria). Without formally rejecting this interpretation, Sterrer, Mainitz and Rieger (1985) point out that there are similarities between gnathostomulids and other groups, such as gastrotrichs (monociliary epithelium and structure of protonephridia) and rotifers. Another interpretation of the affinities of Gnathostomulida is given by Nielsen (1985), who regards them as highly specialized articulates, close to annelids. Accordingly, the monociliate condition and coelomic reduction are interpreted as specializations to mesopsammic life.

7.13 MESOZOANS

Systematic problems are even larger in the Mesozoa. This term, as first proposed by van Beneden in 1882, is still currently used to accommodate two very different and probably unrelated groups of parasitic organisms, Rhombozoa (= Planuloidea = Moruloidea), including dicyemids and heterocyemids, and Orthonectida. Mesozoans have been variously regarded as structurally intermediate between 'protozoa' and 'true Metazoa', or as platyhelminths regressed in relation to their parasitic habits, but to date they still have indeterminate affinities (Stunkard, 1982).

7.14 ASCHELMINTHS

The group Aschelminthes, or Pseudocoelomata, or Nematelminthes, is retained in many recent classifications, sometimes at the rank of a phylum, sometimes as a more or less well defined assemblage of several phyla. The groups (classes, or independent phyla, according to taste) generally assigned to this group include Rotifera, Acanthocephala, Nematoda, Nematomorpha, Kinorhyncha, Gastrotricha and Priapulida, together with the recently discovered Loricifera. Aschelminthes are recognized as a monophyletic group by Andrássy (1976), Coomans (1981) and Nielsen (1985). On the other hand, the group is explicitly rejected as artificial by Clément (1985). Lorenzen (1985) also fails to identify synapomorphies that could justify the monophyly of the group. However, both Clément and Lorenzen recognize the relationship of rotifers with acanthocephalans, whereas most remaining groups (Nematoda, Nematomorpha, Kinorhyncha, Gastrotricha and Priapulida) probably form another monophyletic line (Lorenzen, 1985). The question then is, are these two assemblages sister-groups?

As for nematodes, Wolstenholme *et al.* (1987) discovered unusual sequences in the mitochondrial genes for tRNAs. Brandl, Mann and Sprinzl (1992) have recently developed this line of enquiry, thus suggesting that this group is more closely related to arthropods than to archicoelomates and chordates. In their view (Brandl, Mann and Sprinzl, 1992, p. 325) that implies 'that the pseudocoelomate body plan of nematodes is derived from the coelomate anatomy'. Unfortunately, there are no comparative data for the remaining archicoelomates; in addition, Brandl, Mann and Sprinzl's arguments are simply phenetic and phylogenetically not compelling. The currently available molecular evidence simply confirms the very isolated position of nematodes already suggested by morphology.

Within nematodes, most systematic efforts are directed towards the lower level taxonomy of some groups of major economic interest. Many traditional nematode taxa are widely regarded as artificial, but major overall revisions are not ready.

In recent years, there has been no real progress in the systematic knowledge of rotifers, mainly because of a general shift of interest to other problems, especially ecology (B. Pejler, personal communication).

For the Gastrotricha, Boaden (1985) at variance with most current interpretations has recently suggested that these tiny animals have their closest relatives partly within pseudocoelomates (Nematoda) and partly outside them in Gnathostomulida.

There has been much discussion about the presence of a coelotel (and, hence, of a coelom) in priapulids. However, more recently, Conway Morris (1977), Malakhov (1980) and Kristensen (1983) concur by reaffirming a close relationship of Priapulida with two pseudocoelomate phyla, Nematomorpha and Kinorhyncha. Malakhov (1980) even suggested they be united in a phylum, Cephalorhyncha. Conway Morris based his conclusions on the study of fossil priapulids, particularly from the Cambrian Burgess Shales fauna. Malakhov developed his arguments from embryology and larval morphology. Kristensen regards *Nanaloricus mysticus* (the first described species of his new phylum Loricifera) as a type of missing link between priapulids, nematomorphs and kinorhynchs.

The absence of close relations of priapulids to coelomate groups or to Spiralia in general is confirmed by van der Land and Nørrevang (1985), who fail to find convincing arguments for assessing their affinities.

7.15 POGONOPHORANS

A satisfactory decision on the relationships of pogonophorans has yet to be achieved. Some years ago, different views emerged between those who interpreted

Pogonophora as Protostomia close to Annelida (Nørrevang, 1970) and those who interpreted the segmentation of the posterior part of the metasoma, the setae and the presence of a proliferative zone as simple convergencies between the two phyla. The latter zoologists commonly regarded Pogonophora as closely related to groups of deuterostomes, such as echinoderms, hemichordates and chordates (Ivanov, 1970). The first alternative is currently gaining support in molecular studies. In the molecular phylogenetic tree of Field *et al.* (1988), pogonophorans belong to a group of 'eucoelomate protostomes', together with annelids, molluscs, sipunculans and brachiopods.

It should be noted that in Field *et al.* (1988) Pogonophora are represented by a species belonging to the class Vestimentifera, a group established by Webb (1969) to accommodate *Lamellibrachia barhami*, from off the coast of California. A few additional species of vestimentiferans have been described subsequently, including the exceptional *Riftia pachyptila* from the hydrothermal vents of the East Pacific Rise (Jones, 1981). With the description of these new taxa, the phylogenetic relationships of pogonophorans have become even more difficult to ascertain. Van der Land and Nørrevang (1975) placed both conventional pogonophorans and vestimentiferans, as separate classes, within Annelida, whereas the two groups were retained by Jones (1981, 1985) within an independent phylum, Pogonophora, with two subphyla Perviata and Obturata. A third alternative was proposed by Cutler (1982), who assigned Vestimentifera (Obturata) to Annelida, while retaining Pogonophora as a separate phylum for Perviata. The question of affinities between the two groups seems, however, to be positively settled after the most recent studies. Southward and Southward (1988) simply regard them as two subclasses of a same class.

7.16 ANNELIDS

Contrary to common belief, this phylum, as commonly understood, is probably not a monophyletic taxon. It seems to be futile to look for annelid synapomorphies, as long as we try to retain within the phylum several of the so-called 'archiannelids', for example *Polygordius* and the Lobatocerebridae (Rieger, 1980), or the strange *Aeolosoma*, traditionally placed within the Oligochaeta. Brinkhurst and Jamieson (1971) have excluded this last genus from the Annelida.

Of the traditional taxa, polychaetes are very probably paraphyletic, whereas the clitellates (earthworms and leeches, plus the two small groups of ectoparasitic leech-like forms branchiobdellids and acanthobdellids) are generally regarded as a monophyletic group. There is no support for Sawyer's (1984, 1986) idea of a close relationship between leeches and arthropods. Sawyer identifies a number of arthropod-like characters in leech, including a true haemocoel and compound eyes. Other traits he claims demonstrate a direct link between leeches and insects, such as oogenesis with nurse cells, cephalization and a constant segment number. It is quite possible that all these similarities are nothing but homoplasies. Nevertheless, Sawyer's unconventional views still deserve a detailed examination and discussion.

Despite a widespread view of the artificiality of this taxon, no-one has yet attempted to rearrange the polychaetes in an improved system of annelids.

However, a major improvement has already been achieved by abandoning the traditional distinction between the two subclasses, Errantia and Sedentaria. The current artificial classifications mostly contain 17 (Fauchald, 1977) to 25 orders (Pettibone, 1982) in a linear sequence. Some of these orders have been defined or redefined recently, and may prove to be monophyletic. A few orders within Polychaeta have also been erected to accommodate most of the so-called Archiannelida. After nearly a century of stability in most classifications of annelids, this 'class' has been finally dismantled as artificial. Of the six families still placed until recently in Archiannelida, Westheide (1985) proposes to place Dinophilidae in the polychaete order Eunicida, whereas four additional orders (Polygordiida, Protodrilida, Nerillida and Diurodrilida) are established to accommodate the remaining families in what are possibly monophyletic taxa.

Within clitellates, there is no final agreement for the relationships between Oligochaeta, Branchiobdellida, Acanthobdellida and Hirudinea. Oligochaeta may well prove to be paraphyletic, if the leeches turn out to be the sister-group of Lumbriculidae, as suggested by Michaelsen (1926); an idea which has never been disproved.

Cladistic analyses of the higher taxa within Oligochaeta have been provided by Erséus (1987) and Jamieson (1988) (cf. Appendix 8).

For leeches, an improved knowledge of tropical faunas has demonstrated (Ringuelet, 1954) the artificiality of the distinction between Gnathobdellae and Pharyngobdellae. The two groups should therefore be combined as the Arhynchobdellae, the sister-group of the Rhynchobdellae: similar to Blanchard's (1894) classification.

7.17 MOLLUSCS

Embryological evidence (spiral cleavage) and similarities in the larval stages always suggested close affinities between molluscs and annelids, despite the evidence from adult morphology. Sister-group relationship between the two phyla have been given, for example by Götting (1980) and Wingstrand (1985). Molecular evidence (Field et al., 1988; Ghiselin, 1988a; Lake, 1990) also points in the same direction.

However, there are other strands of evidence which favour an alternative phylogenetic hypothesis. Runnegar (1987) claims that there is 'a general consensus that molluscs were derived from Precambrian animals having the grade of organization and body form found in modern free-living flatworms and nemerteans'. According to Salvini-Plawen (1988, p.396; see also Salvini-Plawen, 1968, 1980), molluscs 'represent a sister-group ... to the common stock of true coelomate organization (i.e. Echiura, Sipuncula, and Annelida) as is expressed by the few synapomorphic characters (Pericalymma larva, cell-junctions?)'. For this hypothesis (also supported by Willmer, 1990), there seems to be some weak molecular evidence (comparisons of 5S rRNA sequences; Kumazaki, Hori and Osawa, 1983a,b).

There is abundant literature speculating about the possible ground-plan of molluscs, but the latest model put forward by Haszprunar (1992a) deserves particular attention. He depicts the first molluscs as small animals, in accordance with some palaeontological evidence (Runnegar, 1987) that Dzik (1991), however, tries to discount as a preservational artefact.

For a systematic arrangement within the molluscan phylum, the detailed phylogenetic scheme developed by Salvini-Plawen (1980) deserves much attention (cf. Appendix 9).

Peel (1991) has recently rejected the molluscan class Monoplacophora, as not monophyletic; he distributes the taxa previously classified as Monoplacophora between two classes, the extinct Helcionelloidea and the still surviving Tergomyia (with *Neopilina* and relatives).

Systematic interrelationships of gastropods are hotly disputed. A new arrangement of this molluscan class has been recently proposed by Haszprunar (1985), who recognizes two subclasses, Prosobranchia and Heterobranchia, but informally admits that the superfamily Rissoelloidea (Rissoellidae and Omalogyridae) represents a 'connecting link' between the two subclasses. The opisthobranchs and the pulmonates of the traditional classifications are united to form the subclass Heterobranchia. To this last taxon is also ascribed the cohort Triganglionata, including the superfamilies Nerineoidea, Architectonicoidea and Pyramidelloidea, generally classified with the Prosobranchia. Haszprunar's scheme, which has received some support from molecular data (the partial sequencing of 28S rRNAs; Tillier *et al.*, 1992b), is summarized in Appendix 10. The phylogenetic interrelationships of major taxa within gastropods, however, do not seem to have been definitely settled, as exemplified by some suggestions deriving from sperm morphology (Healy, 1992) as well as from additional molecular evidence (Tillier *et al.*, 1992a). This last study, extending Tillier *et al.*'s (1992b) investigations on large subunit rRNA sequences, suggests that the first branch splitting from the common stem of gastropods is the Docoglossa, traditionally a subgroup of the Opisthobranchia.

7.18 ARTHROPODS, EXCLUDING INSECTS

Cuvier's (1812) concept of the Articulata, as a major group comprising all metameric invertebrates, is still respected. For instance, Brinkhurst and Nemec (1987, p.70) regard Arthropoda and Annelida as sister-groups. A close relationship between the two phyla is also re-affirmed in recent papers by Nielsen (1985), Westheide (1985) and others. However, there have always been difficulties with this hypothesis, especially at the embryological level. In a recent textbook of comparative embryology, Fioroni (1987, p.170) regrets that the well-established morphological resemblances between annelids and arthropods are not supported by embryology. The search for evidence of spiral cleavage in arthropods has produced unconvincing results.

The first foundation of the current belief in a close relationship between annelids and arthropods, because of their body segmentation, is possibly irrelevant. There is some evidence (Minelli, 1987, 1988b; Minelli and Bortoletto, 1988) suggesting that in the two phyla the segmentation of the body is achieved in completely different ways, primarily involving different embryonic components and under the control of different genetic systems. In annelids, segments arise in serial sequence, by steps punctuated by mitotic cycles, and are primarily mesodermal units. On the contrary, in arthropods, as far as we can understand, particularly from research in *Drosophila*, segments are primarily ectodermal units, arising by iterative splitting (or doubling) of biochemical markings, laid down very early in development, independently from mitotic cycles. The molecular phylogeny elaborated by Field *et al.* (1988) also supports the view of a more distant relationship of arthropods and annelids. In their tree, the annelids cluster with the molluscs, pogonophorans, sipunculids and brachiopods, rather than with arthropods. This fact is also acknowledged by Ghiselin (1988a). Further doubts as to the close relationship between annelids and arthropods are also expressed by Raff and Kaufman (1991) and Delle Cave and Simonetta (1991).

Another much-disputed issue is arthropod monophyly. It is well known that arthropods have been regarded as polyphyletic by Manton, an idea based on her comparative morphological and embryological studies (e.g. Manton, 1973, 1977; Manton and Anderson, 1979). Her views have gained widespread favour, but not by everyone. A polyphyletic concept of arthropods is accepted by some crustaceans specialists (e.g. Schram, 1983b, 1986), but is less accepted by students of insects or myriapods. Schram (1983b) was correct in claiming that myriapod and insect workers mainly believe in the monophyly of Mandibulata (Crustacea plus Tracheata), whereas crustacean students have accepted, together with the dissolution of Arthropoda, Manton's suggestion to reject Mandibulata, because of the structural and functional differences of the jaws of both groups (i.e. crustaceans versus insects and relatives). Personally, I accept arthropod monophyly, although I am not sure whether Crustacea and Tracheata are sister-groups. What I definitely reject (this opinion is also shared by nearly all students of these groups) is a particularly close relationship between tracheate arthropods (i.e. insects and the various myriapod groups) and onychophorans. They were united by Manton to form Uniramia, one of her three or four 'arthropod phyla'. A strong rejection of Uniramia has been developed by Kukalova-Peck (1992), around her re-examination of the ground-plan of pterygote insects; see also Shear (1992).

A new view of arthropod evolution has been developed by Emerson and Schram (Emerson and Schram, 1990; Schram and Emerson, 1991). Their theory is based on the following hypotheses: (1) biramous limbs, as in crustaceans, are not primitive, but evolved through the proximal fusion of uniramous limbs, as in tracheates; (2) a couple of segments of uniramian arthropods (or a diplosegment, as in the trunk of millipedes) are homologous to a single body segment of biramian arthropods; (3) body singularities such as gonopores, occur at fixed positions along the body. This theory allows for a monophyletic concept of arthropods, but a detailed phylogeny of the phylum has yet to be developed from these very interesting premises.

There are problems with the higher-level taxonomy of Chelicerata. There is no consistent placement for Pycnogonida; the group is often excluded from Chelicerata (e.g. in van der Hammen's (1977) system), and probably rightly so. Most of the currently recognized orders seem to be well-founded as monophyletic, except for Acari, the monophyly of which has long been disputed. Van der

Hammen (1977) divides them into two groups with different relations (Actino-trichida closer to Palpigradi, Anactinotrichida closer to Ricinulei). Two cladistic arrangements of the higher taxa of chelicerates have been provided by Weygoldt and Paulus (1979) (cf. Appendix 11) and by Shultz (1990) (cf. Appendix 12). A very good summary of the recent advances in spider systematics, including cladistic analysis of higher taxa, is Coddington and Levi (1991).

Trilobite monophyly has been recently reaffirmed by Fortey and Whittington (1989), Hahn (1989) and Ramsköld and Edgecombe (1991) against a contrary view argued by Lauterbach (1980, 1983, 1989).

In the most recent systems, Crustacea are generally regarded as a superclass (if not a phylum, following Manton; cf. Schram, 1986). Boxshall (1983, p.138), partially incorporating several important contributions of previous authors, divides them into 10 classes: Ostracoda, Branchiopoda, Mystacocarida, Cephalocarida, Phyllocarida, Eumalacostraca, Copepoda, Remipedia, Cirripedia and Branchiura. Enormous differences separate this arrangement from the traditional one. A taxon, Entomostraca, as opposed to Malacostraca, has been rejected for a long time. Phyllocarids are removed from within Malacostraca, thus giving a monophyletic Eumalacostraca. A new class Remipedia (Yaeger, 1981; to accommodate the new species, *Speleonectes lucayensis* from a marine cave in the Bahamas) has been added to the other well-known groups. A further new order-level taxon (Mictacea) was described in 1985 (Bowman *et al.*, 1985) for two unusual peracarid crus-taceans, both described in the same year, *Hirsutia bathyalis* from deep sea waters (around 1000 m depth) (Sanders, Hessler and Garner, 1985) and *Mictocaris halope* from marine caves in Bermuda (Bowman and Iliffe, 1985).

Schram (1983a) points out how rapidly and radically the theories of crustacean phylogeny have changed. A few decades ago, the recently discovered cephalocarids were regarded as the most primitive forms and the system had been revised accordingly. However, this interpretation has subsequently fallen into disrepute, to be substituted by another which regards the biramous paddles of the newly discovered Remipedia as a prototype of ancestral crustaceans.

The affinities of Copepoda are also a matter for discussion. Grygier (1983) accommodates this group within a monophyletic Maxillopoda (rejected by Boxshall, 1983), together with Mystacocarida, Branchiura, Cirripedia (including Rhizocephalia), Ascothoracica, Ostracoda and several strange creatures: (1) the so-called Hansen's Y-larvae; (2) *Basipodella* and *Deoterthron*, two of the smallest crustaceans known so far, copepod-like ectoparasites of other crustaceans; and even (3) pentastomids. The placement of pentastomids within crustaceans had been previously supported by Wingstrand (1972), Riley, Banaja and James (1978), and others. In this context, of some interest is Jamieson's (1991) study of sperm morphology, pointing to the very derived condition of copepods, quite differen-tiated from the primitive flagellate sperm of most maxillopodan groups (Ascotho-racica, Cirripedia, Branchiura, Mystacocarida and Pentastomida); in crustaceans other than Maxillopoda, a flagellate sperm only occurs in Remipedia.

A new arrangement of Branchiopoda is proposed by Fryer (1987), who divides them into 10 orders: besides the traditional Anostraca and Notostraca and two extinct groups (Lipostraca and Kazacharthra), Fryer's system includes two orders (Spinicaudata, Laevicaudata) corresponding to the old Conchostraca and three orders (Anomopoda, Onychopoda, Haplopoda) corresponding to Cladocera.

Within Malacostraca, there are many hotly debated points. Some authors, such as Dahl (1983), or Hessler (1983), do not feel obliged to modify the classical scheme put forward by Calman in 1909, whereas other students propose major modifications. For example, Watling (1983), while refusing to accept that Hoplocarida (Stomatopoda) belong within Eumalacostraca (an opinion also shared by Dahl, 1983), dismembers the traditional superorder Peracarida, by elevating both Isopoda and Amphipoda to the rank of superorders and naming Brachycarida as the superorder comprising the remaining peracaridan groups (Thermosbaenacea, Spelaeogriphacea, Tanaidacea and Cumacea). Further suggestions for new arrangements also occur in the remaining groups and at lower taxonomic levels. Two extensive, but non-concordant cladistic analyses of interrelationships within Isopoda have been performed by Wägele (1989) and Brusca and Wilson (1991).

What characterizes crustacean taxonomy, however, is the very lively debate over higher taxa, which are being continuously revised, not only as far as their mutual relationships are concerned, but also as regards their contents. For instance, whenever we read about Cirripedia, in taxonomic discussion, it should be checked whether Rhizocephala and/or Ascothoracica are included or excluded (Newman, 1987). It is worth noting that the 18S rRNA sequences suggest that Rhizocephala is the sister-group to the Thoracica (barnacles and relatives), whereas the Ascothoracica is less closely related to them (Applegate, Abele and Spears, 1991).

New comprehensive systems of Crustacea, including both extinct and Recent groups, have been devised by Schram (1986) and by Starobogatov (1988). Their views still await more adequate checking by a sound cladistic analysis; however, their classifications are the most useful ground for discussion. I summarize them in Appendices 13 and 14.

The monophyly of Tracheata (or Atelocerata), the group comprising both insects and the different myriapod taxa, is beyond dispute. However, there is no universal agreement as to the major divisions and their mutual relationships. Some authors, such as Boudreaux (1979), insist on Myriapoda monophyly, a point generally dismissed by myriapodologists (Dohle, 1980, 1985; Minelli, 1983). However, there is consensus on the monophyly of several taxa, such as Chilopoda, Diplopoda, Pauropoda, Pauropoda plus Diplopoda, Symphyla and Hexapoda. I regard Chilopoda as the sister-group of all remaining Tracheata (Minelli, 1983), but there are different opinions. The sister-group of Hexapoda is uncertain, although Symphyla seems to be a good candidate.

Within Chilopoda, major progress has been made through the cladistic analysis of Dohle (1980, 1985, 1988), who has definitively clarified the sister-group relations between Notostigmophora (scutigeromorphs) and Pleurostigmophora (all remaining centipedes, i.e. lithobiomorphs, craterostigmomorphs, scolopendromorphs and geophilomorphs). The strange *Craterostigmus* from New Zealand and Tasmania has been finally recognized as a member of an independent clade, probably related to scolopendromorphs plus geophilomorphs (Dohle, 1988, 1990).

Diplopod taxonomy is still in a state of flux; however, major improvements have been accomplished by Hoffman (1979) and Enghoff (1984, 1990), the latter re-evaluating in cladistic terms the systematics of the group.

7.19 INSECTS

The most comprehensive cladistic analyses of phylogenetic interrelationships of insect orders have been published by Hennig (1969) and by Kristensen (1975, 1981, 1991); the evidence discussed in these works can be profitably compared with the palaeontological data summarized and discussed by Kukalova-Peck (1991, 1992).

A main point of concern is the affinities of different apterygote 'orders'. Thysanura has been dismembered, while recognizing that the silverfishes (Zygentoma) are an obvious sister-group of Pterygota, and the bristletails (Archaeognatha) are best regarded as the sister-group of Zygaentoma plus Pterygota. This arrangement was proposed by Hennig (1969; cf. Appendix 15) and has been generally accepted since. Matters are not so straightforward for the remaining primitively wingless groups. For Kristensen (1975), Collembola and Protura are sister-groups, together forming the sister-group of Diplura, all three in turn forming a clade (Entognatha) which is, in turn, the sister-group to Archaeognatha plus Zygaentoma plus Pterygota (Ectognatha); but Kukalova-Peck (1987) regards Diplura alone as the sister-group of Ectognatha.

As far as the huge pterygote group is concerned, major improvements concern the recognition of the paraphyletic nature of a couple of orders.

The relationships within Psocodea are hotly debated; see, for instance, Königsmann (1960), Clay (1970), Kim and Ludwig (1982) and Lyal (1985). Sucking lice (Anoplura) are possibly a monophyletic group, nested within bird lice (Mallophaga) as traditionally understood, and Phthiraptera (Anoplura plus Mallophaga), in turn, are possibly nothing more than specialized Psocoptera, close to the present-day family Lioscelidae.

Within Thysanoptera, one of the two traditional suborders, Tubulifera, appears to be the sister-group to part of one of the seven families of the other suborder, Terebrantia (Mound, Heming and Palmer, 1980).

Heteroptera, as well as the whole of Hemiptera, are universally regarded as a monophyletic unit, whereas many doubts have been cast recently as to the monophyly of Homoptera. Moreover, within Homoptera, the monophyly of both Auchenorrhyncha and Sternorrhyncha is also disputed. Opinions vary among different authors. For instance, according to Hamilton (1988), who relies mostly on evidence from Lower Cretaceous fossils, coccids plus aphids plus psyllids should be considered as the sister-group of cicadas, whereas whiteflies (aleyrodids) are possibly closer to fulgorids.

Lively debates on the internal arrangement of taxa have also been developed by students of Diptera, Lepidoptera and other orders.

Regarding Diptera, a major effort toward a phylogenetic system has been accomplished by McAlpine (1989), Wood and Borkent (1989) and Woodley

(1989). The possibility of further improvements of their arrangement, however, are suggested (e.g. Vossbrinck and Friedman, 1989): the analysis of 28S rRNAs of several cyclorrhaphous Diptera indicates the probably paraphyletic character of acalyptrates.

As for Lepidoptera, a cladistic arrangement of the families has been provided by Kristensen (1976). A great deal of interest has been raised over the true affinities of the Hedyloidea, a small group traditionally placed in the proximity of the geometrid moths, but probably related to the butterflies, perhaps as the sister-group of the Papilionoidea (Scoble, 1986).

Revising the higher level taxonomy of the Coleoptera appears to be a formidable task, so huge as to discourage most efforts. No major comprehensive attempt has been undertaken since that of Crowson (1955, 1960, 1981). Many details within current systems are widely thought to be unsatisfactory. For instance, nearly all of the largest families (30 000 or more species in each when taken in the traditional wide sense) have uncertain, or at least disputed, limits. This is true for Carabidae, Staphylinidae, Scarabaeidae, Chrysomelidae and Curculionidae. In nearly all cases, the traditional circumscriptions of these families are paraphyletic. The problem is most easily overcome in the Carabidae, by inclusion into this huge family of some smaller groups such as the Cicindelidae, Paussidae, Ozaenidae, Trachypachidae, Metriidae and possibly the Rhysodidae (Vigna Taglianti, 1982).

Matters are not so straightforward in the Staphylinidae, where pressures to incorporate some previously segregate groups, such as the Pselaphidae (Newton and Thayer, 1988a) contrasted with trends to dismember the family into several family-level taxa (e.g. Naomi's (1985) proposals and Newton and Thayer's (1988b) reply). Also debated is the Curculionidea, in which Kuschel (1988), for instance, recognizes six families and 30 subfamilies.

The relationships of the Strepsiptera are still uncertain. Analysis of the sperm cells (Baccetti, 1989) suggests that a placement within Coleoptera, as suggested by Crowson (1981), is unwarranted.

For Hymenoptera, it seems to be necessary to abandon the traditional division into Symphyta and Apocrita: the paraphyletic condition of the first group is largely acknowledged, although opinions differ in detail (cf. Gauld and Bolton, 1988). According to G.A.P. Gibson (1985, and personal communication), the basal split within extant Hymenoptera occurs between Xyelidae and the remaining forms, whereas another symphytan family, Orussidae, should be regarded as the extant sister-group of Apocrita.

7.20 ONYCHOPHORANS, TARDIGRADES AND PENTASTOMIDS

These three small groups are often discussed in the context of arthropods and related groups. The Onychophora are traditionally described as a mixture of three kinds of traits, annelid-like, arthropod-like and onychophoran proper. These conditions have often been regarded as proof of the links between the two major articulate phyla, although most authors have regarded onychophorans as more closely related to arthropods than to annelids. These views will need reworking, if annelids and arthropods turn out to be less closely related than previously thought (cf. section 7.18). In this context, the ontogeny of segmentation in onychophorans deserves particular attention.

Onychophora, and their possible links to arthropods, are often debated when discussing the affinities of several segmented fossils of the Cambrian age, particularly those of the well-known Burgess Shale in British Columbia (summaries in Briggs and Fortey, 1989; Conway Morris, 1989; Delle Cave and Simonetta, 1991), or those, of similar age and more or less similar composition, from other sites in Utah (Robison, 1991), or China (Chen and Erdtmann, 1991; Hou and Bergström, 1991; Ramsköld and Hou, 1991). These extraordinary fossil assemblages (cf. Gould, 1989) are full of unusual forms, but erecting high rank taxa to accommodate them is probably unwarranted (Briggs, Fortey and Wills, 1992).

Most authors believe that tardigrades are closely related to arthropods, and Nielsen (1985) even includes them in his Arthropoda s.l.

Regarding Pentastomida, I have already mentioned the increasing preference to place them within Crustacea, although other arrangements (especially, placing them as a separate phylum 'near' Arthropoda) are also widely followed.

7.21 BRYOZOANS, BRACHIOPODS AND PHORONIDS

Bryozoa (Ectoprocta), Brachiopoda and Phoronida have often played a pivotal role in developing hypotheses about animal phylogeny. Most authors treat them as closely related, together forming a group of Tentaculata or Lophophorata. However, in Nielsen's (1985) scheme Bryozoa are removed from within this assemblage and retained within the gastroneuralian (protostomian) radiation, whereas Phoronida and Brachiopoda are placed, together with Echinodermata and Pterobranchia, in one branch of the notoneuralian ('deuterostomian') radiation. Furthermore, Nielsen (1971, 1977, 1985) re-establishes Bryozoa as a taxon including both Ectoprocta (= Bryozoa s. str. of many authors) and Ectoprocta (= Kamptozoa), a group often regarded as quite distant from lophophorates and possibly of pseudocoelomate affinities. In a later study, devoted mainly to the development of the brachiopod *Crania*, Nielsen (1991) reiterates his view that lophophorates or tentaculates are an artificial assemblage: brachiopods and phoronids should be placed with the deuterostomes, without any close relations to bryozoans. This conclusion is puzzling in the light of the molecular evidence which points instead to affinities between brachiopods and annelids plus molluscs (Field *et al.*, 1988).

7.22 DEUTEROSTOMES, EXCLUDING CHORDATES

Schaeffer (1987) has recently reaffirmed deuterostome monophyly, emphasizing that the synapomorphies are almost entirely comprised of embryological characters: the mode of coelom and mesoderm formation (cf. Nielsen, 1985, p.286) and the position of the mouth anterior to the blastopore, which may become the anus. Schaeffer also points to the similar locations of introns of the actin genes in sea urchins, birds and mammals, which contrasts with the pattern in *Drosophila*

(Davidson, Hough-Evans and Britten, 1982). A deuterostomian clade, however, is not present in some of the most recent molecular phylogenetic trees (Field *et al.*, 1988; Ghiselin, 1988a; Christen *et al.*, 1991b).

Echinoderms are a major group where systematics has been largely developed from fossil as well as from extant taxa. Twenty-two classes are now recognized, 16 of which are known only as fossils.

An unexpected addition to the previously known groups has been the recent discovery of *Xyloplax medusiformis*, a small medusa-like echinoderm collected from sunken wood between 1057 and 1208 m off the New Zealand coast. This strange animal has been described by Baker, Rowe and Clark (1986), who have accommodated it into the new class, Concentricycloidea, provisionally placed within Asterozoa, together with the true sea-stars. It is quite possible, however, that *Xyloplax* will turn out to be a type of progenetic (neotenic) sea-star, rather than the representative of a distinct major clade.

Besides this new discovery, the mutual relationships between the extant classes are far from definitively settled. New techniques (radioimmunoassay and analysis of 18S rRNA sequences) suggest a close affinity between Echinoidea and Holothuroidea, at variance with other conventional treatments (Ghiselin and Lowenstein, 1988). The system of Asteroidea has been recently revised by Gale (1987) and by Blake (1987), with contrasting results, largely dependent on opposite assessments of character polarity (cf Smith, 1992). At variance with Blake, Gale tried to supplement neontological evidence with data from fossils.

Arrow-worms (Chaetognatha) have been tentatively related to almost all phyla (Ghirardelli, 1968). Indeed, they are a very isolated group. Nielsen (1985) ascribes them to the Aschelminthes, but they are still currently placed within deuterostomes in most classifications. Once again conclusive evidence is still lacking.

Two small groups of invertebrates of quite different habit, the pterobranchs and the enteropneusts, are often classified within the phylum Hemichordata, as separate classes, together with the fossil graptolites and the poorly-known, possibly neotenic, *Planctosphaera*. Their placement within deuterostomians or, at least, their affinities to echinoderms and chordates are seldom disputed. I have some doubts, however, as to the sister-group relationships of Enteropneusta and Pterobranchia. Enteropneusta are generally regarded as much closer to chordates (e.g. Young, 1981). Sometimes (but this classification is going out of fashion) they have also been included, together with pterobranchs, within the phylum Chordata.

7.23 CHORDATES, EXCLUDING VERTEBRATES

The origins of the chordates are still in dispute, even if one completely disregards the most extravagant hypotheses such as those of van Z. Engelbrecht (1969), contending that chordates arose from annelids, or Gutmann's (1966, 1967) idea that they arose from (not better defined) worm-like organisms with a segmented coelom. Hardly better is Løvtrup's (1977) contention that vertebrates are more closely related to some protostome groups, such as molluscs, than to the remaining deuterostomes. Much more interesting is Jefferies's hypothesis of an origin for chordates from the entirely extinct Calcichordata (Jefferies, 1968, 1979, 1986). Calcichordates are deuterostomes, generally regarded as belonging to the very wide Palaeozoic radiation of echinoderms. Jefferies has elaborated a very detailed re-interpretation of calcichordate structure, suggesting that they are the ancestors of vertebrates. This hypothesis has been rejected by many authors, but is still respectfully discussed by others. Many details 'read' by Jefferies from the fossil remains of his calcichordates are possibly too subtle to be accepted without caution. However, his work has put the search for chordate ancestors within the reasonable context of deuterostome groups. At present, it is difficult to foresee whether palaeontology alone will provide further convincing proof for or against Jefferies's theory. The search for the sister-group relationships of chordates must necessarily rely on broader comparative evidence, particularly on molecular evidence.

Tunicata, Cephalochordata and Vertebrata are mostly classified as three subphyla of Chordata. However, some authors prefer to classify Tunicata as a separate phylum (cf. the chordate classification of Nelson (1969) in Appendix 17) or even to elevate all three groups to the rank of phylum. A cladistic analysis of the inter-relationships of the major chordate clades, based on 320 characters, has been published by Maisey (1986, 1988). His schemes essentially confirm what is currently regarded as established consensus (Figure 7.1). See also Appendix 16 for Hennig's (1985) system of Chordata.

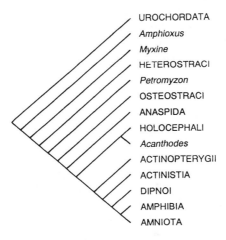

UROCHORDATA
Amphioxus
Myxine
HETEROSTRACI
Petromyzon
OSTEOSTRACI
ANASPIDA
HOLOCEPHALI
Acanthodes
ACTINOPTERYGII
ACTINISTIA
DIPNOI
AMPHIBIA
AMNIOTA

Figure 7.1 The phylogeny of Chordata, according to Maisey (1988), redrawn.

7.24 VERTEBRATES

For decades, conodonts have ranked among the most puzzling of the Problematica, the fossils for which there seems to be no place, in a system essentially built for accommodating the Recent forms. The latest histological investigations into the fine structure of these teeth-like elements has revealed the presence of several tissue types characteristic of the phosphatic skeleton of vertebrates, including cellular bone, two kinds of strongly mineralized enamel and calcified cartilage (Sansom *et al.*, 1992). Hence, conodonts were the teeth of very old vertebrates and their possession of the typical hard tissues of most members of this clade even places them within the vertebrate radiation, above the split separating Myxinoidea (hagfishes) from all other vertebrates (Briggs, 1992). That myxinoids are the sister-group to all remaining vertebrates is also confirmed by Bardack (1991), who has recently described the first fossil hagfish, *Myxinikela siroka* from the Pennsylvanian of Illinois. Stock and Whitt (1992) are of a contrary opinion. They claim that molecular evidence (18S rRNA sequences) supports the monophyly of Recent Agnatha (lampreys plus hagfishes). At any rate, recognizing conodonts as vertebrates has important and unexpected consequences for the diversity of Palaeozoic taxa, so often considered in the search for macroevolutionary trends. With the addition of conodonts, the number of genera of vertebrates from the Cambrian and the Ordovician together rises from five to nearly 150 (Briggs, 1992).

The phylogenetic relationships of lower vertebrates have always been objects of intense investigation. For these groups there is important palaeontological data. Their phylogenetic analysis has become a battlefield for those who deny the relevance of fossils in phylogeny reconstruction versus those who support the opposite view. A comprehensive classification of vertebrates, inclusive of the extinct groups, has been provided by Carroll (1987). Despite its 'eclectic' (i.e. not strictly phylogenetic) nature, this classification seems worthy of inclusion here, in summarized form, as Appendix 19.

Considerable work has been devoted to unravelling the phylogenetic relationships of 'fishes', since the pioneering work of Greenwood *et al.* (1966) and Greenwood, Miles and Patterson (1973). Some classic phylogenetic schemes developed in the works of Nelson (1969), Rosen *et al.* (1981) and Lauder and Liem (1983) are summarized in Appendices 17, 18 and 20.

A sister-group relationship between lungfishes and tetrapods has been proposed by Rosen *et al.* (1981), but subsequently rejected by Holmes (1985). Matters are possibly complicated by the discovery of a fossil form, *Diabolichthys* (Chang and Yu, 1983), perhaps the sister-group of the Dipnoi (Forey, 1987). New debates as to the phylogenetic affinities of *Latimeria* have been recently raised by Gorr, Kleinschmidt and Fricke (1991), who claim that haemoglobin sequences suggest *Latimeria* to be closely related to tetrapods. Stock *et al.* (1991), however, regard this analysis as suspect, because of unequal rates of evolution. Forey (1991) regards the evidence as equivocal.

Tetrapod monophyly was rejected in the past by some students, mostly palaeontologists. Holmgren (1933, 1939, 1949) and Säve-Söderbergh (1945), for instance, regarded the majority of tetrapods as derived from the extinct Rhipidistia, whereas Urodela would have arisen from Dipnoi. Jarvik (1980, 1986) also regarded tetrapods as diphyletic, but contended that Urodela and the remaining tetrapods are independently derived from two separate groups of rhipidistians, Porolepiformes and Osteolepiformes, respectively. Nieuwkoop and Sutasurya (1979) contend that anurans are closer to birds and to some 'reptiles' than to urodeles and caecilians, which are, in turn, closer to other 'reptiles' and mammals. The monophyly of Tetrapoda has been recently reaffirmed by Panchen and Smithson (1988). These authors maintain that tetrapods are the sister-group of the Palaeozoic Osteolepiformes, whereas, among the living groups, they are most closely related to the Dipnoi. Panchen and Smithson (1988, p.4) further regard *Latimeria* as the sister-group of Dipnoa plus Tetrapoda and Actinopterygii as the sister-group of *Latimeria* plus Dipnoi plus Tetrapoda.

Within tetrapods, there is still much debate. Fossils play an important role, but there are also further areas of disagreement. With regard to the comparative anatomy of the heart and circulatory system, for instance, Bishop and Friday (1988, p.33) regard 'the interrelationships of the major groups of tetrapods [as] poorly established'. Panchen and Smithson (1988) tried to identify monophyletic groups within a loose ensemble of 'non-amniote tetrapods', traditionally accommodated within Amphibia, mostly within the paraphyletic (perhaps polyphyletic) Labyrinthodontia. Despite previous disputes, all modern amphibians seem to form a monophyletic unit, Lissamphibia, as reaffirmed by Milner (1988). Caecilians and anurans are regarded by the same author as 'uncontroversially monophyletic', and all living urodeles plus most fossils commonly associated with them also represent another monophyletic unit.

That reptiles are paraphyletic, as traditionally understood, seems no longer to be in dispute. Gauthier, Kluge and Rowe (1988a,b) and Estes (1988) simply equate Reptilia with the conventional reptiles plus the birds. Others also include the mammals within a wider concept of Reptilia. Less certain is the correct placement of Chelonia, although Gaffney and Meylan (1988) now seem to have reached a stable conclusion. They regard turtles as the sister-group of Diapsida, thus reforming the Sauropsida, a taxon established by Goodrich in 1916.

Further problems arise within Diapsida, a major clade within amniotes, comprising birds and most 'reptiles', such as crocodiles, squamates, *Sphenodon* and many extinct groups ('dinosaurs' etc.). Within diapsids, birds, crocodilians, dinosaurs, pterosaurs and a few further fossil groups form the seemingly monophyletic Archosauria.

Regarding Squamata, itself a monophyletic group, it seems probable that lizards are paraphyletic, whereas snakes and amphisbaenians are undisputed monophyletic groups (Rieppel 1988b,c). Rieppel (1988b) regards snakes as 'an example for the origin of a higher taxon obscured by a high degree of character incongruence, indicating commonness of convergence'. Moreover, according to the same author, it is still doubtful whether snakes and amphisbaenians together form a monophyletic taxon.

At the species level, birds are the systematically best known group of animals, but their traditional arrangement into families and orders has been shown to be inadequate and largely erroneous. Cracraft (1988, p.339) notes that 'the phylogenetic relationships of the avian higher taxa are more poorly known than those of any other group of tetrapods' (cf. Cracraft, 1981).

Chatterjee (1991) has recently described *Protoavis texensis*, a new fossil reptile (s.l.) some 225 million years old, claiming that it is a bird, older, indeed, than

Archaeopteryx. The description, however, is incomplete and Ostrom (1991) doubts the true avian nature of the fossil.

There is no problem in recognizing all modern birds (Neornithes) as a monophyletic group. The sister-group is the extinct *Ichthyornis*. Together, they form the Carinatae, with the extinct Hesperornithiformes as sister-group (Cracraft, 1986, 1988). Within Neornithes, Cracraft (1985, 1988) identifies a primary split into Palaeognathae (ratites and tinamous) and Neognathae (all other recent birds). Within neognathes, the same author identifies a few monophyletic groups, for example Anseriformes plus Galliformes, together forming the sister-group of all remaining groups. Loons and grebes are also regarded as sister-groups. When kept together, they are the sister-group of penguins and all three orders together are the sister-group of Pelecaniformes plus Procellariiformes. Other monophyletic groups are possibly Gruiformes plus Charadriiformes plus Columbiformes plus Ardeidae, Falconiformes plus Strigiformes and Caprimulgiformes plus Apodiformes.

It is also possible, however, that the phylogeny of birds is quite different from these conservative views, as strongly suggested by Sibley (review in Sibley, Ahlquist and Monroe, 1988; Sibley and Ahlquist, 1990). For years Sibley has analysed egg-white proteins from a huge number of species. He eventually discovered that the hoatzin (*Opisthocomus hoatzin*) is not related to galliforms but to cuckoos (Sibley and Ahlquist, 1973). Since 1980, Sibley has turned to DNA–DNA hybridization (Sibley and Ahlquist, 1983). His studies (mostly in collaboration with Ahlquist) have produced an astonishing amount of evidence, some partly surprising, some partly confirming previously less substantiated hypotheses of bird affinities. In the case of the New World vultures, they are more closely related to storks than to Old World vultures, eagles and falcons, with which they were traditionally ranked. Sibley's methods have often been criticized, for instance by Cracraft (1987b), thus eliciting strong defensive answers (Sibley and Ahlquist, 1987; Sibley, Ahlquist and Sheldon, 1987; Sibley, Ahlquist and Monroe, 1988). A critical discussion of Sibley's work by Sarich, Schmid and Marks (1989) was quoted in Chapter 2.

The new, revolutionary system of recent birds produced by Sibley, Ahlquist and Monroe (1988), and further developed in book form by Sibley and Ahlquist (1990), is reproduced here as Appendix 21. Another example of the lively debate on bird phylogeny is Cooper *et al.*'s (1992) paper claiming, on molecular evidence (12S mtDNA sequences), independent origins for the two groups of flightless birds of New Zealand, the kiwis and the recently extinct moas.

Gardiner (1982) and Løvtrup (1985) have recently revived the Haemothermia, an old systematic concept introduced by Owen in 1866. The Haemothermia consist of all warm-blooded vertebrates, mammals and birds. Løvtrup and Gardiner have based their proposals on much more critical evidence than that of Owen. Their arguments have been carefully re-examined by Kemp (1988c) and by Gauthier, Kemp and Rowe (1988a). Kluge concludes that an alternative phylogenetic tree (with birds closer to crocodiles than to mammals) is slightly more probable than that supporting the monophyly of Haemothermia. Gauthier, Kluge and Rowe more definitively reject this putative taxon, largely on fossil evidence (cf. Gee, 1988).

Within amniotes, mammals are generally regarded as members of the Synapsida, together with several extinct groups. These have been generally classified within two paraphyletic groups, Pelycosauria and Therapsida (Kemp 1988a, p.1). The case of Therapsida, the so-called 'mammal-like reptiles', is particularly intriguing. This is a typical case of a stem group, with all the associated problems of delimitation and ranking. Gosliner and Ghiselin (1984) wisely suggested that many semantic (and systematic) problems could be resolved if our concept of 'mammal' could be slightly relaxed, thus calling these old synapsids 'reptile-like mammals', instead of 'mammal-like reptiles'. This step has been taken by Ax (1984, 1985b, 1988), who redefines mammals to include therapsids. It may be interesting to know that 'mammal-like reptiles', or 'reptile-like mammals', were perhaps still living in the Palaeocene, some 100 million years after the appearance of 'true' mammals: this type of 'living fossil' (in the concept of a Palaeocene zoologist) has been discovered in the Paskapoo Formation at Cochrane, Alberta, and described as *Chronoperates paradoxus* (Fox, Youzwyshyn and Krause, 1992). There seems to be scope, however, for disputing the true affinities of this fossil tetrapod (Sues, 1992).

The monophyly of Recent mammals has been questioned often, but those former doubts are now generally rejected. However, the results are not so simple. When discussing the relationships of several groups of Mesozoic mammals, Kemp (1988b, p.23) considered that it is impossible to split them into therian and non-therian lineages or to divide therians into just two groups, marsupials and placentals. Recently discovered forms suggest that the relationships of Mesozoic mammals are more complex, and include many short-lived clades. Their mutual relationships have yet to be clarified.

For Recent mammals, the relationships between monotremes, marsupials and placentals are not in dispute, but the systematic arrangement of placental orders has yet to be settled satisfactorily. Benton (1988a) briefly reviewed several recent phylogenies, mostly based on molecular evidence, proposed for the orders of placental mammals. Some of the schemes are reproduced in Figure 7.2.

Slightly less sceptical than Benton are Novacek, Wyss and McKenna (1988). The latter partially summarize the evidence in favour of a close relationship of (1) rodents and lagomorphs (Glires), (2) elephants and sirenians (Tethytheria), (3) hyracoids and Tethytheria (Paenungulata), (4) bats and dermopterans. Less convincing is the evidence for grouping together (1) pangolins and edentates, (2) Glires and macroscelids, (3) tree-shrews, primates, bats and dermopterans (Archonta), as well as (4) all Eutheria excluding edentates and pangolins (Epitheria). According to Novacek (1992), however, there is little congruence between molecular and morphological data, except for the recognition of Paenungulata and Tethytheria and for the isolated placement of edentates. Inclusion of lagomorphs and rodents within a cohort Glires is supported by morphology and developmental biology (skull structure, architecture of the ankle joint, foetal membranes, tooth development), but not by molecular evidence (immunology, protein and gene sequence data).

The monophyly of Tethytheria (elephants and sirenians) and their sister-group relation with hyracoids is supported by Tassy and Shoshani (1988).

According to Holmes (1991), different rates of substitution produce different phylogenies of Eutheria: 17 nuclear genes (out of 26 studied) gave highest support to a grouping of artiodactyls with primates rather than with rodents, whereas three

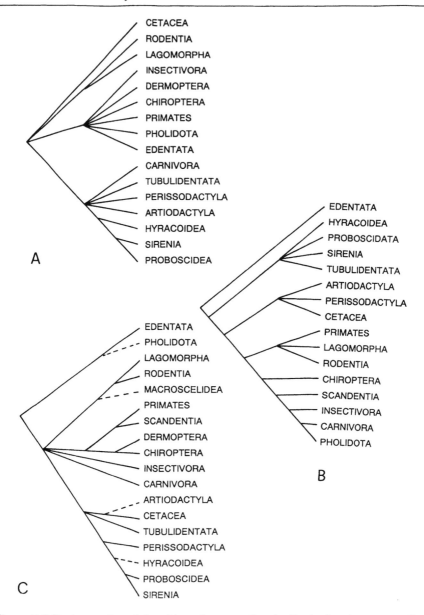

Figure 7.2 Phylogenetic relationships of mammalian (eutherian) orders, according to (A) Simpson (1945); (B) Miyamoto and Goodman (1986); and (C) Novacek (1992), all redrawn.

mitochondrial genes (of the six studied) found a rodents plus artiodactyls grouping best supported. High rates of substitutions in the rodent clade are also reported by Li *et al.* (1990) and by Catzeflis, Aguilar and Jaeger (1992).

The monophyly of the bats (Order Chiroptera) has been recently questioned by Pettigrew (1986, 1991a,b; Pettigrew *et al.*, 1989): the neural anatomy would suggest that microbats (Microchiroptera) and megabats (Megachiroptera) are two independent lines of flying mammals, the latter being particularly close to the primates. Pettigrew's arguments, however, have been rebutted (Wible and Novacek, 1988; Baker, Novacek and Simmons, 1991; Mindell, Dick and Baker, 1991; Simmons, Novacek and Baker, 1991; Bailey, Slightom and Goodman, 1992) on the basis of morphological and molecular evidence, including characters from the skull and the postcranial skeletomuscular system, the cranial vascular system, the foetal membranes and the nervous system, as well as the ε-globin gene, the cytochrome oxidase genes and the mt 12S rRNA.

Rodent monophyly has also been questioned by Graur, Hide and Li (1991), who argue that amino acid sequence data suggest that the guinea-pig (*Cavia*), traditionally a caviomorph, or a hystrichomorph rodent, diverged before the separation of myomorph rodents (the group with mice and rats) and the clade primates plus artiodactyls. According to Hasegawa *et al.* (1992), Allards, Miyamoto and Honeycutt (1991) and Li, Hide and Graur (1992), Graur, Hide and Li's conclusion is perhaps premature and weakly based, but nevertheless possible.

The monophyly of Carnivora is discussed by Flynn, Neff and Jedford (1988). Within Carnivora, they recognize two monophyletic groups, Feliformia (Viverridae, Herpestidae, Hyaenidae and Felidae), and Caniformia (Canidae, Otariidae, Phocidae, Odobaenidae, Ursidae, Procyonidae and Mustelidae). The recent papers on the monophyly of Pinnipedia (Otariidae, Phocidae and Odobaenidae) are very interesting. This view had been discounted for several years in favour of an alternative, diphyletic scheme, with seals originating from weasel- and otter-like forms (Musteloidea) and sea lions and walruses (the 'Otarioidea') originating from bear-like forms (Ursoidea). Janvier (1984) even considered them as a classical example of convergence. However, both molecular and anatomical evidence now seem to favour a hypothesis of pinniped monophyly (Flynn, 1988; Flynn, Neff and Jedford, 1988; Wyss, 1988).

Within carnivores, there seems to be no consensus concerning the relationships of that symbol of endangered species, the giant panda (*Ailuropoda*). Zhang and Shi (1991) have summarized the contradictory evidence gathered to date. Anatomy, immunology, DNA–DNA hybridization, isozyme electrophoresis, karyology and palaeontology, all point to a placement of the giant panda within the bear family (Ursidae); but mitochondrial DNA and haemoglobin, together with ethology, confirm a close relation to the lesser panda (*Ailurus*).

As for the Ungulata, Prothero, Manning and Fischer (1988) regard this group as monophyletic and using morphological (dental and other) characters find a sister-group relation between hyraxes and perissodactyls. Together, they are the sister-group of tethytheres. According to Prothero, Manning and Fischer, tethytheres include the extinct (†) desmostylians, elephants and sirenians. Progressively more distantly related outgroups of this complex are †arsinoitheres, †uintatheres, †phenacodonts plus †meniscotheres, whales plus †mesonychids, †hyopsodonts plus †periptychids, †arctocyonids and eventually artiodactyls. Prothero, Manning and

Fischer (1988) also note that their arrangement is partially supported by molecular analyses that cluster together artiodactyls, perissodactyls and whales, although consistent disagreements emerge as to the affinities of hyraxes. One interesting (indeed, puzzling) consequence of the new view of ungulate phylogeny is the possible aquatic origin of elephants. Janis (1988) suggests that tethytherians are primarily aquatic and within this group developed such aquatic groups as sirenians, desmostylians and *Moeritherium* and eventually, through a return to terrestrial life, true proboscideans.

Mammal taxonomy is still more problematic at lower taxonomic levels. According to Corbet and Hill's checklist (1986), there is hardly a family with more than six species where problems do not exist at the species level. The largest 'non-problematic' families in that list include Peramelidae (21 species), Phalangeridae (14), Equidae (9) and Suidae (8), but problems at species level occur in several small families such as Chinchillidae (about 6 species), Phocoenidae (about 6), Moschidae (about 5), Platanistidae (about 5), Natalidae (about 4), Tragulidae (about 4) and Thryonomidae (about 2).

7.25 GREEN PLANTS, EXCLUDING ANGIOSPERMS

A general cladistic classification of the green plants (Viridiplantae) was attempted by Bremer and Wanntorp (1981), Bremer (1985) (cf. Appendix 22) and Sluiman (1985). These schemes relate to several other modern arrangements of green plants by splitting bryophytes and pteridophytes into several major taxa.

A major contribution towards a systematization of the green algae is based on the work of Mattox and Stewart (1984). Mishler and Churchill (1985) have discussed the relationships between green algae and bryophytes. Manhart and Palmer (1990) have recently added new molecular evidence in support of the origin of land plants from 'green algae' of the Charaphyceae. In the chloroplast genome of three genera of this last group (*Coleochaete, Nitella* and *Spirogyra*) they have identified two molecular apomorphies, previously known only from land plants, the presence of characteristic introns in the genes for the tRNAs for alanine and isoleucine.

Many authors disagree as to the monophyly of bryophytes. Some students regard hornworts as independent from liverworts plus mosses, others even regard liverworts and mosses as having originated independently. There is also disagreement as to their phylogenetic relationships with the 'higher' land plants. Whereas a majority of authors still regard them as leading to the true vascular plants, others regard bryophytes as independent groups of algal origin and a few even maintain that they are derived from vascular plants by progressive simplification of the sporophytic phase. In my opinion, this last view is indefensible. By reviewing these disparate phylogenetic hypotheses, Smith (1986) advocates a phenetic system, as being more stable than one founded on disputable phylogenetic analyses. Despite the large phenetic distances between any two of the three major bryophyte groups, several misplaced taxa have been moved around. For example, *Takakia*, originally described as a moss, has subsequently been transferred to liverworts, but it has been considered that it should be placed in hornworts or even in a fourth, independent group (cf. Engel, 1982; Smith, 1986). The position of *Haplomitrium* has also been disputed between liverworts (Calobryales? Jungermanniales?) and mosses (Schuster, 1966; Engel, 1982). The system of both liverworts and mosses is full of difficult issues. A contribution towards their cladistic arrangement has been offered by Mishler and Churchill (1984).

A basal split between lycopods and the remaining pteridophytes is suggested by the gene order in the chloroplast DNA, because lycopods do not share an apomorphic gene inversion common to the other groups (Raubeson and Jansen, 1992).

For pteridophytes, there is a continuing dispute about relationships and ranks of major groups and the perennial problem of obtaining a satisfactory arrangement at family level (cf. Pichi Sermolli, 1977; Kramer, 1990). There also seems to be some problem about the definition of the ferns as a group and the problem is particularly difficult when fossil taxa are taken into account. Kato (1988) has objected to the living Ophioglossaceae being placed within ferns, suggesting that these plants are perhaps the only survivors of the extinct 'progymnosperms'. In that case, their closest living relatives could be cycads, instead of ferns. Progymnosperms, however, are probably paraphyletic (cf. Doyle and Donoghue, 1987a).

In the last few years, the discovery of new fossil plants and increasingly detailed studies of structure and more careful analyses of characters have led to some new interpretations of relationships.

Of the few studies dealing with the origin of the flowering plants and with the relationships between their major clades, I cite Doyle (1987), Hill and Crane (1982), Crane (1985a,b), Doyle and Donoghue (1987a,b), Donoghue and Doyle (1989), Loconte and Stevenson (1990); for gymnosperms see also Meyen (1984); for angiosperms, Dahlgren (1983) and Dahlgren and Bremer (1985).

7.26 ANGIOSPERMS

One of the major issues within the system of living organisms is the relationships of higher taxa within angiosperms. After a long acceptance of Engler's system (cf. Melchior, 1964), and even the older Bentham and Hooker (1862–1883), the last three decades or so have seen several major efforts to revise the familial, ordinal and superordinal arrangement of flowering plants. The most important systems include: Thorne (1968, 1976, 1983), Takhtajan (1969, 1980, 1983, 1987), R. Dahlgren (1975, 1980, 1983), Cronquist (1981, 1982, 1988), Dahlgren, Clifford and Yeo (1985, for monocotyledons), Goldberg (1986, 1989) and G. Dahlgren (1989a,b). Dahlgren's latest system is reproduced in Appendix 23.

A comparison of the most recent angiosperm systems is given in Table 7.1. A few families are monotypic. In Cronquist (1982), for instance, these are five out of 65 in monocots and 35 out of 318 in dicots.

Goldberg (1986, 1989) has provided a very useful comparative survey of the angiosperm families, as circumscribed by the major plant systematists. However, the table is of little use for understanding major similarities or differences between the different classifications, or for the reasons for their continuous (not always clearly justified) changes.

Table 7.1 Number of orders and families of angiosperms in three recent classifications

	Orders			Families		
	Dicots	Monocots	Total	Dicots	Monocots	Total
Cronquist (1988)	64	19	83	322	66	388
Takhtajan (1987)	128	38	166	429	104	533
Dahlgren (1989a,b)	86	25	111	371	104	475

Cronquist's classification departs less from the traditional schemes than the others. Cronquist rejects a cladistic approach to classification, and overtly defends the use of paraphyletic taxa (cf. Cronquist (1987) and the replies by Donoghue and Cantino (1988) and Humphries and Chappill (1988)). Dahlgren has explicitly embraced cladistics as a major key to develop his system. Some years ago, at a less advanced stage in the development of these major classifications of flowering plants, there seemed to be several important similarities between Cronquist's and Takhtajan's schemes, on one hand, and Thorne's and Dahlgren's, on the other. The subsequent changes have made those of Thorne and Takhtajan more similar.

In 1976, when presenting the second version of his system, Thorne remarked that many higher-level taxa, traditionally accepted within flowering plants, and still 'widely recognized up to date, although sometimes masquerading under other names', are certainly polyphyletic. Examples include Amentiferae, Monochlamydeae, Dialypetalae and Sympetalae. This fact seems to have been sufficiently recognized (and overcome) by all recent authors, although their classifications still include many paraphyletic taxa.

It is now widely acknowledged that monocotyledons have their sister-group within dicotyledons, and Huber (1982) showed that dicotyledons are paraphyletic. The main groups he recognized within angiosperms coincide with those identified by Wenzel and Hemleben (1982) on the basis of DNA reassociation studies.

According to the cladistic analysis of Loconte and Stevenson (1991), the first basal clade splitting within angiosperms corresponds to the Calycanthales (a new order they formally recognize for Calycanthaceae plus Idiospermaceae), followed, in that order, by Magnoliales, Laurales, Illiciales and Lactoridales. The remaining clade, comprising the most Recent angiosperm species, splits into the 'Tricolpates' (Ranunculales plus the 'Hamamelideae', i.e. most of the old dicots) and the 'Paleoherbs' (= Piperales, Aristolochiales, Nymphaeales) within which are nested the 'Alismatidae' (i.e. the old monocots), with the Nymphaeales as the sister-group.

At a lower taxonomic level, it is probable that Lamiaceae (Labiatae) is a monophyletic subgroup of Verbenaceae, as Asclepiadaceae are with respect to Apocynaceae, Apiaceae (Umbelliferae) to Araliaceae and Brassicaceae (Cruciferae) to Capparaceae. If these views are correct, Verbenaceae, Apocynaceae, Araliaceae and Capparaceae must be regarded as paraphyletic. However, these families have been known and recognized for a long time. Therefore, we can expect resistance to change the traditional taxa for the newly recognized clades.

Instances of the difficulties in circumscribing plant families and assessing their affinities may be obtained through a perusal of Mabberley's (1987) *Plant-Book*. For example, Mabberley acknowledges that the Acanthaceae is 'closely allied to Scrophulariaceae with no meaningful distinction between them: Nelsonioideae are perfectly intermediate and are here by tradition'. The limits between Scrophulariaceae and Bignoniaceae are also uncertain, as exemplified by the problematic placement of *Paulownia* and the tribe Schlegeliae. Loganiaceae also pose some problems, because the tribe Buddlejeae (Buddlejaceae *sensu* Cronquist) is 'intermediate between Scrophulariales and Gentianales', whereas another tribe of the same family (Retzieae, also treated as an independent family by Cronquist) has 'some similarities with Verbenaceae' and further tribes, such as Plocospermeae, 'have some similarities with Apocynaceae and Rubiaceae'. The limits of the Euphorbiaceae are unclear with regards to a number of genera currently segregated into independent families, whereas the bulk of the family itself is possibly diphyletic, according to karyological evidence. At the other extreme, the Passifloraceae and Flacourtiaceae can hardly be separated from one another.

Within flowering plants, only one major clade, the Monocotyledones, has been the object of a cladistic analysis at the family level and even this effort (Dahlgren, Clifford and Yeo, 1985) cannot be qualified as definitive.

Of the controversies surrounding the correct placement of individual plant families, simply consider three well-known ones, Cucurbitaceae, Oleaceae and Paeoniaceae. According to Mabberley's (1987) summary, the Cucurbitaceae are 'now placed near Passifloraceae, but seed-structure differs markedly from that in Passifloraceae or Caricaceae and they might be best put near Grossulariaceae (Jeffrey) or even Sapindaceae; they have no members 'intermediate' between Cucurbitaceae and another family'. For the Oleaceae the placement is uncertain between Scrophulariales (as in Cronquist's system) or in Gentianales, as is done traditionally. Paeoniaceae: 'the position of this family is controversial, some characters ... like Dilleniaceae but other features like Ranunculaceae: the forcing of Paeoniaceae into either order demonstrates the artificiality of the system ... The group is probably best treated separately from other orders in the present system as the seed suggests affinities with Cucurbitaceae and Convolvulaceae (Corner), not Dilleniaceae, but as this is controversial ..., it is maintained in its place in the Cronquist system for convenience'.

Many of these problems are simply the result of an unwillingness to reclassify angiosperms using phylogenetic systematics. Things are changing, however.

Interviews on the daily work of systematists: problems and trends

8

il n'est pas possible dans la plupart des 32 Embranchements (Protozoaires compris) du Règne Animal, de reconnaître avec certitude le niveau spécifique; ce n'est que dans seulement quelques Embranchements privilégiés, dont parmi les plus étudiés les Vertébrés, les Arthropodes, les Polychètes et les Echinodermes – groupes qui se prêtent mieux à l'expérimentation et à l'usage de critères diagnostiques variés – qu'il est possible de circonscrire une espèce sans ambiguïté.

J.-L. d'Hondt (1988, p.7)

For most of the 32 phyla of the Animal Kingdom (Protists included) species are not recognizable with any certainty. Species are circumscribed without ambiguity in a few privileged phyla only: as for the best studied ones, in Vertebrates, Arthropods, Polychaetes and Echinoderms, all groups better amenable to experimentation and to the use of diverse diagnostic criteria.

The material for this chapter came from information obtained from some 140 colleagues, who were kind enough to provide me with a picture of the problems they currently face in their specialist groups. In addition, I relied on my personal acquaintance with some taxonomic groups, such as centipedes and leeches, as well as on the literature.

8.1 SPECIALIST GROUPS AS NATURAL GROUPS

A first question was: do taxonomists usually work on natural groups? In so far as this is true, the state of affairs actually gives a sense to their efforts of working at a systematic arrangement of subtaxa within their groups. However, there is sometimes positive knowledge, or at least a 'qualified feeling' that one is currently working on paraphyletic taxa. Despite that, it is still considered that these groups are operationally useful units to work on. Examples include polychaete annelids, oligochaetes, cirripeds, myriapods, symphytes (saw-flies and relatives), butterflies ('Rhopalocera'), lygaeid bugs and staphylinid beetles. Artificial taxa, such as lichens and fossil spores, are also considered acceptable as reasonable fields of investigation. Problems, however, arise not only in the analysis of data, but also in presenting them. Fiedler (1991), when delivering his well-documented volume on fishes in Kaestner's *Lehrbuch der Speziellen Zoologie*, was obliged to specify that there is no natural taxon embracing all fishes, and fishes only, because the old Pisces is but a paraphyletic group. I welcome this acknowledgement of systematic advance in a major treatise the original layout of which was defined before the development of cladistics.

What most seriously affects current systematic work is not so much the (infrequent) artificiality of taxa but the fact that most systematists work on geographically limited sections of a natural taxon. Such samples of species are very often random collections of dubiously related subgroups. Any arrangement built on such a limited scale must be regarded as provisional and potentially artificial, unless confirmed by worldwide revisionary work. Many specialists feel that their groups are placed unsatisfactorily within a wider systematic context and this problem arises at all taxonomic levels.

8.2 GENERA

A list of groups where the generic arrangement is variously perceived as more or less satisfactory is given in Table 8.1. I hope that the reader will take the context of the table at face value, i.e. as opinions, mainly expressed by one specialist for each group, in respect of the current state of affairs.

According to R. Poggi (personal communication) the generic relationships of pselaphid beetles are satisfactorily known for the Palaearctic fauna, but not for the other regions. Analogous problems with non-Palaearctic and especially with Neotropical faunas (and floras) are very widespread.

In many groups there are a lot of monospecific genera (Table 8.2). In fact many authors regard current genera as excessively split: that is the reaction (all personal communications) of C.D. Michener for Apoidea, S. Ruffo for Amphipoda, and S.C. Schell for Trematoda ('many currently recognized genera are probably nothing more than species'). A more satisfactory arrangement after too many years of splitting is much sought for by students of Cactaceae. Several meetings of an *ad hoc* committee have been recently held (Anonymous, 1986) in order to establish a consensus list of genera. It seems that good agreement has already been reached to identify and retain only monophyletic taxa. However, there are still disagreements, as far as the rank to be given to the recognized taxa. 'None of us, it is said in the 1986 report, is so reactionary as to want to return to Schumann's 20 genera or retain *Cereus sensu lato* as a hold-all (as Benson has done), nor so radical as to advocate acceptance of a Backebergian figure of over 200 genera, but a wide spectrum of opinions does exist. The figure of 86 reached at our third meeting in 1985 reflects the majority

Table 8.1 A summary of the answers given by several specialists (all personal communication) to the following question: Are genera satisfactory, as currently recognized in the group you study?

Taxon	Specialist

GENERA SATISFACTORY

Taxon	Specialist
Nematoda	Siddiqi; Tarjan
Cladocera	Margaritora
Crust. Copepoda Cyclopoida	Pesce
Crust. Copepoda (parasitic)	Kabata
Crust. Thermosbaenacea	Wagner
Crust. Mysidacea	Pesce
Plecoptera	Zwick
Orthoptera	Baccetti
Anoplura	Kim
Diptera Chironomidae	Rossaro
Coleoptera Carabidae	Kavanaugh
Hymenoptera Symphyta	Goulet
'Acari' Oribatida	Bernini

GENERA MOSTLY SATISFACTORY

Taxon	Specialist
Anthozoa Alcyonaria	Williams
Cestoda	Schmidt
Rotatoria	Pejler; Wallace; but too lumped for Shiel
Gastrotricha	Tongiorgi; Balsamo
Tardigrada	Bertolani
Oligochaeta (aquatic)	Brinkhurst
Pseudoscorpionida	Gardini
Crust. Amphipoda	Ruffo; Bousfield
Odonata	Brookes; Davies
Thysanoptera	zur Strassen
Homoptera Psylloidea	Hodkinson; Burkhardt; but for the large genus *Trioza*
Diptera Syrphidae	Daccordi
Hymenoptera Apoidea	Michener; but for excessive splitting
Bryozoa	d'Hondt
Ascidiacea	Monniot
Serpentes	Underwood
Laboulbeniales	Tavares
Leguminosae	Isely
Cucurbitaceae	Jeffrey
Malvaceae	Fryxell
Compositae	Nordenstam; Bremer
Gramineae	Clayton; except for bamboos
Cyperaceae	Koyama

Table 8.1 (*contd*)

Taxon	Specialist

GENERA PARTLY UNSATISFACTORY

Porifera	Sarà
Hydrozoa	Boero
Gnathostomulida	Sterrer
Crust. Cirripedia	Relini
Crust. Isopoda	Argano
Embiidina	Ross
Heteroptera generally	Kerzhner
Heteroptera Coreidae	Brailowsky
Heteroptera Lygaeidae	Slater
Diptera Cecidomyidae	Gagné
Lepidoptera Papilionidae	Brown; Racheli
Coleoptera Staphylinidae	Thayer; Frank; Zanetti
Coleoptera Scarabaeoidea	Zunino
Hepaticopsida	Grolle
Compositae	Jones

GENERA QUITE UNSATISFACTORY

Trematoda	Schell
Nemertea	R. Gibson
Polychaeta	Fauchald
Crust. Copepoda (subterranean)	Rouch
Homoptera Aleyrodidae	Mound
Diptera Phoridae	Disney
Lepidoptera Zygaenidae	Racheli
Hymenoptera Vespidae	Carpenter
Opiliones	Martens
'Lichenes'	Nimis
Cactaceae	E.F. Anderson
Sapotaceae	Pennington

viewpoint that generic boundaries should normally be drawn only where there are substantial morphological discontinuities...' This last sentence justifies the exclusive concerns of some of our cactus colleagues for phylogenetic matters; however, owing to the conventional nature I too recognize in all supraspecific taxa, including genera, I cannot disagree (but for the suggestion concerning subgenera) with this comment from the Cactaceae committee: 'Lengthy synonymies for innumerable cactus species are testimony to the argument that the adoption of narrow generic limits is disadvantageous to nomenclatural stability. ... The desirable balance between stability and flexibility in our dual-purpose system of naming is more likely to be achieved by greater use of infrageneric hierarchies within broadly conceived genera'. That is what Crisp and Fogg (1988) recently called for (Chapter 5).

Table 8.2   Percentage of monospecific genera in some groups

6–9%

Cestoda
Lepidoptera Papilionoidea
Crust. Cladocera

10–15%

Nematoda Tylenchidae
Gastrotricha
Crust. Amphipoda
Dermaptera
Diptera Chironomidae
Coleoptera Scolytidae
Pseudoscorpionida
Hepaticopsida
Convolvulaceae

15–25%

Nemertea
Oligochaeta (aquatic)
Crust. Mysidacea
Plecoptera
Diptera Cecidomyidae
Diptera Syrphidae
Coleoptera Nitidulidae
Tunicata Ascidiacea
Sapotaceae
Malvaceae

>25%

Tardigrada
Crust. Copepoda Cyclopoida
Odonata
Anoplura
Psocoptera
Homoptera Aleyrodidae
Coleoptera Cicindelidae
Coleoptera Staphylinidae
Coleoptera Histeridae
Coleoptera Lucanidae
Diptera Stratiomyidae
Lepidoptera Zygaenidae
Begoniaceae
Gramineae
Cucurbitaceae
Ascomycetes Laboulbeniales

8.3 SPECIES

How frequently do specialists believe that species in their group are well circumscribed, natural and easy to identify? The answer to this question is often independent from estimates of the current state of affairs at higher taxonomic levels within the same group. A summary of viewpoints is given in Table 8.3.

Of great interest are the opinions of systematists specializing in enormous genera, such as *Stenus* (Coleoptera Staphylinidae), with some 1800 species described so far, or *Acacia* (Leguminosae), with some 1500 described species. According to V. Puthz (personal communication) and B.R. Maslin (personal communication), in these huge genera there are no major problems at the species level, provided that sufficient materials are available for comparisons. A traditionally difficult very large genus is *Astragalus* (Leguminosae) (D.J.G. Podlech, personal communication), as are *Rhododendron* (approximately 800 species; M.N. Philipson and W.R. Philipson, personal communications) and *Oenothera* (about 130 species; W. Dietrich, personal communication). Many problems derive, in plants, from extensive hybridization, as in *Salix, Eucalyptus, Lotus* or *Rubus*. The worst comments on this question, however, have been those by B. Pejler for rotifers, by R.H. Disney for phorid flies, by F. Monniot on ascidians and by E.F. Anderson on cactuses: according to them, species recognition in these groups is more often difficult than not.

The most difficult problems concern those taxa described in the past without the subtleties of modern research. Old descriptions are poor, the type specimens are either not preserved or very damaged and hardly suitable for re-description, and citations in the literature are unreliable. To obtain a definitive interpretation of an old name might require more time than describing 20 new species in groups not overburdened by old descriptions. In many cases, the problems of old descriptions and old names are acute when dealing with species from Europe or the Mediterranean basin, simply because they have a large history of study.

This is one cause of an apparently puzzling but widespread situation, that within some groups it is often much easier to identify a tropical species, than a European one. There are two problems with our temperate fauna and flora. On one hand, animals and plants from temperate zones have been more thoroughly investigated than most tropical groups. Many details of intraspecific variation, many instances of natural hybridization and other 'subtle' matters have been studied and are widely mirrored in the literature. This is less so for tropical groups. On the other hand, it may also be true that complex variation is more common in temperate areas than in the tropics, partly as a consequence of century-long intensive habitat disturbance, giving rise to secondary contacts, with hybridization between previously isolated populations.

The five volumes of *Flora Europaea* (Tutin *et al.*, 1964–1980) provide us with a good example of this state of affairs. At least 15% of the 1500 or so genera covered by *Flora Europaea* are said to have problems at the species level. Difficult genera are not always the largest, and *vice versa*. There are species problems with *Ranunculus* (131 species) as with *Pulsatilla* (9 species), *Helleborus* (11 species), or

Table 8.3 A summary of the answers given by several specialists (all personal communication) to the following question: Are species satisfactory, as currently recognized in the group you study?

Taxon	Specialist

SPECIES SATISFACTORY

Taxon	Specialist
Gnathostomulida	Sterrer
Nematoda Tylenchidae	Siddiqi
Crust. Mysidacea	Pesce
Crust. Amphipoda	Ruffo; Bousfield
Crust. Isopoda	Argano
Odonata	Brookes; Davies
Anoplura	Kim
Psocoptera	Smithers
Homoptera Psylloidea	Hodkinson; Burkhardt
Diptera Cecidomyidae	Solinas
Diptera Chironomidae	Rossaro
Diptera Sciomyzidae	Munari
Diptera Sepsidae	Munari
Diptera Sphaeroceridae	Munari
Diptera Syrphidae	Daccordi
Diptera Scatopsidae	Haenni
Lepidoptera Heliconiinae	Brown
Coleoptera Carabidae	Kavanaugh
Coleoptera Pselaphidae	Poggi
Coleoptera Dryopidae	Olmi
Coleoptera Elmidae	Olmi
Coleoptera Alticidae	Biondi
Hymenoptera Symphyta	Goulet
Hymenoptera Apoidea	Michener
Hymenoptera Vespidae	Carpenter; but for social ones
Bryozoa	d'Hondt
Cucurbitaceae	Jeffrey
Compositae	Jones; not quite so for Bremer and Nordenstam
Gramineae	Clayton
Cyperaceae	Koyama

PROBLEMS WITH SPECIES, ESPECIALLY WITH TROPICAL FORMS

Taxon	Specialist
Anthozoa Alcyonacea	Williams
Trematoda	Schell
Cestoda	Schmidt
Nemertea	R. Gibson
Gastrotricha	Tongiorgi; Balsamo
Tardigrada	Bertolani
Oligochaeta (aquatic)	Brinkhurst
Crust. Copepoda (subterranean)	Rouch
Plecoptera	Zwick

Table 8.3 (*contd*)

Taxon	Specialist
Orthoptera	Baccetti
Dermaptera	Sakai
Thysanoptera	Mound; zur Strassen
Heteroptera: several families	Heiss; Hasan; Péricart; Slater
Lepidoptera Papilionidae	Brown
Diptera Stratiomyidae	Rozkosny
Coleoptera Scarabaeoidea	Zunino
Coleoptera Scolytidae	Wood
Serpentes	Underwood
Ascomycetes Laboulbeniales	Tavares
'Lichenes'	Nimis
Hepaticopsida	Grolle
Leguminosae	Isely

SPECIES FREQUENTLY ILL-DEFINED

Porifera	Sarà
Hydrozoa	Boero
Mollusca	Giusti
Pseudoscorpionida	Gardini
Homoptera Aleyrodidae	Mound
Coleoptera Histeridae	Vomero
Coleoptera Lucanidae	Bartolozzi
Coleoptera Curculionoidea	Osella
Hymenoptera Dryinidae	Olmi

Caltha (1 species), to give an example within one family (Ranunculaceae). Problems also arise in other genera credited with one species (*Pteridium, Kernera, Ledum, Empetrum, Swertia, Dactylis*) or two species (*Viscum, Nuphar, Capparis, Erophila, Hymenolobus, Hedera, Arbutus, Glechoma, Veratrum, Phippsia, Aeluropus*). On the contrary, several large genera do not seem to be particularly difficult, for example *Teucrium* (49 species), *Pedicularis* (54), *Anthemis* (62), *Artemisia* (57), *Senecio* (67), or *Stipa* (42). Speciose genera such as *Astragalus* (133), *Trifolium* (99), *Euphorbia* (105), *Veronica* (62), *Campanula* (144, but for the group of *C. rotundifolia*) appear to be well understood at the species level. That these uneven difficulties are not simply vagaries of taxonomy is easily confirmed by their idiosyncratic taxonomic distribution: problems are very widespread in families such as Rosaceae or Cruciferae, much less so in Leguminosae and Umbelliferae.

Besides the contrast between temperate and tropical groups, another contrast is readily perceived between freshwater and marine groups. For taxa represented in both major environments, larger problems are encountered with freshwater forms, particularly at the species level. This circumstance is common, for example, in fishes, isopods and amphipods.

8.4 INFRASPECIFIC TAXA

The amount of ink used to propose, defend or contrast the most idiosyncratic attitudes in this subject area is hardly justified. There are many groups where infraspecific taxa are never or only seldom recognized (Table 8.4).

In other groups, infraspecific taxa are recognized and named, but this usage is regarded as unjustified and/or useless by leading specialists, such as J. Golding for begonias, D.J.G. Podlech for *Astragalus*, or W. Dietrich for *Oenothera* (all personal communications).

It is necessary to cite a few arguments in some detail. According to S.C.H. Barrett (personal communication), the varieties currently recognized in *Pontederia* (water plants known as the pickerel weeds or wampees) may well justify species rank. According to R. Poggi (personal communication), those so far described in pselaphid beetles are a type of provisional means to identify something that might in future be recognized as 'full species'. According to a systematist working on a group of large horticultural interest, such as begonias (J. Golding, personal communication), subspecific distinctions are 'only moderately useful to horti-culturists who grow and propagate clones of a particular variety or form'. R. Shiel (personal communication), an Australian rotifer systematist, observes that infraspecific taxa are more commonly recognized and named in developing countries, using a classic (i.e. European) reference as source, and adds: 'Given that recent experimental evidence shows morphotypic variation in response to environmental variables, e.g. predators, pH, temperature, naming of infraspecific taxa is neither justified nor useful. Much of the systematic difficulty in Rotifera is a direct result of the proliferation of form, var., ssp. categories without population measurements, occasionally from single individuals or aberrant forms, in all a reflection of poor taxonomy'. In this context it is fair to say that other systematists of the same group (e.g. B. Pejler and C. Ricci) consider the identification of ecotypes within rotifer species as justifiable. Even biologists who study extremely variable and complex genera of flowering plants such as *Oenothera* (W. Dietrich, personal communication; also Raven, Dietrich and Stubbe (1979)) refuse to recognize infraspecific taxa.

Naturally, there are defenders of subspecies and even of infrasubspecific taxa. Infraspecific distinctions are accepted in groups such as: dragonflies (decreasingly; S. Brookes, D.A.L. Davies), crickets (B. Baccetti), butterflies (K. Brown, Jr), bryozoans (J.-L. d'Hondt), eumenid wasps (A. Giordani Soika; J.M. Carpenter, however, dissents), as well as in several families of beetles (e.g. Carabidae), and especially in birds, mammals and in many families of flowering plants. The infraspecific entities recognized in different groups are different from each other, as are the primary interests (scientific or economic) consequently served. Similarly, the nomenclatural treatments of infraspecific taxa currently recognized.

In my view, those groups where infraspecific taxa are most commonly recognized and named conform with one or more of the following properties. They are organisms with obvious, and sometimes discontinuous, seasonal or otherwise eco-phenotypic changes such as rotifers, cladocerans and butterflies, probably also

Table 8.4 Critical attitudes of several specialists against recognizing infraspecific taxa

Taxon	Specialist

INFRASPECIFIC TAXA NEVER OR SELDOM RECOGNIZED IN THE GROUP

Taxon	Specialist
Porifera	Sarà
Hydrozoa	Boero
Anthozoa Alcyonacea	Williams
Trematoda	Schell
Cestoda	Schmidt
Nemertea	R. Gibson
Gastrotricha	Tongiorgi; Balsamo
Polychaeta	Fauchald
Tardigrada	Bertolani
Pseudoscorpionida	Gardini
Crust. Copepoda (subterranean)	Rouch
Crust. Copepoda (parasitic)	Kabata
Crust. Mysidacea	Pesce
Crust. Amphipoda	Ruffo; Bousfield
Mantodea	La Greca
Plecoptera	Zwick
Anoplura	Kim
Thysanoptera	Mound; zur Strassen
Heteroptera	Heiss; Péricart; Slater; Brailowsky
Homoptera Psylloidea	Hodkinson; Burkhardt
Homoptera Aleyrodidae	Mound
Diptera Cecidomyidae	Solinas
Diptera Phoridae	Disney
Diptera Sciomyzidae	Munari
Diptera Sepsidae	Munari
Diptera Sphaeroceridae	Munari
Lepidoptera Exoporia	Nielsen
Coleoptera Gyrinidae	Beutel
Coleoptera Staphylinidae	Frank; Thayer
Coleoptera Pselaphidae	Poggi
Coleoptera Histeridae	Vomero
Coleoptera Dryopidae	Olmi
Hymenoptera Apoidea	Michener
Opiliones	Martens
'Acari' Oribatida	Bernini
Tunicata Ascidiacea	Monniot
Hepaticopsida	Grolle
Gramineae	Clayton
Pontederiaceae	Barrett

diatoms and many families of flowering plants. They are organisms, mostly animals, with limited dispersal power and discontinuous distributions allowing for recognition of more or less differentiated populations without demonstrable sexual

Table 8.4 (contd)

Taxon	Specialist

INFRASPECIFIC DISTINCTIONS OUTRIGHT REJECTED, SOMETIMES AGAINST CURRENT PRACTICE

Cestoda	Schmidt
Nemertea	R. Gibson
Nematoda	Tarjan
'Polychaeta'	Fauchald
Crust. Amphipoda	Bousfield
Crust. Isopoda	Argano
Anoplura	Kim
Heteroptera	Péricart
Homoptera Psylloidea	Burkhardt; Hodkinson
Homoptera Aleyrodidae	Mound
Diptera Chironomidae	Rossaro
Diptera Cecidomyidae	Gagné
Lepidoptera	Balletto; Nielsen
Coleoptera Histeridae	Vomero
Coleoptera Nitidulidae	Audisio
Coleoptera Lucanidae	Bartolozzi
Hymenoptera Vespidae	Carpenter

A CAREFUL USE OF INFRASPECIFIC TAXA IS RECOMMENDED

Anthozoa Alcyonaria	Williams
Mollusca	Giusti
Crust. Cirripedia	Relini
Crust. Copepoda Cyclopoida	Pesce
Mantodea	La Greca
Heteroptera	Kerzhner
Diptera Syrphidae	Daccordi

isolation, especially when the phenotypic differences are less than the average between recognized species in the same genus: Botta's pocket gopher (*Thomomys bottae*), a North American subterranean rodent, has some 215 named 'races' in California only (Hewitt, 1988). They are very showy organisms, where phenotypic differences, although characteristic and without definite geographic or ecological distribution, are nonetheless identified easily. Such taxa are particularly apparent in animal breeding, in agriculture and horticulture and in the collections of amateurs (e.g. butterflies, several groups of beetles such as tiger-beetles and *Carabus*, birds and several groups of flowering plants). Special problems with domesticated animals and cultivated plants are considered in Chapter 10.

Within such groups, however, there are signs of emerging trends, for example, for recognizing and naming infraspecific taxa. Things seem to be different now from what was current some decades ago, when, for instance, Bright and Leeds (1938) cited 733 named aberrations for the chalkhill blue butterfly (*Lysandra coridon*). Today, many systematists acknowledge that it is important to study the diversity between species adequately. However, at the same time, they reject the practice of giving recognizable populations formal, Linnaean names (e.g. R.H. Disney, phorid flies; E.S. Ross, Embiidina; J.A. Slater, Lygaeidae). Other systematists accept conventional infraspecific taxon names because a name usefully describes variation, calling attention to it and facilitates bibliographic research. Such a view has been repeatedly noted in the literature (e.g. Stace, 1986) and is not easily discounted. In my view it would not be too difficult to devise an informal system of reference to these aspects of infraspecific variability. It would not burden the formal system with a number of additional names not corresponding to natural taxa (cf. Burtt, 1970; and several papers in Styles, 1986). Other students (mostly entomologists) justify the recognition (and naming) of subspecies, but only for strictly allopatric, well-differentiated populations or groups of populations. This opinion is expressed, among others, by students of ground beetles (D.H. Kavanaugh), of staphylinid beetles (A. Zanetti), of blister beetles (M. Bologna), of stratiomyid flies (R. Rozkosny) and of snakes (G. Underwood). Most flowering plant systematists with whom I have corresponded have no problems with the current practice of recognizing infraspecific taxa. More interesting are V. Puthz's comments (personal communication) on the staphylinid genus *Stenus*, where 'the polytypic species concept works for instance in the Mediterranean, in the Oriental region (Sunda Islands) and in the Ethiopian region (disjunctions in montane forests etc.)', less so elsewhere. I am reminded of Aspöck's (1986) observation, that in snake flies (Raphidioptera) several Nearctic species show a particularly high degree of intraspecific variation, not found elsewhere. In a similar vein, Pulawski (1988) remarks on the difficulties in recognizing the North American species of the wasp genus *Tachysphex* through the usual characters from male genitalia universally valuable elsewhere, both in the genus and in its relatives.

Overall, we witness today a definite trend towards the reduction of infraspecific taxa. I favour this trend, which is certainly supported by our increasing knowledge of complex, reticulated patterns of variability within and between populations, sometimes even between 'species'. At the same time, it would be very stupid to return to the old Linnaean dictum: *varietates minores non curat botanicus* (the botanist takes no care of minor varieties). A detailed appreciation of intraspecific variability is one of the most important ingredients of our current knowledge of the living world and its diversity and evolution. Steps should be taken, however, to improve our current linguistic conventions for communicating about infraspecific variability, without interfering with the already cumbersome nomenclatural apparatus we need for the phylogenetic system.

The future of infraspecific categories will probably also depend on the place that infraspecific names will find in the abstracting services databases. For plants, infraspecific taxa are recorded in the *Index Kewensis* only since 1970. What has become of the previous names? Burtt (1970, p.238) refers to them as 'the muck-heap of two centuries of unindexed and inadequately described epithets' and suggests that we 'leave it undisturbed so that it quietly rots down'. This is not a universal viewpoint, however.

8.5 CHARACTERS

I will not deal with biochemical and molecular evidence here, not only because of the more detailed account provided in Chapter 3, but also because of the very limited impact, if any, of biochemical and molecular studies on the daily work of most systematists. Moreover, I confine myself to animals, while referring to Stuessy (1990) for a very well-documented survey of the different types of data gathered and discussed by plant systematists.

Many systematists acknowledge that no major improvement has occurred in the current systematics of their groups. The reason are varied, but often depend on inadequate working conditions, few specialists, limited availability of material owing to inadequate field sampling, a lack of specialized techniques for collecting and a lack of financial support. In many other cases things are not so bad, although the most refined techniques have generally been applied, in nearly all cases, to limited subgroups of particular taxa.

For many groups, recent improvements mainly depend on a more careful and extensive examination of traditional, morphological characters. In sponges, for instance, increasing attention is being paid to structures of the aquiferous system and to the skeletal architecture. In cnidarians, a good description should now include the 'nematocyst formula', illustrating the types and locations of stinging cells. In flatworms and nemerteans, detailed analyses of internal morphology is now routine practice, despite the tedious histological preparations or stained mounts which have to be prepared. In some groups, microscopic analysis may be pushed to unexpectedly subtle details, as in rotifers, where the usefulness of counting the (nuclei of) body cells is sometimes acknowledged.

For each group, however, there seems to be an asymptotic limit to the richness of possible morphological information. It is quite probable that such asymptotic values have never been approached, but it is unlikely that nematode morphology will ever offer as many characters as the morphology of mammals or birds. The morphological terms available for these two vertebrate classes are five times as numerous as those for nematodes, indicating their different degrees of morphological complexity (Schopf et al., 1975; Raff and Kaufman, 1991).

Adult morphology is not always the only or even the best source of taxonomic characters. In whiteflies (aleyrodids), taxonomy is mainly built on the characters of pupal cases (which, incidentally, are easy to collect, to preserve and to study) and adults are hardly identifiable at the species level without the additional information provided by the structure of the case from which they have emerged.

In gall-midges, as in other gall-forming insects, host specificity and gall shape are often the primary clues to identification; however, increasing attention is being paid to both larval and adult morphology. The importance of larval characters is also increasingly recognized in other groups, as in butterflies, where caterpillars sometimes display lesser individual variation than do adults. Even eggs provide useful taxonomic characters, especially when they are encased within a sculptured chorion which can be examined by scanning electron microscopy, as is the case with tardigrades and phasmids. In the latter group, differences in the sculpturing of egg chorion have even been claimed to enable recognition of subspecies.

Data from host–parasite relationships are currently utilized whenever they apply. It should be remarked, however, that an excessive trust has often been put in these studies. Parasites are often polyphagous and occasional 'jumps' from one kind of host to another, not closely related to the first, may occur. The much disputed problem of coevolution between hosts and parasites, according to the so-called Fahrenholz's rule (Fahrenholz, 1913) has been widely discussed in recent times. The studies collected in the volumes edited by Futuyma and Slatkin (1983) and by Stone and Hawksworth (1986) should warn against too precipitous phylogenetic deductions from this type of evidence. This evidence is still largely taken into account, both for animal-parasitic animals, as for plant-parasitic animals, for parasitoids, or for plant-parasitic fungi. Fine examples are provided by Ackery (1988) for the host–plant relationships of nymphalid butterflies and by Schaefer (1988; cf. Schaefer and Mitchell, 1983; Ahmad and Schaefer, 1987; Schaefer and Ahmad, 1987) for the host–plant relationships of many families of bugs (Heteroptera).

Further comparative evidence is provided by behaviour. Coddington's (1986) paper on the monophyletic origin of the orb web must be cited here, as an exemplary phylogenetic treatment of ethological evidence; and this compares with the very different treatments of Kullmann (1964) and Eberhard (1990). Another interesting study is that of Carpenter (1989), where wasp social behaviour is considered against a cladogram of the family derived from morphology.

Developmental biology is gaining increasing attention among systematists, not only at higher levels, but also at the genus and species levels. This trend (Humphries, 1989) is occurring for several possible reasons. One is the widespread (but not universal) opinion, that understanding the developmental origin of morphological traits facilitates recognition of character polarity (cf. Chapter 2). Another is the rejuvenation of the nineteenth century approach, partly stimulated by Gould's (1977) book on *Ontogeny and Phylogeny*. Identification of paedomorphic traits is now fairly common in the literature, both palaeontological and neontological. McKinney and McNamara (1991; see also McNamara, 1982, 1983, 1986a,b, 1988) have provided very fine examples of heterochrony, together with a revision of the associated concepts and terms. Knowledge of morphogenesis helps in identifying homoplasies and also, sometimes, in getting a first, rough guess as to the genetic backgrounds. This last aspect, however, is still in its infancy, in spite of the enormous progress witnessed by developmental biology in these last years. It should be noted that developmental biology is still based on the study of a few species: what we know of developmental mechanisms in animals is derived from the study of only a few dozen species (Schaeffer, 1987). Practical problems obviously prohibit the study of development in a wider array of species. However, progress at the borders of developmental and systematic biology are also slowed down by lack of an adequate theoretical background (Arthur, 1984, 1988).

In the systematic treatments of some groups, it seems that systematists are still tied to methods that Adanson (1763) criticized more than two centuries ago. This consists of choosing one or a few differential characters for distinguishing taxa at each taxonomic level of the Linnaean hierarchy. For example, Kabata (1979) considered that suprafamilial taxa in parasitic Copepoda should be primarily identified according to the types of mandibles, families by the retained body segmentation and the number and structure of natatory appendages. Similarly, the conventional families of geophilomorph centipedes are primarily distinguished on the characters of the mandible. This character is not used at lower taxonomic levels, except for the genera and species of one family, Mecistocephalidae.

The role of cytogenetics is varied in the different groups. The differences are only partly explained by tradition. In fact, karyological characters, like all others, show all possible levels of invariance or plasticity, often within the same lower group. Chromosome studies have become compulsory for the current study of many species complexes, for example in the mosquito genus *Aedes*.

Mating calls provide useful taxonomic characters in several groups, not only in crickets (e.g. Gwynn and Morris, 1986; Otte, 1989; Benedix and Howard, 1991) or frogs (e.g. Hillis, 1981; Frost and Platz, 1983), but also in other groups, such as leafhoppers (e.g. Claridge, 1985a,b, 1986; Claridge, den Hollander and Morgan, 1985) and lacewings (e.g. Martinez Wells and Henry, 1992).

8.6. FROM FIELD WORK TO MONOGRAPH

The dependence of systematic progress on more accurate and broader descriptions of specimens is acknowledged by specialists of most groups. However, considerable progress has been achieved, in many instances, only because of technical improvements.

Important progress has been made in collecting techniques. The most spectacular developments are probably those of insect collecting from the canopies of tropical forests, as described by Mitchell (1986). Stork (1988) also describes how insects can be collected from the tree canopies up to 70 m above ground level by spraying or fogging with insecticides. The canopy studies have stimulated a revision of our estimates of species diversity in tropical biotas (cf. Chapter 6).

Another realm we are now beginning to investigate with more adequate means is the deep sea. For new studies of deepsea biotas, both with respect to the discovery of new animal groups and with respect to speciation see Marshall (1979) and Wilson and Hessler (1987). Species diversity in deepsea communities is unexpectedly high. A series of samples from depths of 155–2500 m off New Jersey contained some 900 different species, mostly belonging to polychaetes (47 families), peracarid crustaceans (39 families) and molluscs (43 families). In the same study, half of the species in many new samples proved to be undescribed and the total number of animal species from most single deepsea sites turned out to be in the thousands for areas of about 1 km^2 (Grassle, 1989).

The fauna of underground waters are an apparently unexhausted source of new, often exceptional, taxa. This is exemplified by the discovery of the first representatives of new major crustacean groups such as Bathynellacea, Mystacocarida, Thermosbaenacea, Spelaeogriphacea and Remipedia.

However, a much better understanding of the systematics of a given group is often simply derived from the availability of more abundant material, the outcome of intensive and more specialized collecting. Efficient collecting stems from a sound knowledge of autoecology. Nocturnal light-trapping, a traditional collecting strategy used by lepidopterologists for a long time, allows for the collection of exceptionally fine and extensive series of insects of different orders.

As for new laboratory techniques, transmission electron microscopy (TEM) and scanning electron microscopy (SEM) have had a widespread impact upon taxonomic practice. Improved light microscopy (e.g. Nomarski interferential microscopy) is also widely recognized as a major help in current taxonomic work, for example in gastrotrichs, or small oligochaetes.

The SEM is now a routine working tool for many taxonomists, particularly in entomology, malacology and protozoology, and many descriptions of new species are now illustrated with SEM micrographs.

By contrast, the TEM is seldom used by workers whose primary interest lies with systematics or phylogenetics, but for those working with protists. Most research on ultrastructure is carried out by students of cytology and related disciplines. However, several discoveries obtained with a TEM have modified our phylogenetic concepts more in depth than any information gathered through SEM. As for the relationships between the major groups of living organisms, TEM data are rivalled only by molecular evidence.

Progress sometimes requires only a small technical improvement. For example, preservation in spirit and mounting between slide and cover-slide is current practice for many invertebrate groups, including most non-insect arthropods, but has been largely avoided for most insect orders. Adopting these procedures, however, offers enormous possibilities for adding new characters to those currently recognized. This type of progress has been witnessed in the case of phorid flies and staphylinid beetles, although this practice is still very far from universal, even in these families.

Practising taxonomists do not necessarily look for revolutionary improvements for their work. They mainly look for the availability of more specimens, especially from the tropics, or for a greater availability of modern techniques (SEM, TEM, protein electrophoresis, DNA studies). For many taxa, there is also a requirement for descriptions of whole life histories or, at least, for morphological information of early ontogenetic stages.

The need to work on a world scale is widely recognized, but this is often unattainable when several systematists are involved in the study of one group. E.S. Ross (personal communication) has collected insects of the order Embiidina in all parts of the world, amassing a huge collection of some 300 000 specimens, including a number of undescribed genera and species. With his collection, we are able to obtain an adequate idea of embiid diversity for the first time. However, there is a problem. The correct description and discussion of all new species is far too great a job for one person. This is just one well-documented case of the inadequacies in human resources currently involved in taxonomic research. While writing, I remember the huge rows of bottles of unidentified spiders which I saw a few years ago in the Arachnid Section of the Natural History Museum (London). Who will ever study them? Active research on spiders has been discontinued in London as has research on whales and coelenterates (Bourne *et al.*, 1988;

Hadlington, 1988). The situation is seldom better, and often much worse, in the other centres of taxonomic research.

Against such a background, the most sensible request of some systematists I have consulted is to have effective acknowledgement of the importance of taxonomic work by the major research funding agencies. They also want a more substantial recruitment of professional taxonomists within museums, universities and other centres (Heywood and Clark, 1982).

As for international cooperation, things are rapidly improving. A major work such as the five volume *Flora Europaea* (Tutin *et al.*, 1964–1980) is the magnificent result of the cooperative efforts of dozens of botanists. To develop that project it was necessary to create an official publicly funded structure and to assure its continuity for 27 years (Tutin *et al.*, 1964–1980, pp.XVII–XX of Vol. 5). This is a method of raising the standards of systematic work.

The unequal distribution of taxonomic diversity

9

Il est donc évident par les faits, que les Genres en général ne peuvent être tous naturels dans aucune Métode artificiele ou arbitrère; et tous les axiomes qui ont été fondés pour l'établissement des Genres naturels, sont sensiblement faux, parce que leurs Auteurs n'aient point une idée juste de la Métode naturele, les rendoient relatifs aux principes abstractifs des Métodes artificieles.

M. Adanson (1763, p.cvii)

Facts also demonstrate, that no artificial or arbitrary Method may be wholly comprised of natural genera; and that all principles underlying natural genera are plainly false, because their Authors had no right idea of the natural Method, but relied instead on the abstract principles of artificial methods.

9.1 THE VERY LARGE GENERA

Raikow (1986) feels that only the vagaries of taxonomy explain the higher number of species belonging to the passerine birds, compared with the non-passerine birds (but see the later comments by Fitzpatrick (1988), Vermeij (1988) and Kochmer and Wagner (1988) and Raikow's (1988) reply). I cannot agree with Raikow's views, as it is difficult to disprove the fact that there are more types of cats than there are types of elephants and more types of cone shells than there are types of nautilus. Despite the vagaries of systematists, there are, in a taxonomic sense, many dense clusters of biotic diversity. In this chapter I briefly discuss some clusters currently recognized as the largest genera. I am not advocating a justification for the genus that I have denied a few pages ago. I simply accept the opinions of systematists working on different groups, who recognize the 'close proximity' of a sometimes huge numbers of species. The meaning of 'large' is quite indeterminate. *Crocidura* with 149 species (Honacki, Kinman and Koeppl, 1982) is enormous within mammals, as is *Eleutherodactylus* with 405 species within amphibians

(Frost, 1985). On the other hand, a genus such as *Viola*, with some 500 species, is not particularly large within flowering plants, where more than 2000 (Mabberley, 1987) or perhaps nearly 3000 species (D.J.G. Podlech, personal communication) are recognized within *Astragalus*.

There seem to be two types of large genera. There are the so-called waste-basket genera, each of them containing an heterogeneous, sometimes overtly polyphyletic assemblage of forms, which await reclassification into natural genera. This is frequently the case with old (e.g. Linnaean) genera, originally embracing what is now a whole family, or even an order, in modern classifications. For example, *Ichneumon* Linnaeus, 1758 or *Geophilus* Leach, 1815, once embraced, more or less, what are now family Ichneumonidae and family Geophilidae (or order Geophilomorpha), respectively. While recognizing the opportunities of splitting these traditional genera into a number of smaller ones, many less well-known species have been left in the old genera (eventually becoming labelled as *species inquirendae*). Consequently, the limits of these genera remain fuzzy and their size may well be unjustifiably large, as in the case of *Ichneumon*, with about 1000 species.

Other waste-basket genera are of more recent origin or, at least, have arisen during the splitting of traditional genera. Examples include *Atheta* within staphylinid beetles and *Onthophagus* within scarabaeids. Both genera include huge numbers of species, 2000–3000, or even more. No recent checklist is available, 1393 species of *Onthophagus* were listed by Boucomont and Gillet in 1927 and 1175 species of *Atheta* by Bernhauer and Scheerpeltz in 1926. As currently understood, *Onthophagus* is very probably artificial (M. Zunino, personal communication), but a core group of several hundreds of species will almost certainly remain under this name, until someone accomplishes the formidable task of systematizing the group. As for *Atheta*, there have already been some proposals to split it into many natural genera, but the task has yet to be accomplished on a world scale. Even with the divisions, the new taxa include one or two groups of respectable size (A. Zanetti, personal communication).

Despite problematic groups, there are many large genera which seem to be good holophyletic groups of closely interrelated species. In a few cases, a 'good' synapomorphy unites all members. This is the case for *Stenus*, a genus of staphylinid beetle with about 1800 species described so far from all over the world except New Zealand. V. Puthz (personal communication) argues that there are at least another 1000 species still to be described. These beetles are characterized by the extraordinary structure of their labium, which works as a protrusible device for capturing prey, in a way strongly suggestive of the 'mask' of the dragonfly nymphs. All species so far described within this genus are well characterized and distinguishable; problematic species groups do not seem to exist. There is no case for splitting the genus into smaller genera. Within the very large family Staphylinidae there is only one genus obviously related to *Stenus*. This is *Dianous*, with some 200 species described so far. *Dianous* lacks the conspicuous apomorphy of the labium found in *Stenus*, but that is insufficient to characterize it as a holophyletic genus. In fact, nobody has been able to identify a synapomorphy for all the species currently ascribed to *Dianous* and it may even be paraphyletic, with *Stenus* representing a specialized offshoot from within it (Puthz, 1981).

For fungi, the largest genera cited by Hawksworth, Sutton and Ainsworth (1983) are listed in Table 9.1. The large figure of 3000 species for *Puccinia*, although impressive, is quite probably misleading, only representing the ecotypes of a relatively few species, described from different host plants. As for the largest genera of lichen-forming fungi, consider some comments from P.L. Nimis (personal communication). In *Usnea*, the high number of 'species' derives from phenomena of vegetative reproduction and probably hybridization. *Parmelia* has

Table 9.1 The largest genera of fungi. Data from Hawksworth, Sutton and Ainsworth (1983). L = lichen-forming genus

Number of species	Genus
3000	*Puccinia*
1270	*Cercospora*
1000	*Meliola*
1000	*Septoria*
600	*Psathyrella*
600	*Rhodophyllus*
600	*Uromyces*
600	L *Usnea*
600	*Aecidium*
550	L *Parmelia*
500	L *Arthonia*
500	*Laboulbenia*
500	*Mycosphaerella*
500	*Uredo*
450	L *Caloplaca*
400	L *Bacidia*
400	L *Buellia*
400	*Cortinarius*
400	L *Lecanora*
400	L *Lecidea*
350	*Ascochyta*
350	L *Cladonia*
350	*Marasmius*
300	L *Graphis*
300	L *Opegrapha*
300	*Ramularia*
300	*Ustilago*
300	L *Verrucaria*
271	*Russula*
270	*Arthopyrenia*
270	L *Graphina*
250	*Appendiculella*
250	*Asteridiella*
250	*Clitocybe*
250	L *Pertusaria*
250	*Poria*

Table 9.1 (*contd*)

Number of species	Genus
226	*Pseudocercospora*
222	*Pestalotia*
200	L *Acarospora*
200	*Agaricus*
200	*Amanita*
200	*Asterina*
200	*Chaetomium*
200	*Galerina*
200	*Leptostroma*
200	*Marasmiellus*
200	*Mycena*
200	*Nectria*
200	L *Ocellularia*
200	L *Phaeographis*
200	*Pleospora*
200	L *Porina*
200	L *Pseudocyphellaria*
200	L *Ramalina*
200	L *Rhizocarpon*
200	L *Rinodina*
200	*Stagonospora*
200	L *Sticta*

been split recently into several genera (e.g. Esslinger, 1977), even 20 or more genera according to some authors. *Bacidia* could possibly follow the same fate. Many genera have also been segregated from *Lecidea*, and the same process has also started in *Lecanora*. However, several very large genera remain, as *Arthonia*, *Caloplaca* (which could also be enlarged to include *Xanthoria* and *Teloschistes*), *Buellia*, *Cladonia*, *Graphis* and *Opegrapha*. In *Verrucaria*, as in *Usnea*, there seems to be much synonymizing still to do.

Several large genera are also known in 'protozoans'. Levine (1962) once calculated that theoretically there could be at least 2 654 736 different kinds of oocysts (and hence, the same number of structurally different species) within one genus of sporozoan. This calculation is little more than a joke. However, in the largest genus of sporozoans (*Eimeria*) some 1000 species have been described.

Within diatoms, more than 1000 species have been described in *Navicula*, and several hundreds in *Nitzschia* (Ross, 1982). However, as discussed in Chapter 4, can it be said that there are species in diatoms? A similar concern should be levelled at the 500 species of *Chlamydomonas* (Leedale and Hibberd, 1985) within Chlorophyta, or the 800 species of *Cosmarium* within desmids.

Several enormous genera have been described in bryophytes, as with the liverworts *Plagiochila* (1600 species) and *Frullania* (800), and the mosses *Bryum* (800) and *Fissidens* (800). As for vascular plants, the largest genera are listed in Table 9.2.

Table 9.2   The largest genera of vascular plants, listed according to the species numbers given by Mabberley (1987). The list includes the genera comprising at least 400 species according to Mabberley and/or Willis (1973)

Mabberley (1987)	Willis (1973)	Genus
2000	2000	Astragalus
1600	2000	Euphorbia
1500	2500	Senecio[a]
1400	1700	Solanum
1400	700	Psychotria
1200	900	Bulbophyllum
1200	775	Acacia
1000	2000	Piper
1000	1750	Carex
1000	1000	Eugenia
1000	1000	Peperomia
1000	1000	Vernonia
1000	700	Miconia
900	1400	Dendrobium
900	1000	Pleurothallis
900	900	Begonia
900	700	Salvia
850	550	Impatiens
850	550	Rhododendron
800	800	Ficus
750	750	Croton
700	700	Indigofera
700	700	Selaginella[b]
700	550	Anthurium
700	450	Allium
665	500	Erica
650	650	Asplenium
600	600	Cyathea
600	600	Dioscorea
600	600	Pandanus
600	600	Phyllanthus
600	550	Crotalaria
600	550	Cyperus
600	500	Panicum
600	450	Quercus
600	400	Cousinia
535	550	Cassia
500	800	Oxalis
500	700	Helichrysum
500	600	Habenaria
500	550	Polygala

[a] Several satellite genera (Brachyglottis, Cineraria, Crassocephalum, Delairea, Gynura, Kleinia, Lachanodes, Ligularia, Othonna, Parasenecio, Pericallis, Pladaroxylon, Roldana, Sinacalia, Sinosenecio) have been recently isolated from Senecio, but their limits towards this genus are not clear.
[b] Some 400 species only are recognized by Jermy (1990).

Table 9.2 (contd)

Mabberley (1987)	Willis (1973)	Genus
500	500	Ipomoea
500	500	Potentilla
500	500	Silene
500	500	Syzygium
500	500	Viola
500	400	Epidendrum
485	485	Paepalanthus
475	500	Diospyros
450	600	Centaurea
450	500	Eucalyptus
450	450	Berberis
450	350	Vaccinium
450	300	Jasminum
430	450	Acalypha
420	300	Justicia
412	350	Oncidium
411	500	Tillandsia
400	500	Primula
400	474	Mimosa
400	400	Clerodendron
400	400	Diplazium
400	400	Elaphoglossum
400	400	Eriocaulon
400	400	Galium
400	400	Hyptis
400	400	Ilex
400	400	Ixora
400	400	Litsea
400	300	Tephrosia
370	400	Hypericum
350	500	Passiflora
350	500	Pedicularis
350	400	Pavetta
300	500	Salix
300	450	Desmodium
300	400	Artemisia
300	400	Gentiana
300	400	Geranium
300	400	Saussurea
250	500	Myrcia
250	500	Aster
250	400	Ardisia
250	400	Pilea
250	400	Ranunculus
200	400	Garcinia
150	400	Medinilla

Table 9.2 (contd)

Mabberley (1987)	Willis (1973)	Genus
?	1200	*Eupatorium*[c]
?	1000	*Hieracium*[d]
?	450	*Lycopodium*[e]
1	600	*Loranthus*

[c] The old genus *Eupatorium* has been dismembered in numerous small size genera, mostly by R.M. King and H. Robinson (King and Robinson, 1970, 1987; Robinson and King, 1977, 1987). This trend is strongly contrasted by other botanists, e.g. Stuessy (1990, pp.203–204).
[d] This genus comprises a small number of sexually reproducing species plus some 250–260 groups of 'microspecies'. According to Stace (1989) more 'species' have been described in *Hieracium* (about 10 000) than in any other genus in the world. Two further genera comprised of a few sexually reproducing species together with a very large number of 'microspecies' (more than 2000 each) are *Rubus* and *Taraxacum*.
[e] Most species traditionally ascribed to *Lycopodium* belong now in *Huperzia* (approximately 300 species according to Øllgaard, 1990).

Comparison of the figures given by Mabberley (1987) and Willis (1973) (Figure 9.1) suggests a certain 'stability' for very large genera, such as *Astragalus, Euphorbia, Senecio* and *Solanum*, whereas *Eupatorium, Lycopodium* and *Loranthus* have been heavily pruned when extensive study of extra-European materials has been carried out. As for *Astragalus*, its holophyletism is still unsettled (Sanderson, 1991). For predominantly apomictic genera, such as *Rubus, Hieracium* and *Taraxacum*, the reader is referred to Chapter 4.

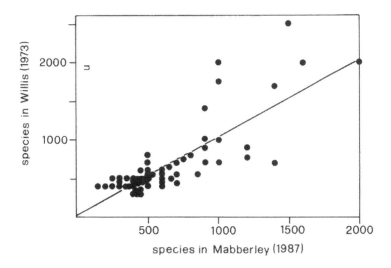

Figure 9.1 Changing attitudes towards the largest genera of vascular plants. The data (N=85) are the same as in Table 9.2, but for *Hieracium, Eupatorium, Lycopodium* and *Loranthus,* which are excluded from this plot and from the calculation of the regression line ($p < 0.05$).

Within animals, most large genera occur in insects, sometimes with more than 1000 species each, but a check on their holophyletic status has seldom been accomplished. Consequently, it is impossible to provide a reasonable list. Just to give a few examples, however, consider the following. Within Lepidoptera, *Coleophora* with 998 species (Vives Moreno, 1988); within Coleoptera, in addition to *Ontophagus*, *Atheta*, *Aphodius*, *Athaenius*, *Phyllophaga* and *Anomala*

Table 9.3 The largest genera of birds, including more than 30 species each according to Howard and Moore (1980) (HM) and/or to Sibley and Monroe (1990) (SM). P = Passeriformes

Number of species in		Genus	Systematic position according to Howard and Moore (1980)	Systematic position according to Sibley and Ahlquist (1990)
HM	SM			
75	79	*Nectarinia*	P Nectariniidae	P Nectariniidae Nectariniinae Nectariniini
63	66	*Turdus*	P Muscicapidae Turdinae	P Muscicapidae Turdinae
63	73	*Zosterops*	P Zosteropidae	P Zosteropidae
60	62	*Ploceus*	P Ploceidae	P Passeridae Ploceinae
51	54	*Columba*	Columbiformes Columbidae	Columbiformes Columbidae
49	51	*Ptilinopus*	Columbiformes Columbidae	Columbiformes Columbidae
48	40	*Pycnonotus*	P Pycnonotidae	P Pycnonotidae
48	51	*Garrulax*	P Muscicapidae Timaliinae	P Sylviidae Garrulacinae
47	48	*Tangara*	P Emberizidae Thraupinae	P Fringillidae Emberizinae Thraupini
47	49	*Accipiter*	Falconiformes Accipitridae	Ciconiiformes Accipitridae
45	51	*Parus*	P Paridae	P Paridae
44	53	*Caprimulgus*	Caprimulgiformes Caprimulgidae	Strigiformes Caprimulgidae
41	50	*Cisticola*	P Muscicapidae Sylviinae	P Cisticolidae
41	46	*Phylloscopus*	P Muscicapidae Sylviinae	P Sylviidae Phylloscopinae
41	48	*Coracina*	P Campephagidae	P Corvidae Corvinae Oriolini
41	41	*Francolinus*	Galliformes Phasianidae	Galliformes Phasianidae

Table 9.3 (*contd*)

Number of species in		Genus	Systematic position according to Howard and Moore (1980)	Systematic position according to Sibley and Ahlquist (1990)
HM	SM			
40	42	*Rhipidura*	P Muscicapidae Rhipidurinae	P Corvidae Dicrurinae Rhipidurini
40	11	*Halcyon*	Coraciiformes Alcedinidae	Coraciiformes Alcedinindae
39	43	*Corvus*	P Corvidae	P Corvidae
38	38	*Emberiza*	P Emberizidae Emberizinae	P Fringillidae Emberizinae Emberizini
38	43	*Larus*	Charadriiformes Laridae	Ciconiiformes Laridae
37	33	*Synallaxis*	P Furnariidae	P Furnariidae
37	39	*Falco*	Falconiformes Falconidae	Ciconiiformes Falconidae
36	13	*Meliphaga*	P Meliphagidae	P Meliphagidae
36	38	*Dicaeum*	P Dicaeidae	P Nectariniidae Nectariniinae Dicaeini
36	42	*Anas*	Anseriformes Anatidae	Anseriformes Anatidae
36	36	*Ducula*	Columbiformes Columbidae	Columbiformes Columbidae
36	53	*Otus*	Strigiformes Strigidae	Strigiformes Strigidae
34	36	*Lonchura*	P Estrildidae	P Passeridae Estrildinae
34	45	*Anthus*	P Motacillidae	P Passeridae Motacillinae
33	11	*Picoides*	Piciformes Picidae	Piciformes Picidae
32	45	*Serinus*	P Fringillidae Fringillinae	P Fringillidae Fringillinae
31	32	*Sporophila*	P Emberizidae Emberizinae	P Fringillidae Emberizinae Emberizini
31	31	*Monarcha*	P Muscicapidae Monarchinae	P Corvidae Dicrurinae Monarchini
24	32	*Sterna*	Charadriiformes Laridae	Ciconiiformes Laridae
27	38	*Phalacrocorax*	Pelecaniformes Phalacrocoracidae	Ciconiiformes Phalacrocoracidae
30	31	*Myrmotherula*	P Formicariidae	P Formicariidae

Table 9.3 (*contd*)

Number of species in		Genus	Systematic position according to Howard and Moore (1980)	Systematic position according to Sibley and Ahlquist (1990)
HM	SM			
29	37	*Zoothera*	P Muscicapidae Turdinae	P Muscicapidae Turdinae
24	39	*Hirundo*	P Hirundinidae	P Hirundinidae
27	32	*Acrocephalus*	P Muscicapidae Silviinae	P Silviidae Acrocephalinae
24	31	*Carduelis*	P Fringillidae Fringillinae	P Fringillidae Fringillinae

In general, there is good agreement between the two checklists, but for a slight increase in the number of species level taxa listed by Sibley and Monroe (1990), in respect to Howard and Moore (1980). The increase mostly depends on elevating some former subspecies to the status of allospecies (cf. Chapter 4), or even of full species. A few genera such as *Cisticola, Anthus, Serinus, Zosterops* and *Zoothera* are more consistently enlarged, always for the same reason. In *Caprimulgus*, the increase due to elevating subspecies to (allo)species sums with the effects of a broader concept of the genus, now including a few species previously placed in *Veles* and *Scotornis*; the same in *Phalacrocorax*, enlarged to include *Haliëtor* and *Nannopterum*. Changes in the taxonomic scope of the genus explain other increases, as for *Hirundo*, now including *Petrochelidon*, or *Carduelis*, incorporating *Acanthis*. More complex are the changes in *Otus*. In this genus, in addition to raising several subspecies to allospecies or even full species, Sibley and Monroe record a few recently described taxa: *O. ireneae* Ripley, 1966; O. *mirus* Ripley and Rabor, 1966; *O. koepckeae* Hekstra, 1982; *O. [marshalli] petersoni* Fitzpatrick and O'Neill, 1986; *O. [marshalli] marshalli* Weske and Terborgh, 1981.

In a few instances, genus size is diminished, always because of splitting the old genera into two or more parts: several species of *Pycnonotus* are now in *Andropadus*; of *Picoides*, in *Dendrocopos*; of *Alcyon*, in *Todirhamphus, Caridonax, Actenoides* and *Syma*; of *Meliphaga*, in *Lichenostomus, Guadalcanaria* and *Xanthotis*.

(Scarabaeidae), *Trachys* and *Agrilus* (Buprestidae), *Monolepta* with 597 species (Wilcox, 1973) and *Longitarsus* (Chrysomelidae s.l.), *Baris* and *Otiorrhynchus* (Curculionidae); within Hymenoptera, *Tenthredo* (Tenthredinidae), *Enicospilus* (Ichneumonidae) with more than 1000 species (Townes, 1969–1971) and *Apanteles* (Braconidae), with several hundreds of species; within Diptera, *Drosophila* (but probably paraphyletic, see Grimaldi, 1990) with 1600 species, and *Limonia*, of similar size.

Marine animal genera with more than 100 species are rare. Some examples are the reef-building coral *Acropora*, with more than 200 species; some molluscan genera, such as *Cypraea* (less than 200, against more than 1000 species names), and *Conus* (with some 500 species, regarded by Kohn (1991) as probably the largest genus of marine invertebrates); within fishes, *Sebastes* and *Chaetodon*, both slightly over 100 species. A much larger number of species belong to some genera of continental vertebrates, such as the freshwater fish genus *Haplochromis*

Table 9.4 The largest genera of mammals (30 or more species each). Data from Honacki, Kinman and Koeppl (1982)

Number of species	Genus	Order and family
149	*Crocidura*	Insectivora Soricidae
88	*Myotis*	Chiroptera Vespertilionidae
65	*Rhinolophus*	Chiroptera Rhinolophidae
64	*Sorex*	Insectivora Soricidae
63	*Rattus*	Rodentia Muridae
60	*Pteropus*	Chiroptera Pteropodidae
57	*Oryzomys*	Rodentia Cricetidae
50	*Peromyscus*	Rodentia Cricetidae
48	*Pipistrellus*	Chiroptera Vespertilionidae
47	*Hipposideros*	Chiroptera Rhinolophidae
47	*Marmosa*	Marsupialia Didelphidae
45	*Microtus*	Rodentia Arvicolidae
40	*Mus*	Rodentia Muridae
36	*Spermophilus*	Rodentia Sciuridae
34	*Gerbillus*	Rodentia Cricetidae
33	*Akodon*	Rodentia Cricetidae
33	*Ctenomys*	Rodentia Ctenomyidae
33	*Eptesicus*	Chiroptera Vespertilionidae

(Cichlidae), with more than 300 lacustrine species. As for amphibians, the largest genus, as already noted, is *Eleutherodactylus*, with 405 species, followed by *Rana* (272), *Hyla* (258) and *Bufo* (206). Taken together, these four genera contain some 28% of the amphibian species described so far (calculations after Frost's (1985) checklist). Within reptiles, the largest genus is *Anolis* (Iguanidae), with some 200 species. The largest genera of birds and mammals are listed in Tables 9.3 and 9.4.

9.2 SIZE DISTRIBUTIONS OF HIGHER TAXA

When looking at the size distributions of supraspecific taxa, measured as the number of subtaxa included (e.g. species) an overall pattern seems to emerge. Several authors, beginning with Willis and Yule (1922), have tried to give an analytical description of this pattern and to explain it, either with reference to evolutionary processes, or as a simple expression of our taxonomic practices. Willis and Yule (1922) found the size distributions of genera within plant families or local floras to follow a common pattern of 'hollow curves', well approximated by straight lines in log–log plots. Within the conceptual framework of his 'age and area' hypothesis (Willis, 1922), Willis regarded these distributions as an effect of age differences, younger taxa being less speciose and less widely distributed on earth than older ones.

A different analytical approach was followed by other authors, such as Williams (1947, 1964), Dehalu and Leclercq (1951) and Clayton (1972, 1974). They used the so-called Fisher's log-series to fit the size distributions of taxa. More important, they considered them within different theoretical frameworks. Williams followed an ecological approach and suggested a similarity between the size distribution of genera and the distribution of species abundances within ecosystems. Dehalu and Leclercq regarded genus size as dependent on an initial burst of speciation, followed by long-term lower-rate speciation. Clayton also correlated the size distributions of families and genera with evolutionary phenomena, especially speciation, but also pointed to the unavoidable bias deriving from the subjective circumscription of taxa.

Different analytical approaches were followed by Dial and Marzluff (1989) and Cronk (1989). Dial and Marzluff's method emphasized the few largest 'dominant' taxa, rather than the much more numerous smaller ones, as in previous authors' analyses. Dial and Marzluff explored five 'null' models of diversity-generating processes, checked several size distributions of actual taxa against them and concluded that 'overdominance of an assemblage by one unit is a common and nonrandom feature of distributions of taxonomic diversity'. Finally, Cronk (1989), after checking some size distributions against an expected log-normal distribution, came to the conclusion that small taxa are generally relicts, thus overturning Willis' original explanations.

Contrary to the views of most other authors, Walters (1961, 1986) regarded the size distributions of plant genera and families as taxonomic artefacts.

Size distributions have been examined by Burlando (1990), Green (1991) and Minelli, Fusco and Sartori (1990, 1991). According to them, the size distributions of taxa mirror a fractal structure of diversity, due perhaps (Burlando, 1990) to scale independence of some evolutionary processes. The study by Minelli, Fusco and Sartori rested on recursive analysis of size patterns at different taxonomic levels. Their attention focused on the distributions of species within genera (= 'size' of genera), of genera within families and of families within orders, as well as on distributions of species within families. They tried to fit them with a power law

$$N(x) = a\,x^{-b}$$

where x is the taxon size and $N(x)$ the number of taxa of size x. In most instances the size distributions of taxa followed the power law in statistically satisfactory terms, with the exception of a small tail comprised of the largest taxa. Distributions were not correlated with biological traits such as size of organisms, habitat, diet, mobility or with the very unequal taxonomic standards or traditions relating to the different groups (e.g. birds versus tapeworms). More important, the same pattern resulted when moving upwards along the conventional taxonomic hierarchy, thus comparing the size distributions of families, as measured by the number of included genera with that of orders, as the number of included families. Finally, this pattern proved to be independent of the splitting or lumping of taxa. The consistent behaviour of size distribution of genera within families and families within orders under the power law approximation confirms that the size distributions of taxa exhibit self-similarity, or scale invariance. It is this behaviour that suggests a fractal structure (Mandelbrot, 1982) of biological classifications.

What is the meaning of this pattern of taxonomic diversity? I cannot rule out the burden of taxonomic bias, but I do not think, as Walters (1961, 1986) does, that it explains the whole pattern. Some effect is possibly contributed by the cognitive perceptual biases common to all human beings, as primarily different from the 'professional' biases of taxonomists. Folk taxonomies have developed classifications with up to five hierarchical levels, very close, indeed, to the number of traditional Linnaean ranks. Moreover, the continuity of folk taxonomies into early 'scientific' classifications has been convincingly demonstrated (Cain, 1958; Walters, 1961; Raven, Berlin and Breedlove, 1971; Berlin, Breedlove and Raven, 1973; Atran, 1987). True taxonomic biases, however, are possibly not too large, in so far as the self-similar pattern in the size distribution of taxa emerges from within different groups, with different taxonomic traditions, and is not sensitive to variation in scope of the 'genera' we recognize.

Should we be concerned about the conventional nature of ranking species into genera, families etc., or about the probably nonmonophyletic nature of many, perhaps of most, taxa the size of which has been investigated. The pattern could be the outcome of a partly phenetic approach to classification. Such a concern may be justified, but in a much subtler way than one could reasonably suppose.

It is possible (Minelli, Fusco and Sartori, 1991) that the pattern observed depends in part on the adaptive events that have led to unequal but not necessarily random occupancy of ecospace (Levin, 1989; Sugihara and May, 1990), in the history of the individual groups. These events, however, are not necessarily uncoupled from cladogenesis. It is tempting to speak of speciational trends (Futuyma, 1989), but a discussion of macroevolutionary issues is outside the scope of this book. One may wonder, instead, what would happen, if we could search for size patterns within major sections of the system, rather than within classifications. We simply do not have system's excerpts large enough to begin this work. However, it is already possible to say that we shall work at the highest possible level of resolution. Otherwise, by telescoping a fully resolved system into a shorter form, we do not simply miss part of the structural information of the tree (Tassy, 1988) but, much worse, we subjectively, if unconsciously, suppress the phylogenetic information of certain nodes (O'Hara, 1992), with the most disastrous consequences for the estimation of the relative size of terminal taxa.

Domesticated animals and cultivated plants 10

...there are hardly any domestic races, either amongst animals or plants, which have not been ranked by some competent judges as mere varieties, and by other competent judges as descendant of originally distinct species. If any marked distinction existed between domestic races and species, this source of doubt could not so perpetually recur.

C. Darwin (1859)

The taxonomic and nomenclatural treatment of domesticated animals and cultivated plants is very often controversial and complicated. Many problems I have dealt with in Chapter 9 reappear on a grand scale, when dealing with cats, cattle and cabbages. Theoretical and practical issues seldom contrast so vividly as when circumscribing species within cultivated *Brassica*, and their wild and weedy relatives.

In these matters, tradition and practical requirements deserve more attention than usual in systematics. As for ranks, we simply cannot banish all subspecies and variety names of cultivated plants on some alleged scientific grounds.

Special rules of nomenclature have been devised for dealing with domesticated animals and particularly with cultivated plants. These rules are often regarded as independent from the scientific, Linnaean nomenclature, especially at levels below the species. In the case of plants, the existence and role of an International Code for the Nomenclature of Cultivated Plants is fully acknowledged by the ICBN. In this way, the official nomenclature of cultivated plants meets with the requirements of users, more than with those of taxonomists, and often merges with problems of trade marks and registration of varieties and cultivars. Similar problems begin to emerge for organisms produced by genetic engineering, as foreseen by Harlan and de Wet (1986, p.75).

10.1 TAXONOMY AND NOMENCLATURE OF DOMESTICATED ANIMALS

Guidelines for the taxonomic and nomenclatural treatment of domesticated animals have been proposed by Corbet and Clutton-Brock (1984, p.434). They illustrate the problems arising from the application of the biological species concept to domesticated animals and from an improved knowledge of the pedigree of domesticated forms. 'With a few notable exceptions it can be argued that the domesticated and wild forms constitute at least a potentially interbreeding assemblage.' They argue that these problems may be answered best by abandoning both strict scientific requirements, as in the use of the biological species concept, and strict adherence to the ICZN, at least for animals such as dogs, cats, cattle and horses. They identify three groups of domesticated animals, worthy of different taxonomic and nomenclatural treatment at the species level.

10.1.1 GROUP 1

'Anciently domesticated forms that are distinctive, and rarely bred with their wild ancestors and have well-established scientific names that are normally understood as applying only to the domesticated forms.' Corbet and Clutton-Brock (1984) argue that these domesticated forms are better regarded as derivatives of wild relatives, but not as parts of them. Strictly speaking, this means identifying an ancestor–descendant relationship between two coexisting species, therefore acknowledging paraphyly in the wild species, if treated as specifically distinct from its domesticated descendants. However, this is possibly not the worst aspect of the problem. An additional aspect derives from the fact that the domesticated form itself is not necessarily monophyletic. The domesticated 'species' belonging to Group 1 are not 'orthodox species' and Corbet and Clutton-Brock (1984) suggest that they are better identified by vernacular names than by Latin epithets. However, whenever a Latin name is also required for them, the well-known Linnaean names, such as *Bos taurus* Linnaeus, 1758, may still be used, while retaining different names for the wild ancestors (as *Bos primigenius* Bojanus, 1827, for the aurochs). To remind readers that 'cattle' is not an 'orthodox species', Corbet and Clutton-Brock (1984) suggest the placement of the 'specific' epithet within quotation marks, e.g. *Bos 'taurus'* Linnaeus, 1758. Their list of species is reproduced in Table 10.1.

Corbet and Clutton-Brock's solution for the nomenclature of these domesticated animals is not universally accepted. For instance, Herre and Röhrs (1990), following Bohlken (1961), do not accept any specific distinction between wild and domestic forms, but refuse to take into account, for the purposes of the principle of priority, the names only used for the domesticated forms. In consequence, cattle becomes *Bos primigenius*, like aurochs, while horse must wait until 1881, to be named *Equus przewalskii*, after the discovery of its wild relatives. Is all that not just another reason for fostering a divorce between 'scientific' names and 'names for users'?

10.1.2 GROUP 2

'Distinctive domesticated forms that are readily distinguishable from their wild ancestral species but have not generally been given separate specific names.' Included in this group are many species popularly known and named as wild animals, before they gave rise to domesticated forms, together with species which are traditionally given the same name in their wild and domesticated forms. For these species, Corbet and Clutton-Brock (1984) suggest identification of the domesticated forms by simply adding a vernacular term to the Latin name of the wild form, as in the following example: *Anas platyrhynchos* (domesticated). A list of these species is reproduced in Table 10.1.

Table 10.1 Nomenclature of domesticated animals according to Corbet and Clutton-Brock (1984)

Group 1. 'Anciently domesticated forms that are distinctive, and rarely bred with their wild ancestors and have well-established scientific names that are normally understood as applying only to the domesticated forms' (Corbet and Clutton-Brock, 1984, pp.435–7)

Domesticated form	Probable wild relative (= ancestor)
Bos 'taurus' Linnaeus, 1758 Common cattle (including zebu)	*Bos primigenius* Bojanus, 1827 Aurochs (including *B.indicus* Linnaeus, 1758)
Bos 'frontalis' Lambert, 1804 Gayal, mithan	*Bos gaurus* H. Smith, 1827 Gaur
Bos 'grunniens' Linnaeus, 1766 Domestic yak	*Bos mutus* (Przewalski, 1883) Wild yak
Bubalus 'bubalis' (Linnaeus, 1758) Domestic water buffalo	*Bubalus arnee* (Kerr, 1792) Wild water buffalo, arni
Ovis 'aries' Linnaeus, 1758 Domestic sheep	*Ovis orientalis* Gmelin, 1774 Asiatic mouflon
Capra 'hircus' Linnaeus, 1758 Domestic goat	*Capra aegagrus* Erxleben, 1777 Bezoar goat
Camelus 'dromedarius' Linnaeus, 1758 Arabian camel, dromedary	unknown
Camelus 'bactrianus' Linnaeus, 1758 Domestic Bactrian camel	*Camelus ferus* Przewalski, 1883 Wild Bactrian camel
Lama 'glama' Llama	*?Lama guanicoe* Muller, 1776 Guanaco
Lama 'pacos' (Linnaeus, 1758) Alpaca	unknown
Sus 'domesticus' Erxleben, 1777 Domestic pig	*Sus scropha* Linnaeus, 1758 Wild boar
Equus 'caballus' Linnaeus, 1758 Domestic horse	*Equus ferus* Boddaert, 1785 Wild horse, tarpan
Equus 'asinus' Linnaeus, 1758 Donkey	*Equus africanus* Fitzinger, 1857 *Wild ass*

Table 10.1 (*contd*)

Canis 'familiaris' Linnaeus, 1758 Dog	*Canis lupus* Linnaeus, 1758 Wolf (including *C. dingo* Meyer, 1793)
Felis 'catus' Linnaeus, 1758 Domestic cat	*Felis silvestris* Schreber, 1777 Wild cat
Mustela 'furo' Linnaeus, 1758 Ferret	*Mustela putorius* Linnaeus, 1758 Western polecat
Cavia 'porcellus' (Linnaeus, 1758) Domestic guinea-pig	*Cavia aperea* Erxleben, 1777 Cavy
Bombyx 'mori' (Linnaeus, 1758) Domestic silk moth	*Bombyx mandarina* (Moore, 1872) Chinese wild silk moth

Group 2. 'Distinctive domesticated forms that are readily distinguishable from their wild ancestral species but have not generally been given separate specific names' (Corbet and Clutton-Brock, 1984, pp.437–8)

Wild species	Domesticated form
Bos javanicus d'Alton, 1823	Banteng, Bali cattle
Rangifer tarandus (Linnaeus, 1758)	Reindeer, caribou
Vulpes vulpes (Linnaeus, 1758)	Red fox, Silver fox, etc.
Mustela vison Schreber, 1777	American mink, Ranch mink
Oryctolagus cuniculus (Linnaeus, 1758)	Rabbit
Mesocricetus auratus (Waterhouse, 1839)	Golden hamster, Syrian hamster
Mus domesticus Rutty, 1771	House mouse, Laboratory mouse
Rattus norvegicus (Berkenhout, 1769)	Norway rat, Laboratory rat
Gallus gallus (Linnaeus, 1758)	Red jungle fowl, Domestic fowl
Meleagris gallopavo Linnaeus, 1758	Turkey
Anas platyrhynchos Linnaeus, 1758	Mallard, Domestic duck
Cairina moschata (Linnaeus, 1758)	Muscovy duck
Anser anser (Linnaeus, 1758)	Greylag goose, Domestic goose
Anser cygnoides (Linnaeus, 1758)	Swan goose, Chinese goose
Columba livia Gmelin, 1789	Rock dove, Domestic pigeon
Serinus canarius (Linnaeus, 1758)	Canary
Melopsittacus undulatus (Shaw, 1805)	Budgerigar
Carassius auratus (Linnaeus, 1758)	Goldfish
Apis mellifera Linnaeus, 1758	Honeybee

10.1.3 GROUP 3

'Wild species that are commonly bred or kept in captivity but in which the majority of domesticated individuals are not readily distinguishable as a group from the wild species and are not normally given separate names.' For this group, no special comment seems to be necessary. A few examples include the following:

Elephas maximus (Linnaeus, 1758) (Asian elephant);
Myocastor coypus (Molina, 1782) (coypu, nutria);
Chinchilla laniger (Molina, 1782) (chinchilla);
Numida meleagris (Linnaeus, 1758) (guinea-fowl);
Pavo cristatus Linnaeus, 1758 (peafowl);
Struthio camelus (Linnaeus, 1758) (ostrich).

The case of the New World camelids (llama, alpaca, guanaco and vicuna) provides an opportunity to appreciate the systematic problems often encountered with domesticated animals. All hybrid combinations occur between the four forms (Novoa and Wheeler, 1984). Despite that, some authors regard them as four separate species of the genus *Lama*; others even separate the vicuna into the genus *Vicugna*. At the opposite extreme, 'the four forms have been also described both as subspecies of *L. glama*, and varieties of *L. g. glama*, incorrectly utilizing *forma domestica* for the llama and alpaca' (Novoa and Wheeler, 1984). I cannot guess on what terms the taxonomy and nomenclature of this group will finally be stabilized.

10.2 TAXONOMY AND NOMENCLATURE OF CULTIVATED PLANTS

Generally speaking, systematic problems concerning cultivated plants are much greater than those concerning domesticated animals. We have already seen that it is generally more difficult to circumscribe plant species than animal species. These problems are exaggerated in domesticated plants.

De Wet, Harlan and Brink (1986) observe that post-fertilization mechanisms of genetic isolation are rare, within the complexes comprising wild, weedy and cultivated forms, despite frequent conspicuous differences in morphology. However, as they comment further (1986, p.212), 'hybridization between domesticates and their wild progenitors is common, but extensive gene exchange is prevented by disruptive selection for habitat preference. The wild complex cannot successfully invade the man-made habitat, and the cultivated complex cannot compete successfully for natural habitats. Weedy and cultivated complexes both occupy man-made habitats, with weeds spontaneous in this habitat and domesticates depending on man for seed dispersal'. In a sense, the wild, weedy and domesticated forms of a given complex owe their distinctiveness to the continuing agricultural activities of man.

The problems of infraspecific classification in wild and cultivated plants were debated widely during a recent symposium (Styles, 1986). A few major contributions are summarized here. What is astonishing is the uncertainty of taxonomic treatments and the profound disagreement on details regarding more than 4000 species (Hanelt, 1986, p.142), many of which are of major economic interest. For example, the grain sorghum (*Sorghum bicolor* (L.) Moench) was divided by Snowden (1936, 1954) into 13 wild species, seven species of weedy forms and 28 cultivated species, the latter were further subdivided into 156 taxonomic varieties and 521 forms. At the other extreme, Harlan and de Wet (1971) (see also de Wet, 1978; de Wet, Harlan and Brink, 1986) regard all these taxa as one species with three subspecies. De Wet, Harlan and Brink (1986) acknowledge that these subspecies, and particularly the wild and the cultivated one (the third is the weedy subsp. *drummondii*), exhibit an extensive variability, but they prefer not to identify racial or varietal units formally within them.

Problems appear both at the species level and below. It is customary to use different Linnaean species names for the cultivated plants and their wild relatives (e.g. *Beta vulgaris* L. for cultivated beet and *Beta maritima* L. for wild beet). However, the two 'species' are fully interfertile, and often cross spontaneously.

For cultivated plants, many authors have used and still use conventional categories of the Linnaean hierarchy, but this trend has been abandoned by some, such as Mansfeld (1953), Jirásek (1961) and Jeffrey (1968). They have introduced an array of categories to be used only for cultivated plants. These categories are distinct from, although parallel to, those commonly used for wild plants. Jeffrey (1968) for example, suggested the use of subspecioid, convar (= convarietas, cultiplex etc.), provar (= cultigen, conculta, group, nidus, provarietas, cyclus, sortotypus), and cultivar. According to Jeffrey, there are some objective reasons suggesting this separate treatment of cultivated plants: they mainly lack a true population structure; they do not occupy natural areas; they have a range of variability larger than wild plants; they suffer from a breakdown of isolation barriers, thus favouring hybridization and introgression; and so on. Other authors, such as Hanelt (1986), argue against this view and reject the use of independent categories for cultivated plants.

Another system for the classification of cultivated plants was proposed by Harlan and de Wet (1971), in terms of gene pools: 'The primary gene pool (GP–1) simply represented the well-known concept of the biological species (i.e. those individuals that cross readily to form fertile offspring). The secondary gene pool (GP–2) included individuals that could be crossed with the GP–1, but with distinct genetic barriers. The tertiary gene pool (GP–3) represented individuals that could yield hybrids with GP–1, but that were completely sterile and/or gene exchange was impossible without some radical technique' (Harlan and de Wet, 1986, p.71). The same authors give further examples of the very complex problems of species (or gene pool) delimitation in plants, and especially in grasses. For example, introgression between *Bothriochloa ischaemum* and *B. bladhii*, occurring in recently deforested areas along the front range of the Himalayas and Kara Korum, is a common occurrence. But the type of *B. bladhii* involved in this introgression was further demonstrated to be 'itself, an introgression product combining a more tropical form of *B. bladhii* and the related *Dichanthium annulatum*. ... In Australia, we found that *B. bladhii* introgressed there with *Capillipedium parviflorum* ..., producing an intermediate type that has been given the epithet *B. glabra* ... The latter species has, in turn, introgressed with *C. parviflorum*, producing an intermediate named *C. spicigerum* ... *B. bladhii* also introgresses with *B. ewartiana* ... in northern Australia...'

Harlan and de Wet (1986) further comment upon the so-called compilospecies, absorbing germ plasm from related species. This behaviour has been demonstrated, for example, in *Poa pratensis* (introgression from other sympatric *Poa* species), for *Elymus glaucus* (introgression from both *Sitanion hystrix* and *S. jubatum*), *Dactylis glomerata* (introgression from several diploid grasses) and *Saccharum officinarum* (introgression from several genera of grasses).

Problems with introgression between crop plants and their wild or weedy relatives are currently investigated with the powerful support of molecular techniques. For a good review, see Doebley (1992).

In an effort to produce a causal analysis of differentiation occurring in cultivated plants, Pickersgill (1986, p.191) distinguishes three principal levels of change, 'changes associated with domestication itself; subsequent partition of variation in the cultigen through geographical isolation, ecological adaptation, or human selection for different usages; and differentiation into the cultivars or landraces which constitute the actual field populations'. According to Pickersgill (1986), the infraspecific categories recognized by the Codes of Nomenclature are inadequate to cope with this type of hierarchically arranged infraspecific differences. Instead, she suggests formally recognizing four levels of variation at least:

1   Subspecioid (subspd.), as the taxonomic category expressing the contrast between the wild ancestor and the cultivated form (e.g. *Arachys hypogaea* subsp. *monticola* and subsp. *hypogaea*);
2   Convar (conv.), to recognize 'major geographic agronomic divisions within the cultigen' (e.g. for groundnuts, conv. 'Hypogaea' and conv. 'Fastigiata');
3   Group (provar) and Sub-group (sub-provar) for groups of cultivars or groups of landraces (e.g. for groundnuts, 'Virginia', 'Peruvian', as groups; and 'Virginia Bunch', 'Virginia Runner' as sub-groups);
4   Cultivar or landrace.

That the Codes are inadequate to cope with the infraspecific diversity within cultivated plants is also acknowledged by Stace (1986). Moreover, the problem seems hardly amenable to an easy universal solution. Hawkes (1986, p.5) contends that the cultivar, often recognized as the lowest category for infraspecific classification of cultivated plants, 'may be more or less acceptable in developed countries, but it must be remembered that many developing countries and even some developed countries possess land races and primitive forms in which the cultivar as we envisage it does not exist'.

Most taxonomic approaches to infraspecific variability within cultivated plants more or less explicitly assume that diversity is hierarchical, but this point is convincingly refuted by Stace (1986): horticulturists are interested in naming as 'variety' a number of phenotypes that are liable to appear within different 'subspecies' of species. Stace therefore concludes that a hierarchical classification cannot adequately express these patterns of diversity. A similar point has been also made by Heywood (1973) and by Elkington (1986). Prentice (1986) even contends that many interesting patterns of intraspecific variation are continuous; nevertheless, partitioning this variation into arbitrary segments may often be a practical necessity.

Special concerns with the nomenclature of cultivated plants were recently expressed at an international symposium on the stability of botanical and zoological nomenclature (Brandenburg, 1991; Chauvet, 1991; Gunn, Wiersema and Kirkbride, 1991). The general feeling was that 'users' want increasingly more freedom from the Linnaean standards of the Botanical Code. However, if 'users' can feel authorized to free themselves from the bonds of Linnaean tradition, why (Humphries, 1991; Minelli, 1991a) should 'scientific', phylogenetic systematics feel obliged to look for compromises, only to stay indefinitely within the same tradition?

PART THREE

Epilogue

The relative stability of Angiosperm families must be understood in terms of the unwillingness of taxonomists to change them.

S.M. Walters (1961, p.82)

Some dangerous trends, and a hope for the future

11

The present classification of birds amounts to little more than superstition and bears about so much relationship to a true phylogeny of the class as Greek mythology does to the theory of relativity.

S.L. Olson (1981, p.193)

The preceding chapters should convince the reader that systematists were not uniformly successful until they consciously began developing phylogenetic methods. Unfortunately, there is often a hiatus between the theoreticians who discuss the nature of species and homology, and the specialists working on a given group of animals, plants, or fungi. The gap between the frontiers of pure research and the application of principles is still very large, and there are several reasons for it. Some reasons are of a contingent nature, and can be easily avoided. Others are much more deeply entrenched in our ways of working, and I cannot see an easy way of changing the current state of affairs. In my view, several theoretical papers are developed by people with experience narrowly restricted to a single group and their attempts to generalize are always dangerous. However, still harder to overcome are other, more subtle difficulties, derived from an unconscious and uncritical reliance on traditional methods, which is widespread among systematists. This is possibly the cause of the largest number of problems within systematics.

Many systematists have problems applying new techniques to their study objects. It is not necessary to use mtDNA characters, or allozymes, or data from scanning electron microscope studies, or counting chromosomes, to make many systematists uneasy with recent literature. For many entomologists, for example, to suggest alcohol preservation or mounting of whole specimens on slides, instead of sticking them on pinned pieces of cardboard, is enough to evoke strong adverse reactions. A small staphylinid beetle or a phorid fly looks quite different when observed as dry specimens through a binocular microscope, or as whole-mounts through a conventional microscope. When changing from the pinned specimens to the slide preparations, a few traditional characters, such as those relating to colour or pubescence, may become difficult to observe (or, at least, no longer directly comparable to the conventional descriptions), but the technical improvements are often rewarding, in so far as they allow the observation of many additional characters. Many old descriptions are useless, until typical material is re-described under new standards.

Many workers are accustomed to identify specimens in the field. This is possible for several types of organisms visible to the naked eye, after suitable training, such as birds, butterflies, or orchids, just to mention three groups. This is impossible for many other groups, not only for small organisms such as copepods, mites or moulds, but also for several larger organisms, where identification rests on fine traits requiring preparation and microscope work.

Methods are changing in some groups, where systematists were accustomed to field identification through easily observed characters. For example, the identification of sibling species may require routine examination of chromosomes, such as in some genera of plants and in insects like mosquitoes. This progress is often opposed by people contending that 'good' species are only those identifiable without the 'artifices' of modern technique. According to Hawksworth, Sutton and Ainsworth (1983), Fries' fungal classification, with genera recognized on macroscopic characters of the fruiting body and spore colour, was reluctantly replaced by a modern system, because the older one had the advantage of allowing easy identification of genera and species in the field.

Walters (1961, 1986) has written two delightful papers on the huge heritage of pre-scientific knowledge still preserved in our current classifications of the flowering plants. I have been deeply struck by these papers and I fully acknowledge their influence although I disagree with Walters in some details. As noted by Walters, we generally underestimate the importance of folk taxonomy in shaping our classifications. One of the reasons for this unbalanced view is the scanty attention paid to the main literary sources of Linnaeus. These are the authors of the first modern classifications, such as Tournefort, or Ray. The interest in pre-Linnaean botany and zoology has been indirectly restricted by the requirement of the International Codes of Nomenclature, identifying two Linnaean works as the starting points for the nomenclature of most living organisms, i.e. *Species Plantarum* (1753) for plants and *Systema naturae* (edn. X, 1758) for animals. Accordingly, many systematists have at least a cursory acquaintance with the work of Linnaeus, whereas a familiarity with the pre-Linnaeans is essentially limited to a few historians of natural history. As a consequence, many students fail to perceive the debt they have inherited (through Linnaeus) towards the major pre-Linnaeans and, through them, towards the European folk taxonomy that they have steadily rationalized, embellished and refined in their work. The contrast between woody and herbaceous plants goes back to the ancient Greeks: woody plants were conceived as essentially different from herbaceous plants. In the Renaissance, this contrast persisted in many printed herbals and was even used in the otherwise valuable classification of plants of John Ray. Interestingly, a contrast between 'fundamentally woody' and 'fundamentally herbaceous' plants (i.e. Lignosae and Herbosae) still persisted, in this century, in Hutchinson's system, although the distinction had already been abandoned as unnatural by most botanists. As appropriately argued by Mabberley

(1987), *Averrhoa* and *Sarcotheca*, two woody genera of an otherwise herbaceous family such as Oxalidaceae, have been segregated into an independent family Averrhoaceae, together with the lianoid genus *Dapania*, only because of habit differences.

With the discovery of a new species, authors seldom have the opportunity for revising the whole group to consider where the new taxon should be correctly placed. Instead, 'the taxa initially recognized seem immune from criticism and provide a pattern to which other taxa must conform' (Stevens, 1986, p.329). In many groups, this way of working leads to poor classifications. Many Linnaean genera, established more than two centuries ago for accommodating a small number of more or less closely related species, still continue to be used as 'waste-baskets'. Newly discovered species that do not obviously fit within segregate genera are simply placed in the old genus. As a consequence, these genera established later are often more or less natural, whereas the old Linnaean genus remains ill-defined and probably para- or polyphyletic.

Heywood (1977, 1988) has demonstrated that many genera and most families of flowering plants have grown by accretion around an ancient nucleus of species or genera, without a detailed and deliberate re-synthesis of data concerning old and new taxa. Of course, revisionary work is increasingly difficult, as the size of a taxon increases.

Walters (1961) has demonstrated that most of the largest families of flowering plants were already circumscribed in an essentially modern sense in de Jussieu's (1789) *Genera plantarum secundum ordines naturales disposita*, but this stability is not to be regarded as a proof of their 'correctness'. This is an important point. The revolutions in angiosperm taxonomy in the last three decades, although mostly affecting suprafamilial taxa, decidedly point to a new concept of many families too.

The species retained as *typus generis* for nomenclatural reasons is sometimes 'unbalanced' in view of the heterogeneity of the species originally placed with it in the same genus. For instance, it is often a marginal temperate offshoot of a predominantly tropical genus. Under these circumstances, things often remain fuzzy until the tropical members of the group are better known. At last, when things begin to be more distinct, the type species of the old big genus may turn out to be isolated from most members of the group. Therefore, the old, often large, genus may be drastically reduced to a few species, or even to one, as happened to *Loranthus*, which now only includes the type species *L. europaeus* Jacq. (a relative of the mistletoe), whereas at one time it included up to 600 species (Willis, 1973), now distributed between several genera. Regarding insects, Gauld and Mound (1982) remarked that the classification of groups such as Tenthredinoidea (Hymenoptera) and Aphidoidea (Homoptera), primarily worked out on temperate faunas, can be easily extrapolated to the world fauna, because the diversity centres of these groups are essentially temperate. Things are different, however, with groups such as Psylloidea (Homoptera), Phlaeothripidae (Thysanoptera) and many groups of Ichneumonidae (Hymenoptera), the centres of diversity of which are in the tropics, thus posing major problems when the old taxonomy, elaborated again on the temperate groups, is being extended to a worldwide perspective.

I have repeatedly cited Mayr's (1982a) vision of microtaxonomy and macro-taxonomy as two largely separate fields of enquiry, with different problems and different procedures. While understanding the historical reasons for this contrast, I feel uneasy with the attitudes of some systematists, who attempt to undertake phylogenetic work, say, at the genus, tribe, or family level, without any acquaintance of the problems of the groups at species level. How accurate is their character coding and evaluation, if they fail to appreciate intraspecific variability? How can they 'reconstruct the affinities' between genera, without primarily checking the monophyly of genera or species groups?

When referring to a systematic framework in studies of ecology, biogeography and evolution, we are seldom aware of the intrinsic biases in the data. Some of these problems have been discussed (section 4.1.2), but a few additional comments should be developed here. Let me cite a passage from Stevens (1986, p.330): 'Stebbins (1981) analysed characters used in recognizing families and found that gynoecial characters were more frequently used, especially in orders having an early position in the overall phylogenetic scheme. He then suggested that such characters were therefore being selected for in the early course of angiosperm evolution. Such a conclusion depends on classifications having a cladistic basis, and also one in which the main structure is not so heavily dependent on gynoecial characters. The careful analysis by Gilmartin (1981) shows some of the dangers inherent in comparing taxa at the same hierarchical level from an evolutionary perspective'. That is what we currently do in ecology, in biogeography and in other fields of biological sciences. As Stevens (1986) concludes, 'Biogeography has until recently emphasized counting number of families, genera and species in different areas. Such an approach, which has predominated for close to two centuries, would be problematic even if a cladistic framework were available'. This problem has been well perceived by Mayr (1980, p.101), who has objected that it would be misleading, when comparing different faunas, to give the same weight to semispecies (allospecies) as to 'genuine' species (i.e. those which are not members of superspecies) (cf. section 4.1.2): 'Many superspecies of New Guinea birds, for instance, have a different allospecies on almost every mountain range (for example, *Astrapia, Parotia*). Superspecies are particularly widespread in archipelagoes, like the West Indies or the Solomon Islands, and when the fauna of such island regions are compared with that of the nearest mainland, the unit of comparison must be the superspecies, or more correctly the new category zoogeographic species, composed of superspecies and all those isolated species that are not members of superspecies'. Mayr estimates that the more than 9000 'conventional' species of birds reduce to some 7000 ± 200 zoogeographical species.

'Ranking of taxa can often be traced back to decisions made over a century ago that were explicitly decisions of convenience' (Stevens, 1986, p.328). In my opinion, higher ranks are very often the levels of acknowledged difficulty in identifying homologies, or relationships. Two species are placed in separate phyla (the heirs of Cuvier's embranchements, or groups characterized by rationally irreducible body-plans) whenever their affinities cannot be reasonably established by ordinary means. In this sense, we can understand why the increasing knowledge of prokaryotes and protists gained through electron microscopy and biochemical and molecular investigations has led to proposals of many new phyla. In the same sense, we can understand why animals of problematic affinities such as Pogonophora, Gnathostomulida or Loricifera are generally regarded as members of separate phyla. Coordinating them with other groups or subordinating them to previously established taxa would mean firmly establishing their affinities. The situation remains more flexible as long as they are treated as independent phyla.

A similar perception of the current attitude towards taxa of uncertain affinities was expressed by van Steenis (1969, pp.116–17), when commenting on 'isolated species' that are numerous in many tropical floras. These species often belong to mono- or bi-typic genera and: 'in many cases one can hardly speak of 'next of kin' as there is hardly any, or even none'. In many cases (van Steenis cites *Scyphostegia, Corsia, Alangium, Crypteronia, Daphniphyllum, Sarcosperma, Nyssa* and *Suriana* as examples) these genera have been accommodated within independent families for a long time. The same trend has been followed for other genera, such as *Ctenolophon, Eupomatia, Erythropalum, Cardiopteris, Tetramerista, Saurauia, Schisandra* and *Averrhoa*.

This trend is fairly easy to understand in the context of traditional systematics producing a classification rather than a system. This might be said for phyla such as Gnathostomulida and Pogonophora and for the plant families, such as Scyphostegiaceae and Corsiaceae. I am not suggesting that they are, respectively, natural phyla or natural families, as I regard these ranks of the Linnaean hierarchy plainly as conventions. Provided that we are aware that statements such as: 'Canidae, Carabidae, Rosaceae and Agaricaceae are equivalent groups within the System of Living Beings' are biologically meaningless, placing *Nanaloricus mysticus* into the new phylum Loricifera, or the genus *Eupomatia* into the family Eupomatiaceae may stimulate progress. Isolated taxa invite further study to trace their relationships. Dismembering a traditional heterogeneous taxon into several smaller taxa may facilitate recognition of monophyletic units. For instance, splitting the Orchidaceae into Apostasiaceae, Cypripediaceae and Orchidaceae s.str. has facilitated a reappraisal of the very close relationship existing between Orchidaceae and Liliaceae (Dahlgren, Clifford and Yeo, 1985). Placing flamingoes into a separate order Phoenicopteriformes has facilitated their movement within the system of birds until they have reached their current placement, close to herons (Appendix 17). There are, however, many problems with names. Should every higher group, although informally proposed as a working hypothesis, be formally named?

Names are reference points for concepts and natural entries for indexes. Consequently, a name is more or less necessary in order to capture attention for a 'group'. However, we should also address another question: should formal names be introduced in accordance with the codes of nomenclature; or, should we give preference to vernacular names, if the groups are nothing more than working hypotheses? It would be easy to answer by recommending formal names only for resolved cases. However, how and when can we be sure that our group is something more sound than a working hypothesis? It could be argued that all groupings are working hypotheses.

On a practical note, I believe that the worst formal malpractice is to employ old names for new concepts. New names are always less confusing, despite their ever increasing number. If the concept it identifies proves to be a poor working hypothesis, the name may naturally fall into disuse without necessitating further discussion, except to be recalled to use if the old concept is resuscitated, as in the case of Owen's Haemothermia (cf. section 7.24). Moreover, it may prove erroneous to consider that informal names for suprageneric groupings are better than formal ones until the concepts are finally accepted by the scientific community. It is always easy to find an advocate of formal nomenclature, ready to rename groups, which would then burden the system with two names, both informal and formal.

There are many more problems and prejudices, in the science of biological systematics. For example, a widespread capricious dislike for large genera, or families, that must be dismembered just because of their size, or the unwillingness to think that many living organisms are possibly not part of a species. However, I want to finish these pages with the hope that things are changing, with an increasing number of systematists consciously developing a critical attitude towards the problems and the methods of their discipline.

Appendix 1
Zoological checklists
and catalogues

In addition to the few broad scope works mentioned in the text, I list here a representative sample of zoological works offering worldwide coverage of major groups. The reader will easily appreciate the major gaps waiting to be filled.

I shall begin with insects, that major heap of biological diversity. For the largest insect groups, several multi-authored catalogues have been available for some time. There is a huge and formally complete *Coleopterorum Catalogus*, but most issues are now quite old and revised editions exist only for a few families or subfamilies. Matters are worse with works dealing with butterflies, moths and hymenoptera (*Lepidopterorum Catalogus, Hymenopterorum Catalogus*, two projects stopped short at the beginning). However, a new edition of the *Hymenopterorum Catalogus* was started in 1969 by van der Vecht and Ferrière, as has a companion work on Lepidoptera. Much more advanced, though still incomplete, is *Orthopterorum Catalogus* (Beier, 1962). For the Hemiptera, there is Metcalf *et al.*'s (1927–1968) work. Again, this work does not cover all groups and is out of date. A supplement on the Cicadoidea has been published by Duffels and van der Laan (1985). Diptera have never been covered worldwide within a single major work but the order is comprehensively dealt with in six excellent catalogues, respectively devoted to the Palaearctic (Soós and Papp, 1984), Nearctic (Stone *et al.*, 1965), Afrotropical (Crosskey *et al.*, 1980), Neotropical (Papavero, 1967), Oriental (Delfinado and Hardy, 1973–1977) and Australasian and Oceanian species (Evenhuis, 1989). For several smaller insect orders there are relatively recent catalogues or checklists, such as Salmon (1964–1965) on Collembola, Davies and Tobin (1984–1985) on Odonata, Illies (1966) and Zwick (1973) on Plecoptera, Sakai (1970–1988) and Steinmann (1989) on Dermaptera, Smithers (1967) on Psocoptera, Hopkins and Clay (1952) on Mallophaga, Jacot-Guillarmod (1970–1979) on Thysanoptera, Fischer (1960–1973) on Trichoptera.

Relatively recent monographs also cover several large insect families. Within Hemiptera, for instance, Slater (1964) for Lygaeidae, Carvalho (1957–1960) for Miridae, Drake and Ruhoff (1965) for Tingidae, Kormilev and Froeschner (1987) for Aradidae; within Homoptera, Mound and Halsey (1978) for Aleyrodidae and Eastop and Lambers (1976) for aphids; within Coleoptera, Udayagiri and Wadhii

(1989) for Bruchidae, Mazur (1984) for Histeridae, Didier and Séguy (1953) for Lucanidae; within Diptera, Hull (1962) for Asilidae, Hull (1973) for Bombylidae; within Hymenoptera, Olmi (1984) for Dryinidae, Gordh and Moczar (1990) for Bethylidae, Townes (1969–1971) for Ichneumonidae, Townes and Townes (1981) and Johnson (1992) for Proctotrupoidea, Bohart and Menke (1976) for Sphecidae, Kimsey and Bohart (1990) for Chrysididae and Boucek (1974) for Leucospidae. For groups lacking more detailed treatments, simple checklists of genera are available, such as Blackwelder (1952) on staphylinid beetles, or Seeno and Wilcox (1982) on Chrysomelidae s.l.

For arthropods other than insects, spiders are adequately covered by Roewer's (1942, 1954) catalogue, if used in conjunction with Brignoli's (1983) and Platnick's (1989) excellent supplements. However, there are no supplements to Roewer's (1923, 1934) monographs on harvestmen and solifuges, whereas Beier's (1932a,b) volumes on pseudoscorpions have been aptly supplemented by Harvey's (1990) catalogue. A new comprehensive *Crustaceorum Catalogus* has recently been started by Gruner and Holthuis, whereas family and genus level taxa of millipedes (Diplopoda) have been listed by Jeekel (1971) and Hoffman (1979).

For other invertebrate groups, things are not much better. A few recent world-wide monographs, however, deserve mention: Schmidt (1986) on cestodes; Siddiqi (1986) on tylenchid nematodes; Stephen and Edmonds (1972) on echiurids and sipunculans; Fauchald (1977) and Hartmann (1959–1965) on polychaetes; Reynolds and Cook (1976, 1981), Brinkhurst and Jamieson (1971) and Brinkhurst and Wetzel (1984) on oligochaetes, Sawyer (1986) on leeches, Ramazzotti and Maucci (1983) on tardigrades, and others. The literature on Mollusca is quite extensive. I only cite a recent list of world genera (Cunningham Vaught, 1989); at the species level, however, the lists available cover selected molluscan taxa only.

As for vertebrates, comprehensive and updated checklists of species are available for amphibians (Frost, 1985), birds (Howard and Moore, 1980; and others) and mammals (Honacki, Kinman and Koeppl, 1982; Corbet and Hill, 1986). As for fishes, a most scholarly reference is Eschmeyer's (1990) generic catalogue. For several groups, such as sharks and several families of bony fishes of major economic interest, FAO publishes a series of illustrated catalogues.

For many animal groups, however, world checklists are now in preparation. I am aware, for instance, of new projects on snakes, amphipods, neuropterans, scolytid beetles and others.

Appendices 2–23

Appendices 2–23 include excerpts from major systematic works of the last few years. This choice of schemes is not offered as the draft of a possible system of living organisms. Instead, it may provide the reader with a first general reference to the current understanding of relationships within groups with which he or she is less familiar. Moreover, it exemplifies several contrasting attitudes and procedures widely occurring in the systematic literature: for instance, the use of non-conventional ranks in a Linnaean hierarchy (Appendix 21), or the refusal to adopt a Linnaean hierarchy altogether (Appendices 6, 17, 20); or the adoption of Nelson's sequential ranking procedure (Appendix 17, 20) and the use of Patterson and Rosen's (1977) 'plesion' convention (Appendix 18).

The taxa marked [†] are extinct.

Appendix 2

MÖHN'S (1984) GENERAL CLASSIFICATION OF LIVING ORGANISMS

Superkingdom Prokaryonta
  Suprakingdom Archaebacteria
    Kingdom Archaebacteriobionta
      Infrakingdom Methanobacteriobionta
        Phylum Methanococcacea
        Phylum Methanospirillacea
        Phylum Methanosarcinacea
        Phylum Methanobacteriacea
      Infrakingdom Halobacteriobionta
        Phylum Halobacteriacea
        Phylum Halococcacea
      Infrakingdom Caldariabionta
        Phylum Sulfolobacea
        Phylum Thermoplasmatacea
  Suprakingdom Neobacteria
    Kingdom Bacteriobionta
      Infrakingdom Eubacteria
        Superphylum Grambacteria
          Phylum Streptococcacea
          Phylum Clostridiacea
          Phylum Bacillacea
          Phylum Mycoplasmatacea
          Phylum Chlamydiacea
          Phylum Micrococcacea
          Phylum Actinomycetacea
        Superphylum Agrambacteria
          Phylum Desulfovibrionacea
          Phylum Rhodobacteriacea
          Phylum Thiobacillacea
          Phylum Azotobacteriacea
          Phylum Pseudomonadacea
          Phylum Photobacteriacea
          Phylum Enterobacteriacea
      Infrakingdom Spirochaetae
        Phylum Spirochaetacea
      Infrakingdom Myxobacteria
    Kingdom Cyanobionta
        Phylum Cyanophyta
        Phylum Prochlorophyta
Superkingdom Eukaryonta

Suprakingdom Aconta
  Kingdom Rhodocyanobionta
    Phylum Cyanidiophyta
  Kingdom Erythrobionta
    Phylum Rhodophyta
Suprakingdom Contophora
  Kingdom Chlorobionta
    Phylum Prasinophyta
    Phylum Charophyta
    Phylum Ulvaphyta
    Phylum Chlorophyta
  Kingdom Flagelloopalinida
    Phylum Protomonada
    Phylum Opalinidea
  Kingdom Euglenophytobionta
    Phylum Euglenophyta
  Kingdom Eumycota
    Phylum Opisthomastigomycota
    Phylum Amastigomycota
  Kingdom Dinophytobionta
    Infrakingdom Dinophytea
      Phylum Dinophyta
      Phylum Granuloreticulosa
      Phylum Acanthiolaria
      Phylum Polannulifera
    Infrakingdom Ciliata
      Phylum Ciliophora
  Kingdom Cryptophytobionta
    Phylum Cryptophyta
  Kingdom Colponemata
    Phylum Colponemaria
  Kingdom Chloromonadophytobionta
    Phylum Chloromonadophyta
  Kingdom Chromophytobionta
    Infrakingdom Chromobionta
      Phylum Xanthophyta
      Phylum Pantonemomycota
      Phylum Proteomyxidea
      Phylum Labyrinthomorpha
      Phylum Chrysophyta
      Phylum Hydraulea
      Phylum Trichomycetea
      Phylum Mycetozoidea
      Phylum Pedinellaphyta
      Phylum Bacillariophyta
      Phylum Phaeophyta
    Infrakingdom Eustigmatobionta
      Phylum Eustigmatophyta
    Infrakingdom Choanobionta
      Phylum Craspedophyta
    Infrakingdom Haptophytobionta
      Phylum Haptophyta
Suprakingdom Cormobionta
Suprakingdom Animalia (Metazoa)

I have translated in the following way Möhn's German names for the individual hierarchical ranks: Superkingdom for Überreich, Suprakingdom for Oberreich, Kingdom for Reich, Infrakingdom for Unterreich, Superphylum for Überstamm, Phylum for Stamm.

Bacteriobionta (without Cyanobionta), Chlorobionta (without Cormobionta) and Chromophytobionta (without Animalia) are paraphyletic (Möhn, 1984, p.79).

Appendix 3

'PROVISIONAL CLASSIFICATION' OF THE PROTISTA, ACCORDING TO CORLISS (1984, 1986A,B, 1987)

Rhizopods
  Karyoblastea
  Amoebozoa
  Acrasia
  Eumycetozoa
  Plasmodiophorea
  Granuloreticulosa
  Xenophyophora, *incertae sedis*
Mastigomycetes
  Hyphochytridiomycota
  Oomycota
  Chytridiomycota, *incertae sedis*
Chlorobionts
  Chlorophyta
  Prasinophyta
  Conjugatophyta
  Charophyta
  Glaucophyta, *incertae sedis*
Chlorarachniophytes
  Chlorarachniophyta
Euglenozoa
  Euglenophyta
  Kinetoplastidea
  Pseudociliata, *incertae sedis*
Rhodophytes
  Rhodophyta
Cryptomonads
  Cryptophyta
Choanoflagellates
  Choanoflagellata
Chromobionts
  Chrysophyta
  Haptophyta
  Bacillariophyta
  Xanthophyta
  Eustigmatophyta
  Phaeophyta
  Pteromonadea, *incertae sedis*

Bicosoecidea, *incertae sedis*
Heterochloridea, *incertae sedis*
Rhaphidophyceae, *incertae sedis*
Labyrinthomorphs
  Labyrinthulea
  Thraustochytriacea
Polymastigotes
  Metamonadea
  Parabasalia
Paraflagellates
  Opalinata
Actinopods
  Heliozoa
  Taxopoda
  Acantharia
  Polycystina
  Phaeodaria
Dinoflagellates
  Peridinea
  Syndinea
  Ebriidea, *incertae sedis*
  Ellobiophyceae, *incertae sedis*
  Acritarcha, *incertae sedis*
Ciliates
  Ciliophora
Sporozoa
  Sporozoa
  Perkinsida, *incertae sedis*
Microsporidia
  Microsporidia
Haplosporidia
  Haplosporidia
Myxosporidia
  Myxosporidia
  Actinomyxidea, *incertae sedis*
  Marteiliidea, *incertae sedis*
  Paramyxidea, *incertae sedis*

In this provisional scheme, Corliss recognizes a conventional kingdom Protista, with about 50 formal phyla distributed within 19 informally named supraphyletic assemblages.

Appendix 4

PHYLA AND CLASSES OF THE PROTOCTISTA (CORLISS' PROTISTA) ACCORDING TO MARGULIS *ET AL.* (1990)

Phylum Rhizopoda
   Class Lobosea
   Class Filosea
Phylum Haplosporidia
   Class Haplosporea
Phylum Paramyxea
   Class Paramyxidea
   Class Marteiliidea
Phylum Myxozoa
   Class Myxosporea
   Class Actinosporea
Phylum Microspora
   Class Rudimicrosporea
   Class Microsporea
Phylum Acrasea
   Class Acrasids
Phylum Dictyostelida
   Class Dictyostelids
Phylum Rhodophyta
   Class Rhodophyceae
Phylum Coniugatophyta
   Class Coniugatophyceae
Phylum Xenophyophora
   Class Psamminida
   Class Stannominida
Phylum Cryptophyta
   Class Cryptophyceae
Phylum Glaucocystophyta
   Class Glaucocystophyceae
Phylum Karyoblastea
   Class Karyoblastea
Phylum Zoomastigina
   Class Amebomastigota
   Class Bicoecids
   Class Choanomastigota
   Class Diplomonads
   Class Pseudociliata
   Class Kinetoplastids
   Class Opalinids
   Class Proteromonads

   Class Parabasalia
   Class Retortomonads
   Class Pyrsonymphida
Phylum Euglenida
   Class Euglenophyceae
Phylum Chlorarachnida
   Class Chlorarachniophyceae
Phylum Prymnesiophyta
   Class Prymnesiophyceae
Phylum Raphidophyta
   Class Rhaphidophyceae
Phylum Eustigmatophyta
   Class Eustigmatophyceae
Phylum Actinopoda
   Class Polycystina
   Class Phaeodaria
   Class Heliozoa
   Class Acantharia
Phylum Hyphochytriomycota
   Class Hyphochytrids
Phylum Labyrinthulomycota
   Class Labyrinthulids
   Class Thraustochytrids
Phylum Plasmodiophoromycota
   Class Plasmodiophorids
Phylum Dinomastigota
   Class Dinophyceae
   Class Syndiniophyceae
Phylum Chrysophyta
   Class Chrysophyceae
   Class Synurophyceae
   Class Pedinellophyceae
   Class Dictyochophyceae
   Class Silicomastigota
Phylum Chytridiomycota
   Class Chytridiomycetes
Phylum Plasmodial slime moulds
   Class Protostelids
   Class Myxomycetes
Phylum Ciliophora

Class Karyorelictea
Class Spirotrichea
Class Protostomatea
Class Litostomatea
Class Phyllopharyngea
Class Nassophorea
Class Oligohymenophorea
Class Colpodea
Phylum Granuloreticulosa
Class Athalamea
Class Foraminifera
Phylum Apicomplexa
Class Gregarinia
Class Coccidia
Class Hematozoa
Phylum Bacillariophyta
Class Cosciniodiscophyceae

Class Fragilariophyceae
Class Bacillariophyceae
Phylum Chlorophyta
Class Prasinophyceae
Class Chlorophyceae
Class Ulvophyceae
Class Charophyceae
Phylum Oomycota
Class Saprolegniomycetidae
Class Peronosporomycetidae
Phylum Xanthophyta
Class Xanthophyceae
Phylum Phaeophyta
Class Phaeophyceae
Phylum: *incertae sedis*
Class Ebriids
Class Ellobiopsids

For several taxa, only vernacular names are suggested, mostly because of troubles with the 'ambiregnal' nomenclature of these groups, officially covered by the International Code of Zoological Nomenclature as well as by the International Code of Botanical Nomenclature.

Appendix 5

MÖHN'S (1984) CLASSIFICATION OF ANIMALS

Suprakingdom Animalia (Metazoa)
  Midkingdom Parazoa
    Kingdom Porifera
       Infrakingdom Calcarosponga
         Phylum Calcara
       Infrakingdom Silicosponga
         Phylum Demospongea
         Phylum Hexactinellidea
    Kingdom Archaeata
       Infrakingdom Archaeozoa
         Phylum Archaeocyatha
    Kingdom Placozoomorpha
         Phylum Placozoa
  Midkingdom Eumetazoa
    Kingdom Radiata
         Phylum Cnidaria
         Phylum Ctenophora
    Kingdom Bilateralia
      Interkingdom Aproctophora
       Infrakingdom Acoelomata
         Superphylum Parenchymata
            Phylum Plathelminthes
         Superphylum Mesozoia
            Phylum Mesozoa
            Phylum Cnidospora
      Interkingdom Proctophora
       Infrakingdom Nemertinea
         Superphylum Nemertina
            Phylum Nemertini
       Infrakingdom Neoproctophora
         Superphylum Pseudocoelia
            Midphylum Nematelminthes
            Phylum Priapulida
            Phylum Acanthocephala
            Phylum Kinorhyncha
            Phylum Rotatoria
            Phylum Gastrotricha
            Phylum Nematoda

Phylum Nematomorpha
Midphylum Entoprocta
Phylum Kamptozoa
Infrakingdom Coelomata
Superphylum Mollusca
Phylum Aplacophora
Phylum Conchifera
Superphylum Eucoelomata
Supraphylum Articulata (Teloblastomata)
Midphylum Annelidea
Phylum Annelida
Midphylum Proarthropoda
Phylum Onychophora
Midphylum Pentastomidea
Phylum Pentastomida
Midphylum Tardigradea
Phylum Tardigrada
Midphylum Euarthropoda
Phylum Amandibulata
Phylum Mandibulata
Supraphylum Monocoelomata
Phylum Myzostomida
Phylum Sipunculida
Phylum Echiurida
Supraphylum Pogonophorea
Phylum Pogonophora
Supraphylum Chaetognathea
Phylum Chaetognatha
Supraphylum Enterocoelomata
Midphylum Tentaculata
Phylum Phoronidea
Phylum Bryozoa
Phylum Brachiopoda
Midphylum Deuterostomia
Phylum Hemichordata
Phylum Echinodermata
Phylum Chordata

In addition to the names translated in the way explained in Appendix 2, I have used Midkingdom for Mittelreich, Interkingdom for Zwischenreich, Supraphylum for Oberstamm and Midphylum for Mittelstamm. Besides a few unconventional names, the arrangement of taxa is mainly traditional.

Appendix 6

NIELSEN'S (1985) SYSTEM OF THE ANIMALIA

Kingdom Animalia
               Phylum Choanoflagellata
    Metazoa
            Phylum Porifera
            Phylum Placozoa
      Gastraeozoa (Eumetazoa)
            Phylum Cnidaria
     Trochaeozoa
      Gastroneuralia
       Spiralia
        Articulata
           Phylum Annelida *sensu lato*
           Phylum Echiura
           Phylum Gnathostomulida
           Phylum Onychophora
           Phylum Arthropoda *sensu lato*
           Phylum Mollusca
           Phylum Sipuncula
        Parenchymia (Acoelomata)
           Phylum Nemertini
           Phylum Platyhelminthes (Turbellaria)
       Bryozoa
           Phylum Entoprocta
           Phylum Ectoprocta
      Aschelminthes (Nematelminthes, Pseudocoelomata)
           Phylum Rotifera
           Phylum Acanthocephala
           Phylum Chaetognatha
           Phylum Nematoda
           Phylum Nematomorpha
           Phylum Kinorhyncha
           Phylum Loricifera
           Phylum Priapulida
           Phylum Gastrotricha
     Protornaeozoa
            Phylum Ctenophora
      Notoneuralia
       Brachiata
           Phylum Pterobranchia

        Phylum Phoronida
        Phylum Brachiopoda
        Phylum Echinodermata
    Cyrtotreta
        Phylum Enteropneusta
    Chordata
        Phylum Urochordata
        Phylum Cephalochordata
        Phylum Vertebrata

Several ranks intermediate between the kingdom and the phylum are left unnamed.

Appendix 7

EHLERS' S (1985) SYSTEM OF THE PLATHELMINTHES

Plathelminthes
  Catenulida
  Euplathelminthes
    Acoelomorpha
      Nemertodermatida
      Acoela
    Rhabditophora
      Macrostomida
      Trepaxonemata
        Polycladida
        Neoophora
          Lecithoepitheliata, *incertae sedis*
          N.N.1
            Prolecithophora, *incertae sedis*
            N.N.2
              Seriata
                Proseriata
                Tricladida
              Rhabdocoela
                'Typhloplanoida' (including Kalyptorhynchia)
                Doliopharyngiophora
                  'Dalyellioidea' (including Temnocephalida and Udonellida)
                  Neodermata
                    Trematoda
                      Aspidobothrii
                      Digenea
                    Cercomeromorphae
                    Monogenea
                    Cestoda
                      Gyrocotylidea
                      Nephroposticophora
                        Amphilinidea
                      Cestoidea
                        Caryophyllidea
                        Eucestoda

Ranks are not named according to a Linnaean hierarchy. Taxa N.N.1 and N.N.2 have received no formal name, because of the uncertain relationships to their postulated sister-groups.

Appendix 8

JAMIESON'S (1988) SYSTEM OF THE OLIGOCHAETA

Class Oligochaeta (Octogonadia?)
1. Subclass Randiellata
    Order Randiellida
        Family Randiellidae
2. Subclass Tubificata
    Order Tubificida
        Family Tubificidae
        Family Enchytraeidae
        Family Capilloventridae
        Family Phreodrilidae
        Family Dorydrilidae
        Family Opisthocystidae
3a. Subclass Lumbriculata
    Order Lumbriculida
        Family Lumbriculidae
3b. Subclass Diplotesticulata
    Superorder Haplotaxidea
        Family Haplotaxidae
        Family Tiguassuidae
    Superorder Metagynophora
    Order Moniligastrida
        Family Moniligastridae
    Order Opisthopora
        Suborder Alluroidina
            Superfamily Alluroidoidea
            Family Alluroididae
            Family Syngenodrilidae
        Suborder Crassiclitellata
            Cohort Aquamegadrili
                1. Superfamily Sparganophiloidea
                    Family Sparganophilidae
                2a. Superfamily Biwadriloidea
                    Family Biwadrilidae
                2b. Superfamily Almoidea
                    Family Almidae
                    Family Lutodrilidae
            Cohort Terrimegadrili
                1. Superfamily Ocnerodriloidea
                    Family Ocnerodrilidae
                2. Superfamily Eudriloidea
                    Family Eudrilidae

3a. Superfamily Lumbricoidea
Family Kynotidae?
Family Komarekionidae
Family Ailoscolecidae
Family Microchaetidae
Family Hormogastridae
Family Glossoscolecidae
Family Lumbricidae
3b. Superfamily Megascolecoidea
Family Megascolecidae

A fundamentally cladistic approach coexists here with the use of formal 'Linnaean' ranks, as well as with a limited use of numerical prefixes, as in Hennig's systems (Appendices 15, 16).

Appendix 9

SALVINI-PLAWEN'S (1980) CLASSIFICATION OF THE PHYLUM MOLLUSCA

Subphylum Scutopoda
    Classis Caudofoveata
        Ordo Chaetodermatida
Subphylum Adenopoda
   Infraphylum/Superclassis Heterotecta
     Classis Solenogastres
        Superordo Aplotegmentaria
          Ordo Pholidoskepia
          Ordo Neomeniamorpha
        Superordo Pachytegmentaria
          Ordo Sterrofustia
          Ordo Cavibelonia
     Classis Placophora
      †Subclassis Heptaplacota
        Ordo Septemchitonida
      Subclassis Loricata
        †Ordo Chelodida
        †Ordo Scanochitonida
        Ordo Lepidopleurida
        Ordo Chitonida
   Infraphylum/Superclassis Conchifera
     Classis Galeroconcha
        Ordo Tryblidiida (=Monoplacophora)
        †Ordo Bellerophontida
     Classis Gastropoda
      Subclassis Prosobranchia
        Ordo Archaeogastropoda
        Ordo Caenogastropoda
      Subclassis Pulmonata
        Ordo Archaeopulmonata
        Ordo Basommatophora
        Ordo Stylommatophora
      Subclassis Gymnomorpha
        Ordo Onchidiida
        Ordo Soleolifera = Veronicellida
        Ordo Rhodopida
      Subclassis Opisthobranchia
        Ordo Pyramidellimorpha

   Ordo Cephalaspidea
   Ordo Anaspidea
   Ordo Saccoglossa
   Ordo Notaspidea
   Ordo Nudibranchia
   Ordo Anthobranchia
 Classis Bivalvia
  Subclassis Pelecypoda
  Superordo Ctenidiobranchia
   Ordo Nuculida
  Superordo Palaeobranchia
   Ordo Solemyida
   †Ordo Praecardiida
  Superordo Autobranchia
   Ordo Pteriomorpha
   Ordo Palaeoheterodonta
   Ordo Heterodonta
   Ordo Anomalodesmata
  Superordo Septibranchia
   Ordo Poromyida
  †Subclassis Rostroconchia
   Ordo Ribeiriida
   Ordo Ischyriniida
   Ordo Conocardiida
 Classis Scaphopoda
   Ordo Dentaliida
   Ordo Siphonodentaliida
 Classis Siphonopoda = Cephalopoda
  †Subclassis Orthoceratoida
   Ordo Ellesmerocerida
   Ordo Orthocerida
   Ordo Ascocerida
   Ordo Discosorida
   Ordo Endocerida
   Ordo Actinocerida
  Subclassis Nautiloida
   †Ordo Oncocerida
   Ordo Nautilida
   †Ordo Tarphycerida
  †Subclassis Ammonoida
   Ordo Bactritida
   Ordo Goniatitida
   Ordo Ammonitida
  Subclassis Coleoida
   †Ordo Aulacocerida
   †Ordo Belemnitida
   Ordo Sepiida
   †Ordo Phragmoteuthida
   Ordo Teuthida
   Ordo Vampyromorpha
   Ordo Octobrachia = Octopoda

A well-worked traditional classification, with Linnaean ranks, including both Recent and extinct taxa.

Appendix 10

HASZPRUNAR'S (1986) CLASSIFICATION OF GASTROPODS

Subclass Prosobranchia
    Order Archaeogastropoda
    Order Caenogastropoda
    Suborder Architaenioglossa
    Suborder Neotaenioglossa
    Suborder Heteroglossa
    Suborder Stenoglossa
Connecting link
    Superfamily Rissoelloidea
Subclass Heterobranchia
  Cohors Triganglionata
    Superorder Allogastropoda
  Cohors Pentaganglionata
    Superorder Architectibranchia
    Superorder Tectibranchia
      Order Bullomorpha
      ?Order Acochlidiomorpha

      Order Aplysiomorpha
      Order Saccoglossa
      Order Thecasomata
      ?Order Gymnosomata
    Superorder Eleutherobranchia
      Order Notaspidea
      Order Nudibranchia
      Order Anthobranchia
      ?Order Smeagolida
    Superorder Gymnomorpha
      Order Onchidiida
      Order Soleolifera
      ?Order Rhodopida
    Superorder Pulmonata
      Order Archaeopulmonata
      Order Branchiopulmonata
      Order Stylommatophora

The scheme incorporated the author's new concept of Heterobranchia, with a revised placement of Pulmonata and the disappearance of a traditional paraphyletic Opisthobranchia. Several problems still appear to be open.

Appendix 11

WEYGOLDT AND PAULUS'S (1979) SYSTEM OF THE CHELICERATA

†Aglaspida
Euchelicerata
  Xiphosurida
  Metastomata
    †Eurypterida
  Arachnida
      Ctenophora: Scorpiones
      Lipoctena
        Megaoperculata
          Uropygi
          Labellata
            Amblypygi
            Araneae
      Apulmonata
        Palpigradi
        Holotracheata
          Haplocnemata
            Chelonethi
            Solifugae
          Cryptoperculata
            Acannomorpha
              Ricinulei
              Acari
            Opiliones

The placement of Pantopoda within Chelicerata is regarded by Weygoldt and Paulus as doubtful. Ranks are left unnamed. To be compared with Appendix 12.

Appendix 12

SHULTZ'S (1990) SYSTEM OF THE CHELICERATA

Arachnida
  Micrura
    Megoperculata
      Palpigradi
        Tetrapulmonata
          Araneae
          Pedipalpi
            Amblypygi
            Uropygi
              Thelyphonida
              Schizomida
    Acaromorpha
      Ricinulei
      Acari
  Dromopoda
  Novogenuata
    Scorpiones
    Haplocnemata
      Pseudoscorpiones
      Solifugae

To be compared with Appendix 11.

Appendix 13

SCHRAM'S (1986) CLASSIFICATION OF THE CRUSTACEA

Class Remipedia
  †Order Enantiopoda
  Order Nectipoda
Class Malacostraca
  Subclass Hoplocarida
    †Order Aeschronectida
    †Order Palaeostomatopoda
    Order Stomatopoda
      †Suborder Archeostomatopodea
      Suborder Unipeltata
  Subclass Eumalacostraca
    Order Syncarida
      Suborder Bathynellacea
      Suborder Anaspidacea
      †Suborder Palaeocaridacea
    †Order Belotelsonidea
    Order Euphausiacea
    Order Amphionidacea
    Order Decapoda
      Suborder Dendrobranchiata
      Suborder Eukyphida
      Suborder Euzygida
      Suborder Reptantia
    †Order Waterstonellidea
    Order Mysida
    Order Lophogastrida
    †Order Pygocephalomorpha
    Order Myctacea
    Order Edriophthalma
      Suborder Isopoda
      Suborder Amphipoda
    Order Thermosbaenacea
    Order Hemicaridea
      Suborder Cumacea
      Suborder Tanaidacea
      Suborder Spelaeogriphacea
Class Phyllopoda
  Subclass Phyllocarida
    †Order Archaeostraca
    †Order Canadaspidida
    †Order Hymenostraca

    †Order Hoplostraca
    Order Leptostraca
  Subclass Cephalocarida
    Order Brachypoda
    †Order Lipostraca
  Subclass Sarsostraca
    Order Anostraca
  Subclass Calmanostraca
    Order Notostraca
    †Order Kazacharthra
    Order Conchostraca
Class Maxillopoda
  Subclass Tantulocarida
    Order Tantulocaridida
  Subclass Branchiura
    Order Arguloida
    ?Order Pentastomida
  Subclass Mystacocarida
    Order Mystacocaridida
  Subclass Ostracoda
    †Order Bradoriida
    †Order Phosphatocopida
    †Order Leperditicopida
    Order Palaeocopida
    Order Myodocopida
    Order Podocopida
  Subclass Copepoda
    Order Calanoida
    Order Misophrioida
    Order Harpacticoida
    Order Mormonilloida
    Order Siphonostomatoida
    Order Monstrilloida
    Order Cyclopoida
    Order Poecilostomatoida
  Subclass Thecostraca
    Order Facetotecta
    Order Rhizocephala
    Order Ascothoracica
    Order Cirripedia

Schram regards Crustacea as an independent phylum (see section 7.18). This classification incorporates all higher taxa of Recent and fossil Crustacea.

Appendix 14

STAROBOGATOV'S (1988) CLASSIFICATION OF THE CRUSTACEA

Supraclass Crustacea
  Class Branchipodiodes
    Subclass Polyphemiones
      Infraclass Speleonectioni (=Remipedia)
        Superorder Speleonectiformii
          Order Speleonectiformes
        [†]Superorder Tesnusocaridiformii
          Order Tesnusocaridiformes (=Enantiopoda)
      Infraclass Hutchinsonellioni (=Cephalocarida)
        Order Hutchinsonelliformes
      Infraclass Polyphemioni (=Brachypoda)
        [†]Superorder Lepidocaridiformii
          Order Lepidocaridiformes (=Lipostraca)
        Superorder Polyphemiformii
          Order Polyphemiformes (=Onychopoda)
          Order Leptodoriformes (=Haplopoda)
    Subclass Branchipodiones
      Infraclass Triopioni (=Calmanostraca)
        [†]Superorder Odariiformii
          Order Odariiformes
        Superorder Triopiformii
          Order Triopiformes (=Notostraca)
        [†]Superorder Ketmeniiformii
          Order Ketmeniiformes (=Kazacharthra)
      Infraclass Branchipodioni
        [†]Superorder Leanchoiliformii
          Order Leanchoiliformes
          Order Yohoiiformes
        Superorder Branchipodiformii
          Order Branchipodiformes (=Anostraca)
    Subclass Daphniiones
        Superorder Daphniiformii
          [†]Order Protocaridiformes
          Order Daphniiformes (=Ctenopoda plus Adenopoda)
        Superorder Limnadiiformii
          Order Lynceiformes (=Laevicaudata)
          Order Limnadiiformes (=Spinicaudata)
  Class Carcinoides (=Malacostraca)
    Subclass Carciniones (=Eumalacostraca)
      Infraclass Squillioni (=Hoplocarida)
      Infraclass Carcinioni (=Eumalacostraca *sensu* Schram 1982)
    Subclass Nebaliiones (=Phyllocarida)
  Class Ascothoracioides
    Subclass Ascothoraciones

Order Ascothoraciformes
Subclass Cyclopiones
  Infraclass Derocheilocaridioni
          Order Derocheilocaridiformes (=Mystacocarida)
  Infraclass Calanioni (=Gymnoplea)
      Superorder Basipodelliformii (=Tantulocarida)
          Order Basipodelliformes (=Tantulocarida)
      Superorder Calaniformii
          Order Calaniformes
          Order Platycopiiformes (=Progymnoplea)
  Infraclass Cyclopioni (=Podoplea)
      Superorder Misophriiformii
          Order Misophriiformes
      Superorder Mormonilliformii
          Order Mormonilliformes
      Superorder Monstrilliformii
          Order Monstrilliformes
      Superorder Harpacticiformii
          Order Longipediiformes (=Polyarthra)
          Order Harpacticiformes
      Superorder Lernaeiformii (=Gnathostoma =Cyclopoida *sensu* Kabata)
          Order Cyclopiformes
          Order Enterocoliformes
          Order Lernaeiformes
      Superorder Philichthyiformii (=Poecilostomata)
          Order Lichomolgiformes
          Order Philichthyiformes
          Order Chondracanthiformes (=Andreinidea)
          Order Sarcotacideiformes
      Superorder Caligiformii (=Siphonostoma)
          Order Caligiformes
          Order Asterocheiriformes
          Order Lernaeopodiformes
          Order Dichelestiformes
          Order Herpylobiformes
          Order Choinostomatiformes
Class Halicynoides
  Subclass Halicyniones
          †Superorder Halicyniformii
          Order Cycliformes
          Order Halicyniformes
      Superorder Arguliformii
          Order Arguliformes
  Subclass Cypridiones
  Infraclass Lepadioni
          †Superorder Concavicaridiformii (=Thylacocephala)
          Order Concavicaridiformes
      Superorder Sacculiniformii
          Order Sacculiniformes (=Rhizocephala Kentrogonida)
      Superorder Lepadiformii (Thoracica plus Acrothoracica plus Rhizocephala
      Akentrogona)
  Infraclass Cypridioni (=Ostracoda)

The most unusual feature of Starobogatov's classification probably consists in his replacement of many traditional names with new ones, formed according to his idiosyncratic proposal for naming higher taxa (Starobogatov, 1991).

Appendix 15

HENNIG'S (1969) SYSTEM OF THE INSECTA

1. Entognatha
   1.1. Diplura
   1.2. Ellipura
   1.2.1. Protura
   1.2.2. Collembola
2. Ectognatha
   2.1. Archaeognatha (Microcoryphia)
   2.2. Dicondylia
   2.2.1. Zygentoma
   2.2.2. Pterygota
   2.2.2.1. Palaeoptera
   2.2.2.1..1. Ephemeroptera
   2.2.2.1..2. Odonata
   2.2.2.2. Neoptera
   2.2.2.2..1. Plecoptera
   2.2.2.2..2. Paurometabola
   2.2.2.2..2.1. Embioptera
   2.2.2.2..2.2. Orthopteromorpha
   2.2.2.2..2.2..1. Blattopteriformia
   2.2.2.2..2.2..1.1. Notoptera (Grylloblattodea)
   2.2.2.2..2.2..1.2. Dermaptera
   2.2.2.2..2.2..1.3. Blattopteroidea
   2.2.2.2..2.2..1.3.1. Mantodea
   2.2.2.2..2.2..1.3.2. Blattodea (including termites)
   2.2.2.2..2.2..2. Orthopteroidea
   2.2.2.2..2.2..2.1. Ensifera
   2.2.2.2..2.2..2.2. Caelifera
   2.2.2.2..2.2..2.3. Phasmatodea
   2.2.2.2..3. Paraneoptera
   2.2.2.2..3.1. Zoraptera
   2.2.2.2..3.2. Acercaria
   2.2.2.2..3.2..1. Psocodea (including lice)
   2.2.2.2..3.2..2. Condylognatha
   2.2.2.2..3.2..2.1. Thysanoptera
   2.2.2.2..3.2..2.2. Hemiptera
   2.2.2.2..3.2..2.2.1. Heteropteroidea
   2.2.2.2..3.2..2.2.1.1. Coleorrhyncha
   2.2.2.2..3.3..2.2.1.2. Heteroptera
   2.2.2.2..3.2..2.2.2. Sternorrhyncha
   2.2.2.2..3.2..2.2.3. Auchenorrhyncha
   2.2.2.2..4. Holometabola
   2.2.2.2..4.1. Neuropteroidea
   2.2.2.2..4.1..1. Megaloptera

2.2.2.2..4.1..2. Raphidioptera
2.2.2.2..4.1..3. Planipennia
2.2.2.2..4.2. Coleoptera
2.2.2.2..4.3. Strepsiptera
2.2.2.2..4.4. Hymenoptera
2.2.2.2..4.5. Siphonaptera
2.2.2.2..4.6. Mecopteroidea
2.2.2.2..4.6..1. Amphiesmenoptera
2.2.2.2..4.6..1.1. Trichoptera
2.2.2.2..4.6..1.2. Lepidoptera
2.2.2.2..4.6..2. Antliophora
2.2.2.2..4.6..2.1. Mecoptera
2.2.2.2..4.6..2.2. Diptera

Notice the idiosyncratic use of numerical prefixes and several unresolved polytomies.

Appendix 16

HENNIG'S (1985) SYSTEM OF THE CHORDATA

[A. Echinodermata]
B. Chordata
B.1 Tunicata
B.11 Ascidiacea
B.12 Pelagotunicata
B.12.1 Pyrosomida
B.12.2 Larvacea
B.12.3 Cyclomyaria
B.12.4 Desmomyaria
B.2 Vertebrata
B.21 Acrania
B.22 Craniota
B.22.1 Agnatha
B.22.2 Gnathostomata
B.22.21 Chondrichthyes
B.22.21.1 Elasmobranchia
B.22.21.2 Holocephali
B.22.22 Osteognathostomata
B.22.22.1 Actinopterygii s. l.
B.22.22.1.C.1 Brachyopterygii
B.22.22.1.C.2 Actinopterygii s. str.
B.22.22.1.C.21 Palaeopterygii
B.22.22.1.C.22 Neopterygii
B.22.22.1.C.22.1 Rhomboganoidea
B.22.22.1.C.22.2 Teleostei s. l.
B.22.22.1.C.22.2.D.1 Cycloganoidea
B.22.22.1.C.22.2.D.2 Teleostei s. str.
B.22.22.2 Choanatae
B.22.22.21 Dipnoi
B.22.22.22 Kinocrania
B.22.22.22.1 Coelacanthi
B.22.22.22.2 Tetrapoda
B.22.22.22.2.C.1. Amphibia
B.22.22.22.2.C.11 Gymnophiona
B.22.22.22.2.C.12 Caudata
B.22.22.22.2.C.13 Salientia (Anura)
B.22.22.22.2.C.2 Amniota
B.22.22.22.2.C.21 Lepidosauria
B.22.22.22.2.C.21.D.1 Rhynchocephalia
B.22.22.22.2.C.21.D.2 Squamata
B.22.22.22.2.C.22 Testudines

B.22.22.22.2.C.22.D.1 Pleurodira
B.22.22.22.2.C.22.D.2 Cryptodira
B.22.22.22.2.C.23 Archosauromorpha
B.22.22.22.2.C.23.D.1 Crocodilia
B.22.22.22.2.C.23.D.2 Aves
B.22.22.22.2.C.24 Mammalia
B.22.22.22.2.C.24.D.1 Prototheria
B.22.22.22.2.C.24.D.2 Theria
B.22.22.22.2.C.24.D.22 Eutheria (Placentalia)

As in Appendix 14, notice the use of numerical prefixes and a few unresolved polytomies. Acrania (amphioxus) is placed here in Vertebrata, whereas Vertebrata *Auctorum* is entered in this system as Craniota.

Appendix 17

THE MAJOR GROUPS OF CHORDATA ACCORDING TO NELSON (1969)

Phylum Chordata
  Subphylum Acrania
  Subphylum Vertebrata
    Superclass Cyclostomata
    Superclass Gnathostomata
      Class Elasmobranchiomorphi
        †Subclass Acanthodii
        Subclass Holocephali
        Subclass Elasmobranchii
      Class Teleostomi
        Subclass Actinopterygii
          Infraclass Chondrostei
          Infraclass Neopterygii
            Division Holostei
            Division Teleostei
              Cohort Osteoglossomorpha
              Cohort Clupeomorpha
              Cohort Elopomorpha
              Cohort Euteleostei
                Superorder Protacanthopterygii
                Superorder Ostariophysi
                Superorder Neoteleostei
                  Series Atherinomorpha
                  Series Myctophiformes
                  Series Paracanthopterygii
                  Series Acanthopterygii
        Subclass Sarcopterygii
          Infraclass Brachiopterygii
          Infraclass Coelacanthini
          Infraclass Dipnoi
          Infraclass Choanata
            †Division Rhipidistia
            Division Batrachomorpha
            Division Reptilomorpha
              Cohort Sauropsida
                Superorder Chelonia
                Superorder Lepidosauria
                  Series Rhynchocephalia
                  Series Squamata
                Superorder Archosauria

Series Crocodylia
Series Aves
Cohort Mammalia
Superorder Prototheria
Superorder Theria
Series Metatheria
Series Eutheria

In this system, the tunicates are not regarded as belonging to the phylum Chordata. Notice the use (it was 1969) of formal Linnaean ranks.

Appendix 18

ROSEN *ET AL.'*S (1981) CLASSIFICATION OF GNATHOSTOME VERTEBRATES

Superclass Gnathostomata
   plesion [†] *Acanthodes bronni* Agassiz
   class Chondrichthyes
      Subclass Selachii
      Subclass Holocephali
   Class Osteichthyes
      Subclass Actinopterygii
         Infraclass Cladistia
         Infraclass Actinopteri
            Series Chondrostei
            Series Neopterygii
               Division Ginglymodi
               Division Halecostomi
      Subclass Sarcopterygii
      Sarcopterygii *incertae sedis* [†]Porolepiformes
   plesion [†]*Eusthenopteron foordi*
      Infraclass Actinistia
      Infraclass Choanata
         Series Dipnoi
         Series Tetrapoda

Notice the use of formal ranks for Recent groups with plesions for extinct taxa.

Appendix 19

CARROLL'S (1987) CLASSIFICATION OF VERTEBRATES, INCLUDING BOTH EXTINCT AND LIVING FORMS

Class Agnatha
 †Subclass Pteraspidomorphi (Diplorhina)
                     Order Heterostraci (Pteraspidiformes)
                     Order Thelodontida
    Subclass Cephalaspidomorpha
                         †Order Osteostraci
                         †Order Galeaspida
                         †Order Anaspida
                          Order Petromyzontiformes
      Agnatha *incertae sedis*
                          Order Myxiniformes
†Class Placodermi
                     Order Stensioellida
                     Order Pseudopetalichthyda
                     Order Rhenanida
                     Order Ptyctodontida
                     Order Acanthothoraci
                     Order Petalichthyida
                     Order Phyllolepida
                     Order Arthrodira
                     Order Antiarchi
Class Chondrichthyes
    Subclass Elasmobranchii
                         †Order Cladoselachida
                         †Order Coronodontia
                         †Order Symmoriida
                         †Order Eugeneodontida (Edestida, Helicoprionida)
                         †Order Orodontida
                         †Order Squatinactida
                         †Order Ctenacanthiformes
                         †Order Xenacanthida
                          Order Galeomorpha
                          Order Squalomorpha
                          Order Batoidea
    Subclass Holocephali
                         †Order Chondrenchelyiformes
                         †Order Copodontiformes
                         †Order Psammodontiformes
                         †Order? (5 suborders *incertae sedis*)
                          Order Chimaeriformes

Chondrichthyes, subclass?, *incertae sedis*
†Order Iniopterygiformes
†Order Petalodontida
†Class *incertae sedis* Acanthodii
Order Climatiiformes
Order Ischnacanthiformes
Order Acanthodiformes
Class Osteichthyes
Subclass Actinopterygii
Infraclass Chondrostei
†Order Paleonisciformes
†Order Haplolepiformes
†Order Dorypteriformes
†Order Tarrasiiformes
†Order Ptycholepiformes
†Order Pholidopleuriformes
†Order Luganoiiformes
†Order Redfieldiiformes
†Order Perleidiformes
†Order Peltopleuriformes
†Order Phanerorhynchiformes
†Order Saurichthyiformes
Order Polypteriformes (Cladistia)
Order Acipenseriformes
Infraclass Neopterygii
Order Lepisosteiformes (Ginglymodi)
†Order Semionotiformes
†Order Pycnodontiformes
†Order Macrosemiiformes
Order Amiiformes
†Order Pachycormiformes
†Order Aspidorhynchiformes
Division Teleostei
†Order Pholidophoriformes
†Order Leptolepiformes
†Order Ichthyodectiformes
Order Osteoglossiformes
Order Elopiformes
Order Anguilliformes
Order Notacanthiformes
†Order Ellimmichthyiformes
Order Clupeiformes
Subdivision Euteleostei
†Order?, *incertae sedis* (*Pharmacichthys*)
Order Salmoniformes
†Superorder undescribed (*Pyrenichthys*)
Superorder Ostariophysi
Order Gonorhynchiformes
†Order?, *incertae sedis* (*Chanoides*)
Order Characiformes
Order Cypriniformes
Order Siluriformes
Superorder Stenopterygii
Order Stomiiformes

                        Order Aulopiformes
                Superorder Scopelomorpha
                        Order Myctophiformes
                †Superorder undesigned
                        Order Pattersonichthyiformes
                        Order Ctenothrissiformes
                Superorder Paracanthopterygii
                        Order Percopsiformes
                        Order Batrachoidiformes (Haplodoci)
                        Order Gobiesociformes (Xenopteri)
                        Order Lophiiformes (Pediculati)
                        Order Gadiformes
                        Order Ophidiiformes
                Superorder Acanthopterygii
                        Series Atherinomorpha
                        Order Atheriniformes s. l.
                        Series Percomorpha
                        Order Beryciformes
                        Order Zeiformes
                        Order Lampriformes
                        Order Gasterosteiformes
                        Order Syngnathiformes
                        Order Synbranchiformes
                        Order Indostomiformes
                        Order Pegasiformes
                        Order Dactylopteriformes
                        Order Scorpaeniformes
                        Order Perciformes
                        Order Pleuronectiformes (Heterosomata)
                        Order Tetraodontiformes
        Subclass Sarcopterygii
                        Order Crossopterygii
                        Order Dipnoi
Class Amphibia
        †Subclass Labyrinthodontia
                        Order Ichthyostegalia
                        Order?, *incertae sedis* (several taxa)
                        Order Temnospondyli
                        Order Anthracosauria
        †Subclass?, *incertae sedis*
                        Order Diadectida
        †Subclass Lepospondyli
                        Order Aistopoda
                        Order Nectrida
                        Order Microsauria
                        Order Lysorophia
                        Order?, >*incertae sedis* (several taxa)
        Subclass not stated
                        Order Urodela
                        †Order Proanura
                        Order Anura
Class Reptilia
        †Subclass Anapsida
                        Order Captorhinida

Order Mesosauria
Order?, *incertae sedis* (*Eunotosaurus*)
Subclass Testudinata
Order Chelonia
Subclass Diapsida
†Order Araeoscelida
†Order?, *incertae sedis* (several taxa)
†Order Choristodera
†Order Thalattosauria
Infraclass Lepidosauromorpha
†Order Eosuchia
Superorder Lepidosauria
Order Sphaenodonta
Order Squamata
†Superorder Sauropterygia
Order?, *incertae sedis* (*Claudiosaurus*)
Order Nothosauria
Order Plesiosauria
Order Protorosauria
Order Trilophosauria
Order Rhynchosauria
Superorder Archosauria
†Order Thecodontia
Order Crocodylia
†Order Pterosauria
†Order Saurischia
†Order Ornithischia
†Diapsida *incertae sedis*
Order Placodontia
†Subclass Ichthyopterygia
Order Ichthyosauria
†Subclass Synapsida
Order Pelycosauria
Order Therapsida
Class Aves
†Subclass Archaeornithes
Order Archaeopterygiformes
Subclass Neornithes
†Order not stated (Ambiortidae)
†Superorder Odonthognathae
Order Hesperornithiformes
Order Ichthyornithiformes
†Superorder?, *incertae sedis*
Order Gobipterygiformes
Order Enantiornithiformes
Superorder Palaeognathae
†Order unnamed (Lithornidae)
Order Tinamiformes
Order Struthioniformes
Order Rheiformes
Order Casuariiformes
Order Aepyornithiformes
Order Dinornithiformes
Order Apterygiformes

Superorder Neognathae
Order Cuculiformes
Order Falconiformes
Order Galliformes
Order Columbiformes
Order Psittaciformes
†Order?, *incertae sedis* (Zygodactylidae)
Order Coliiformes
Order Coraciiformes s. l.
Order Strigiformes
Order Caprimulgiformes
Order Apodiformes
Order Bucerotiformes
Order Piciformes
Order Passeriformes
Order Gruiformes
Order Podicipediformes
†Order Diatrymiformes
Order Charadriiformes
Order Anseriformes
Order Ciconiiformes
Order Pelecaniformes
Order Procellariiformes
Order Gaviiformes
Order Sphenisciformes
Class Mammalia
  Subclass Prototheria
Order Monotremata
†Order Triconodonta
†Order Docodonta
  †Subclass Allotheria
Order Multituberculata
  Subclass Theria
    †Infraclass Trituberculata
Order Symmetrodonta
Order?, *incertae sedis* (*Shuotherium*)
Order Eupanthotheria
    †Theria *incertae sedis* (several taxa)
    Infraclass Metatheria
Order Marsupialia
    Infraclass Eutheria
†Order?, *incertae sedis* (several taxa)
†Order Apatotheria
†Order Leptictida
†Order Pantolesta
Order Scandentia
Order Macroscelidea
Order Dermoptera
Order Insectivora
†Order Tillodontia
†Order Pantodonta
†Order Dinocerata
†Order Taeniodontia
Order Chiroptera

        Order Primates
        †Order Creodonta
        Order Carnivora
        †Order Anagalida
        Order Rodentia
        Order Lagomorpha
        †Order Condylarthra
        Order Artiodactyla
        †Order Mesonychia (Acreodi)
        Order Cetacea
        Order Perissodactyla
        Order Proboscidea
        Order Sirenia
        †Order Desmostylia
        Order Hyracoidea
        †Order Embrithopoda
        Order Tubulidentata
        †Order Notoungulata
        †Order Astrapotheria
        †Order Litopterna
        †Order Xenungulata
        †Order Pyrotheria
        Order Xenarthra
        †Order?, *incertae sedis* (suborder Palaeanodonta)
        Order Pholidota
†Mammalia *incertae sedis* (several taxa)

The arrangement is eclectic and often less than satisfactory. The usefulness of Carroll's classification is with its comprehensive coverage of Recent and extinct taxa, with a checklist of all extinct genera.

Appendix 20

LAUDER AND LIEM'S (1983) CLASSIFICATION OF LIVING BONY FISHES

Actinopterygii
  Actinopteri
    Chondrostei
    Neopterygii
      Ginglymodi
      Halecostomi
        Halecomorphi
        Teleostei
          Osteoglossomorpha
          Elopomorpha
          Clupeomorpha
          Euteleostei
            Ostariophysi
            'Protacanthopterygii'
            Neoteleostei
              Stomiiformes
              Aulopiformes
              Myctophiformes
              Paracanthopterygii
              Acanthopterygii
                Atherinomorpha
                Percomorpha

Cladistic, following in part Nelson's sequential principle.

Appendix 21

SIBLEY AND AHLQUIST'S (1990) CLASSIFICATION OF BIRDS

Class Aves
  [Subclass Archeornithes: *Archeopteryx*]
  Subclass Neornithes
    Infraclass Eoaves
      Parvclass Ratitae
        Order Struthioniformes
          Suborder Struthioni
            Infraorder Struthionides
              Family Struthionidae
            Infraorder Rheides
              Family Rheidae
          Suborder Casuarii
            Family Casuariidae
              Tribe Casuariini
              Tribe Dromaiini
            Family Apterygidae
        Order Tinamiformes
          Family Tinamidae
    Infraclass Neoaves
      Parvclass Galloanserae
        Superorder Gallomorphae
        Order Craciformes
          Suborder Craci
            Family Cracidae
          Suborder Megapodii
            Family Megapodiidae
         Order Galliformes
          Parvorder Phasianida
            Superfamily Phasianoidea
            Family Phasianidae
            Superfamily Numidoidea
            Family Numididae
          Parvorder Odontophorida
            Family Odontophoridae
        Superorder Anserimorphae
        Order Anseriformes
          Infraorder Anhimides
            Superfamily Anhimoidea
            Family Anhimidae
            Superfamily Anseranatoidea

       Family Anseranatidae
     Infraorder Anserides
       Family Dendrocygnidae
       Family Anatidae
        Subfamily Oxyurinae
        Subfamily Stictonettinae
        Subfamily Cygninae
        Subfamily Anatinae
         Tribe Anserini
         Tribe Anatini
Parvclass Turnicae
 Order Turniciformes, *incertae sedis*
       Family Turnicidae
Parvclass Picae
 Order Piciformes
    Infraorder Picides
       Family Indicatoridae
       Family Picidae
    Infraorder Rhamphastides
      Superfamily Megalaimoidea
       Family Megalaimidae
      Superfamily Lybioidea
       Family Lybiidae
      Superfamily Rhamphastoidea
       Family Rhamphastidae
        Subfamily Capitoninae
        Subfamily Rhamphastinae
Parvclass Coraciae
    Superorder Galbulimorphae
   Order Galbuliformes
    Infraorder Galbulides
       Family Galbulidae
    Infraorder Bucconides
       Family Bucconidae
   Superorder Bucerotimorphae
   Order Bucerotiformes
       Family Bucerotidae
       Family Bucorvidae
   Order Upupiformes
    Infraorder Upupides
       Family Upupidae
    Infraorder Phoeniculides
       Family Phoeniculidae
       Family Rhinopomastidae
   Superorder Coraciimorphae
   Order Trogoniformes
       Family Trogonidae
        Subfamily Apaloderminae
        Subfamily Trogoninae
         Tribe Trogonini
         Tribe Harpactini
  Order Coraciiformes
   Suborder Coracii
     Superfamily Coracioidea

                        Family Coraciidae
                        Family Brachypteraciidae
                     Superfamily Leptosomoidea
                        Family Leptosomidae
            Suborder Alcedini
               Infraorder Alcedinides
                  Parvorder Momotida
                     Family Momotidae
                  Parvorder Todida
                     Family Todidae
                  Parvorder Alcedinida
                        Family Alcedinidae
                  Parvorder Cerylida
                     Superfamily Dacelonoidea
                        Family Dacelonidae
                     Superfamily Ceryloidea
                        Family Cerylidae
               Infraorder Meropides
                        Family Meropidae
Parvclass Coliae
   Order Coliiformes
                        Family Coliidae
                           Subfamily Coliinae
                           Subfamily Urocoliinae
Parvclass Passerae
   Superorder Cuculimorphae
      Order Cuculiformes
            Infraorder Cuculides
               Parvorder Cuculida
                     Superfamily Cuculoidea
                        Family Cuculidae
                     Superfamily Centropodoidea
                        Family Centropodidae
               Parvorder Coccyzida
                        Family Coccyzidae
            Infraorder Crotophagides
               Parvorder Opisthocomida
                        Family Opisthocomidae
               Parvorder Crotophagida
                        Family Crotophagidae
                           Tribe Crotophagini
                           Tribe Guirini
               Parvorder Neomorphida
                        Family Neomorphidae
   Superorder Psittacimorphae
      Order Psittaciformes
                  Family Psittacidae
   Superorder Apodimorphae
      Order Apodiformes
                  Family Apodidae
                  Family Hemiprocnidae
      Order Trochiliformes
                  Family Trochilidae
                     Subfamily Phaethornithinae

Subfamily Trochilinae
Superorder Strigimorphae
Order Musophagiformes
Family Musophagidae
Subfamily Musophaginae
Subfamily Criniferinae
Order Strigiformes
Suborder Strigi
Parvorder Tytonida
Family Tytonidae
Parvorder Strigida
Family Strigidae
Suborder Aegotheli
Family Aegothelidae
Suborder Caprimulgi
Infraorder Podargides
Family Podargidae
Family Batrachostomidae
Infraorder Caprimulgides
Parvorder Steatornithida
Superfamily Steatornithoidea
Family Steatornithidae
Superfamily Nyctibioidea
Family Nyctibiidae
Parvorder Caprimulgida
Superfamily Eurostopoidea
Family Eurostopodidae
Superfamily Caprimulgoidea
Family Caprimulgidae
Subfamily Chordeilinae
Subfamily Caprimulginae
Superorder Passerimorphae
Order Columbiformes
Family Raphidae
Family Columbidae
Order Gruiformes
Suborder Grui
Infraorder Eurypygides
Family Eurypygidae
Infraorder Otidides
Family Otididae
Infraorder Gruides
Parvorder Gruida
Superfamily Gruoidea
Family Gruidae
Family Heliornithidae
Tribe Aramini
Tribe Heliornithini
Superfamily Psophioidea
Family Psophiidae
Parvorder Cariamida
Family Cariamidae
Family Rhynochetidae
Suborder Ralli

Family Rallidae
Suborder Mesornithi, *incertae sedis*
Family Mesornithidae
Order Ciconiiformes
Suborder Charadrii
Infraorder Pteroclides
Family Pteroclidae
Infraorder Charadriides
Parvorder Scolopacida
Superfamily Scolopacoidea
Family Thinocoridae
Family Pedionomidae
Family Scolopacidae
Subfamily Scolopacinae
Subfamily Tringinae
Superfamily Jacanoidea
Family Rostratulidae
Family Jacanidae
Parvorder Charadriida
Superfamily Chionidoidea
Family Chionididae
Superfamily Charadrioidea
Family Burhinidae
Family Charadriidae
Subfamily Recurvirostrinae
Tribe Haematopodini
Tribe Recurvirostrini
Subfamily Charadriinae
Superfamily Laroidea
Family Glareolidae
Subfamily Dromadinae
Subfamily Glareolinae
Family Laridae
Subfamily Larinae
Tribe Stercorariini
Tribe Rynchopini
Tribe Larini
Tribe Sternini
Subfamily Alcinae
Suborder Ciconii
Infraorder Falconides
Parvorder Accipitrida
Family Accipitridae
Subfamily Pandioninae
Subfamily Accipitrinae
Family Sagittariidae
Parvorder Falconida
Family Falconidae
Infraorder Ciconiides
Parvorder Podicipedida
Family Podicipedidae
Parvorder Phaethontida
Family Phaethontidae
Parvorder Sulida

Superfamily Suloidea
Family Sulidae
Family Anhingidae
Superfamily Phalacrocoracoidea
Family Phalacrocoracidae
Parvorder Ciconiida
Superfamily Ardeoidea
Family Ardeidae
Superfamily Scopoidea
Family Scopidae
Superfamily Phoenicopteroidea
Family Phoenicopteridae
Superfamily Threskiornithoidea
Family Threskiornithidae
Superfamily Pelecanoidea
Family Pelecanidae
Subfamily Balaenicipitinae
Subfamily Pelecaninae
Superfamily Ciconioidea
Family Ciconiidae
Subfamily Cathartinae
Subfamily Ciconiinae
Superfamily Procellarioidea
Family Fregatidae
Family Spheniscidae
Family Gaviidae
Family Procellariidae
Subfamily Hydrobatinae
Subfamily Procellariinae
Subfamily Diomedeinae
Order Passeriformes
Suborder Tyranni
Infraorder Acanthisittides
Family Acanthisittidae
Infraorder Eurylaimides
Superfamily Pittoidea
Family Pittidae
Superfamily Eurylaimoidea
Family Eurylaimidae
Family Philepittidae, *incertae sedis*
Infraorder Tyrannides
Parvorder Tyrannida
Family Tyrannidae
Subfamily Tyranninae
Subfamily Tityrinae
Tribe Schiffornithini
Tribe Tityrini
Subfamily Cotinginae
Subfamily Piprinae
Parvorder Thamnophilida
Family Thamnophilidae
Parvorder Furnariida
Superfamily Furnarioidea
Family Furnariidae

Subfamily Furnariinae
Subfamily Dendrocolaptinae
Superfamily Formicarioidea
Family Formicariidae
Family Conopophagidae
Family Rhinocryptidae
Suborder Passeri (Oscines)
Parvorder Corvida
Superfamily Menuroidea
Family Climacteridae
Family Menuridae
Subfamily Menurinae
Subfamily Atrichornithinae
Family Ptilonorhynchidae
Superfamily Meliphagoidea
Family Maluridae
Subfamily Malurinae
Tribe Malurini
Tribe Stipiturini
Subfamily Amytornithinae
Family Meliphagidae
Family Pardalotidae
Subfamily Pardalotinae
Subfamily Dasyornithinae
Subfamily Acanthizinae
Tribe Sericornithini
Tribe Acanthizini
Superfamily Corvoidea
Family Eopsaltriidae
Family Irenidae
Family Orthonychidae
Family Pomatostomidae
Family Laniidae
Family Vireonidae
Family Corvidae
Subfamily Cinclosomatinae
Subfamily Corcoracinae
Subfamily Pachycephalinae
Tribe Neosittini
Tribe Mohouini
Tribe Falcunculini
Tribe Pachycephalini
Subfamily Corvinae
Tribe Corvini
Tribe Paradisaeini
Tribe Artamini
Tribe Oriolini
Subfamily Dicrurinae
Tribe Rhipidurini
Tribe Dicrurini
Tribe Monarchini
Subfamily Aegithininae
Subfamily Malaconotinae
Tribe Malaconotini

Tribe Prionopini
Family Callaeatidae, *incertae sedis*
Parvorder *incertae sedis*
Family Picathartidae
Parvorder Passerida
Superfamily Muscicapoidea
Family Bombycillidae
Tribe Dulini
Tribe Ptilogonatini
Tribe Bombycillini
Family Cinclidae
Family Muscicapidae
Subfamily Turdinae
Subfamily Muscicapinae
Tribe Muscicapini
Tribe Saxicolini
Family Sturnidae
Tribe Sturnini
Tribe Mimini
Superfamily Sylvioidea
Family Sittidae
Subfamily Sittinae
Subfamily Tichodromadinae
Family Certhiidae
Subfamily Certhiinae
Tribe Certhiini
Tribe Salpornithini
Subfamily Troglodytinae
Subfamily Polioptilinae
Family Paridae
Subfamily Remizinae
Subfamily Parinae
Family Aegithalidae
Family Hirundinidae
Subfamily Pseudochelidoninae, *incertae sedis*
Subfamily Hirundininae
Family Regulidae
Family Pycnonotidae
Family Hypocoliidae
Family Cisticolidae
Family Zosteropidae
Family Sylviidae
Subfamily Acrocephalinae
Subfamily Megalurinae
Subfamily Garrulacinae
Subfamily Sylviinae
Tribe Timaliini
Tribe Chamaeini
Tribe Sylviini
Superfamily Passeroidea
Family Alaudidae
Family Nectariniidae
Subfamily Promeropinae
Subfamily Nectariniinae

Tribe Dicaeini
Tribe Nectariniini
Family Melanocharitidae
Tribe Melanocharitini
Tribe Toxorhamphini
Family Paramythiidae
Family Passeridae
Subfamily Passerinae
Subfamily Motacillinae
Subfamily Prunellinae
Subfamily Ploceinae
Subfamily Estrildinae
Tribe Estrildini
Tribe Viduini
Family Fringillidae
Subfamily Peucedraminae
Subfamily Fringillinae
Tribe Fringillini
Tribe Carduelini
Tribe Drepanidini
Subfamily Emberizinae
Tribe Emberizini
Tribe Parulini
Tribe Thraupini
Tribe Cardinalini
Tribe Icterini

Slightly modified from the previous classification proposed by Sibley, Ahlquist and Monroe (1988), this is by far the most impressive, though discussed, result of DNA–DNA hybridization studies. Taxonomic ranks are intended to represent degrees of molecular distance.

Appendix 22

BREMER'S (1985) CLADISTIC CLASSIFICATION OF GREEN PLANTS

Subkingdom Chlorobionta
  Division Chlorophyta
      Class 'Ulvophyceae'
      Class Pleurastrophyceae
      Class Chlorophyceae
  Division Streptophyta
    Subdivision Chlorokybophytina
      Class Chlorokybophyceae
    Subdivision 'Zygophytina'
      Class 'Klebsormidiophyceae'
      Class Zygophyceae
    Subdivision Chaetosphaeridiophytina
      Class Chaetosphaeridiophyceae
    Subdivision Charophytina
      Class Charophyceae
    Subdivision 'Coleochaetophytina'
      Class 'Coleochaetophyceae'
    Subdivision Embryophytina
     Superclass Marchantiatae
      Class Marchantiopsida
     Superclass Anthocerotatae
      Class Anthocerotopsida
     Superclass Bryatae
      Class Bryopsida
     Superclass Tracheidatae
      Class Psilotopsida
      Class Lycopodiopsida
      Class Equisetopsida
      Class 'Polypodiopsida'
        Subclass Ophioglossidae
        Subclass Marattiidae
        Subclass Polypodiidae
      Class Spermatopsida
        Subclass Cycadidae
        Subclass Pinidae
        Subclass Ginkgoidae
        Subclass Gnetidae
        Subclass Magnoliidae

Following the author, undefined groups (paraclades) are put within inverted commas. Nelson's sequencing convention is adopted, except within Spermatopsida, where sister-group relationships are regarded by the author as uncertain. Formal Linnaean ranks are consequently applied.

Appendix 23

DAHLGREN'S (1989A,B) CLASSIFICATION OF FLOWERING PLANTS

Dicotyledones
Superorder Magnolianae
   Order Annonales
      Family Annonaceae
      Family Myristicaceae
      Family Eupomatiaceae
      Family Canellaceae
      Family Austrobaileyaceae
   Order Aristolochiales
      Family Aristolochiaceae
   Order Rafflesiales
      Family Rafflesiaceae
      Family Hydnoraceae
   Order Magnoliales
      Family Degeneriaceae
      Family Himantandraceae
      Family Magnoliaceae
   Order Lactoridales
      Family Lactoridaceae
   Order Winterales
      Family Winteraceae
   Order Chloranthales
      Family Chloranthaceae
   Order Illiciales
      Family Illiciaceae
      Family Schisandraceae
   Order Laurales
      Family Amborellaceae
      Family Trimeniaceae
      Family Monimiaceae
      Family Gomortegaceae
      Family Calycanthaceae
      Family Lauraceae
   Order Nelumbonales
      Family Nelumbonaceae
Superorder Nympheanae
   Order Piperales
      Family Saururaceae
      Family Piperaceae
   Order Nympheales
      Family Cabombaceae
      Family Nymphaeaceae
      Family Ceratophyllaceae

Superorder Ranunculanae
 Order Ranunculales
  Family Lardizabalaceae
  Family Sargentodoxaceae
  Family Menispermaceae
  Family Kingdoniaceae
  Family Circaeasteraceae
  Family Ranunculaceae
  Family Hydrastidaceae
  Family Berberidaceae
 Order Papaverales
  Family Papaveraceae
  Family Fumariaceae
Superorder Caryophyllanae
 Order Caryophyllales
  Family Molluginaceae
  Family Caryophyllaceae
  Family Phytolaccaceae
  Family Achatocarpaceae
  Family Agdestidaceae
  Family Basellaceae
  Family Portulacaceae
  Family Stegnospermataceae
  Family Nyctaginaceae
  Family Aizoaceae
  Family Halophytaceae
  Family Cactaceae
  Family Didiereaceae
  Family Hectorellaceae
  Family Chenopodiaceae
  Family Amaranthaceae
Superorder Polygonanae
 Order Polygonales
  Family Polygonaceae
Superorder Plumbaginanae
 Order Plumbaginales
  Family Plumbaginaceae
  Family Limoniaceae
Superorder Malvanae
 Order Malvales
  Family Sterculiaceae
  Family Plagiopteridaceae
  Family Bixaceae
  Family Cochlospermaceae
  Family Cistaceae
  Family Sphaerosepalaceae
  Family Sarcolaenaceae
  Family Huaceae
  Family Tiliaceae
  Family Dipterocarpaceae
  Family Bombacaceae
  Family Malvaceae
 Order Urticales
  Family Ulmaceae

Family Moraceae
Family Cecropiaceae
Family Berbeyaceae
Family Cannabaceae
Family Urticaceae
Order Euphorbiales
 Family Euphorbiaceae
 Family Simmondsiaceae
 Family Pandaceae
 Family Aextoxicaceae
 Family Dichapetalaceae
Order Thymelaeales
 Family Gonystylidaceae
 Family Thymelaeaceae
Order Rhamnales
 Family Rhamnaceae
Superorder Violanae
 Order Violales
  Family Flacourtiaceae
  Family Berberidopsidaceae
  Family Aphloiaceae
  Family Physenaceae
  Family Passifloraceae
  Family Dipentodontaceae
  Family Peridiscaceae
  Family Scyphostegiaceae
  Family Violaceae
  Family Turneraceae
  Family Malesherbiaceae
  Family Caricaceae
 Order Cucurbitales
  Family Achariaceae
  Family Cucurbitaceae
  Family Begoniaceae
  Family Datiscaceae
 Order Salicales
  Family Salicaceae
 Order Tamaricales
  Family Tamaricaceae
  Family Frankeniaceae
 Order Capparales
  Family Capparaceae
  Family Brassicaceae
  Family Tovariaceae
  Family Resedaceae
  Family Gyrostemonaceae
  Family Batidaceae
  Family Moringaceae
 Order Tropaeolales
  Family Tropaeolaceae
  Family Limnanthaceae
 Order Salvadorales
  Family Salvadoraceae
Superorder Theanae

Order Dilleniales
    Family Dilleniaceae
Order Paeoniales
    Family Glaucidiaceae
    Family Paeoniaceae
Order Theales
    Family Stachyuraceae
    Family Pentaphylacaceae
    Family Marcgraviaceae
    Family Quiinaceae
    Family Ancistrocladaceae
    Family Dioncophyllaceae
    Family Nepenthaceae
    Family Medusagynaceae
    Family Caryocaraceae
    Family Strasburgeriaceae
    Family Ochnaceae
    Family Chrysobalanaceae
    Family Oncothecaceae
    Family Scytopetalaceae
    Family Theaceae
    Family Bonnetiaceae
    Family Hypericaceae
    Family Elatinaceae
Order Lecythidales
    Family Lecythidaceae
Superorder Primulanae
Order Primulales
    Family Myrsinaceae
    Family Aegicerataceae
    Family Theophrastaceae
    Family Primulaceae
    Family Coridaceae
Order Ebenales
    Family Sapotaceae
    Family Styracaceae
    Family Lissocarpaceae
    Family Ebenaceae
Superorder Rosanae
Order Trochodendrales
    Family Trochodendraceae
    Family Tetracentraceae
Order Cercidiphyllales
    Family Cercidiphyllaceae
    Family Eupteleaceae
Order Hamamelidales
    Family Hamamelidaceae
    Family Platanaceae
    Family Myrothamnaceae
Order Balanopales
    Family Balanopaceae
Order Fagales
    Family Nothofagaceae
    Family Fagaceae

Family Corylaceae
Family Betulaceae
Order Juglandales
    Family Rhoipteleaceae
    Family Juglandaceae
Order Myricales
    Family Myricaceae
Order Casuarinales
    Family Casuarinaceae
Order Buxales
    Family Buxaceae
    Family Daphniphyllaceae
    Family Didymelaceae
Order Geissolomatales
    Family Geissolomataceae
Order Cunoniales
    Family Cunoniaceae
    Family Baueraceae
    Family Brunelliaceae
    Family Davidsoniaceae
    Family Eucryphiaceae
Order Saxifragales
    Family Saxifragaceae
    Family Francoaceae
    Family Greyiaceae
    Family Brexiaceae
    Family Grossulariaceae
    Family Iteaceae
    Family Cephalotaceae
    Family Crassulaceae
    Family Podostemaceae
Order Droserales
    Family Droseraceae
    Family Lepuropetalaceae
    Family Parnassiaceae
Order Rosales
    Family Rosaceae
    Family Neuradaceae
    Family Malaceae
    Family Amygdalaceae
    Family Anisophyllaceae
    Family Crossosomataceae
    Family Surianaceae
    Family Rhabdodendraceae
Order Gunnerales
    Family Gunneraceae
Superorder Proteanae
Order Proteales
    Family Proteaceae
Order Elaeagnales
    Family Elaeagnaceae
Superorder Myrtanae
Order Myrtales
    Family Psiloxylaceae

Family Heteropyxidaceae
Family Myrtaceae
Family Onagraceae
Family Trapaceae
Family Lythraceae
Family Combretaceae
Family Melastomataceae
Family Memecylaceae
Family Crypteroniaceae
Family Oliniaceae
Family Penaeaceae
Family Rhynchocalycaceae
Family Alzateaceae
Order Haloragidales
Family Haloragidaceae
Superorder Rutanae
Order Sapindales
Family Coriariaceae
Family Anacardiaceae
Family Leitneriaceae
Family Podoaceae
Family Sapindaceae
Family Hippocastanaceae
Family Aceraceae
Family Akaniaceae
Family Bretschneideraceae
Family Emblingiaceae
Family Staphylaeaceae
Family Melianthaceae
Family Sabiaceae
Family Meliosmaceae
Family Connaraceae
Order Fabales
Family Mimosaceae
Family Caesalpiniaceae
Family Fabaceae
Order Rutales
Family Rutaceae
Family Ptaeroxylaceae
Family Cneoraceae
Family Simaroubaceae
Family Tepuianthaceae
Family Burseraceae
Family Meliaceae
Order Polygalales
Family Malpighiaceae
Family Trigoniaceae
Family Vochysiaceae
Family Polygalaceae
Family Krameriaceae
Order Geraniales
Family Zygophyllaceae
Family Peganaceae
Family Ledocarpaceae

Family Biebersteiniaceae
Family Dirachmaceae
Family Balanitaceae
Order Linales
Family Linaceae
Family Humiriaceae
Family Ctenolophonaceae
Family Ixonanthaceae
Family Erythroxylaceae
Family Lepidobotryaceae
Family Oxalidaceae
Order Celastrales
Family Stackhousiaceae
Family Lophopyxidaceae
Family Cardiopteridaceae
Family Corynocarpaceae
Family Celastraceae
Order Rhizophorales
Family Rhizophoraceae
Family Elaeocarpaceae
Order Balsaminales
Family Balsaminaceae
Superorder Vitanae
Order Vitales
Family Vitaceae
Superorder Santalanae
Order Santalales
Family Olacaceae
Family Opiliaceae
Family Loranthaceae
Family Medusandraceae
Family Misondendraceae
Family Eremolepidaceae
Family Santalaceae
Family Viscaceae
Superorder Balanophoranae
Order Balanophorales
Family Cynomoriaceae
Family Balanophoraceae
Superorder Aralianae
Order Pittosporales
Family Pittosporaceae
Family Tremandraceae
Family Byblidaceae
Order Araliales
Family Araliaceae
Family Apiaceae
Superorder Asteranae
Order Campanulales
Family Pentaphragmataceae
Family Campanulaceae
Family Lobeliaceae
Order Asterales
Family Asteraceae

Superorder Solananae
  Order Solanales
    Family Solanaceae
    Family Sclerophylacaceae
    Family Goetzeaceae
    Family Convolvulaceae
    Family Cuscutaceae
    Family Cobaeaceae
    Family Polemoniaceae
  Order Boraginales
    Family Hydrophyllaceae
    Family Ehretiaceae
    Family Boraginaceae
    Family Lennoaceae
    Family Hoplestigmataceae
Superorder Ericanae
  Order Bruniales
    Family Bruniaceae
    Family Grubbiaceae
  Order Fouquieriales
    Family Fouquieriaceae
  Order Ericales
    Family Actinidiaceae
    Family Clethraceae
    Family Cyrillaceae
    Family Ericaceae
    Family Empetraceae
    Family Monotropaceae
    Family Pyrolaceae
    Family Epacridaceae
  Order Stylidiales
    Family Stylidiaceae
  Order Sarraceniales
    Family Sarraceniaceae
Superorder Cornanae
  Order Cornales
    Family Garryaceae
    Family Alangiaceae
    Family Nyssaceae
    Family Cornaceae
    Family Roridulaceae
    Family Davidiaceae
    Family Escalloniaceae
    Family Helwingiaceae
    Family Toricelliaceae
    Family Aucubaceae
    Family Aralidiaceae
    Family Diapensiaceae
    Family Phellinaceae
    Family Aquifoliaceae
    Family Paracryphiaceae
    Family Sphenostemonaceae
    Family Symplocaceae
    Family Icacinaceae

    Family Montiniaceae
    Family Columelliaceae
    Family Alseuosmiaceae
    Family Hydrangeaceae
    Family Sambucaceae
    Family Viburnaceae
    Family Menyanthaceae
    Family Adoxaceae
    Family Dulongiaceae
    Family Tribelaceae
    Family Eremosynaceae
    Family Pterostemonaceae
    Family Tetracarpaeaceae
  Order Eucommiales
    Family Eucommiaceae
  Order Dipsacales
    Family Caprifoliaceae
    Family Valerianaceae
    Family Dipsacaceae
    Family Morinaceae
    Family Calyceraceae
Superorder Loasanae
  Order Loasales
    Family Loasaceae
Superorder Gentiananae
  Order Goodeniales
    Family Goodeniaceae
  Order Oleales
    Family Oleaceae
  Order Gentianales
    Family Desfontainiaceae
    Family Loganiaceae
    Family Dialipetalanthaceae
    Family Rubiaceae
    Family Theligonaceae
    Family Gentianaceae
    Family Saccifoliaceae
    Family Apocynaceae
    Family Asclepiadaceae
Superorder Lamianae
  Order Lamiales (=Scrophulariales)
    Family Retziaceae
    Family Stilbaceae
    Family Buddlejaceae
    Family Scrophulariaceae (including Orobanchaceae and Selaginaceae)
    Family Myoporaceae
    Family Globulariaceae
    Family Plantaginaceae
    Family Lentibulariaceae
    Family Pedaliaceae
    Family Trapellaceae
    Family Martyniaceae
    Family Gesneriaceae

Family Bignoniaceae
Family Acanthaceae
Family Verbenaceae
Family Lamiaceae
Family Callitrichaceae
Order Hydrostachyales
Family Hydrostachyaceae
Order Hippuridales
Family Hippuridaceae
Monocotyledones
Superorder Alismatanae
Order Alismatales
Family Aponogetonaceae
Family Butomaceae
Family Hydrocharitaceae
Family Limnocharitaceae
Family Alismataceae
Order Najadales
Family Scheuchzeriaceae
Family Juncaginaceae
Family Najadaceae
Family Potamogetonaceae
Family Zosteraceae
Family Posidoniaceae
Family Cymodoceaceae
Family Zannichelliaceae
Superorder Triuridanae
Order Triuridales
Family Triuridaceae
Superorder Aranae
Order Arales
Family Araceae
Family Acoraceae
Family Lemnaceae
Superorder Lilianae
Order Dioscoreales
Family Trichopodaceae
Family Dioscoreaceae
Family Stemonaceae
Family Taccaceae
Family Trilliaceae
Family Rhipogonaceae
Family Petermanniaceae
Family Smilacaceae
Order Asparagales
Family Philesiaceae
Family Luzuriagaceae
Family Behniaceae
Family Convallariaceae
Family Dracaenaceae
Family Asparagaceae
Family Ruscaceae
Family Herreriaceae
Family Nolinaceae

Family Asteliaceae
Family Dasypogonaceae
Family Calectasiaceae
Family Blandfordiaceae
Family Xanthorrhoeaceae
Family Agavaceae
Family Hypoxidaceae
Family Tecophilaeaceae
Family Lanariaceae
Family Ixioliriaceae
Family Cyanastraceae
Family Johnsoniaceae
Family Phormiaceae
Family Doryanthaceae
Family Eriospermaceae
Family Asphodelaceae
Family Simethidaceae
Family Anthericaceae
Family Aphyllanthaceae
Family Hemerocallidaceae
Family Funkiaceae
Family Hyacinthaceae
Family Alliaceae
Family Amaryllidaceae
Order Liliales
Family Colchicaceae
Family Uvulariaceae
Family Iridaceae
Family Alstroemeriaceae
Family Calochortaceae
Family Liliaceae
Order Melanthiales
Family Melanthiaceae
Family Campynemaceae
Order Burmanniales
Family Burmanniaceae
Family Corsiaceae
Order Orchidales
Family Neuwiediaceae
Family Apostasiaceae
Family Cypripediaceae
Family Orchidaceae
Superorder Bromelianae
Order Velloziales
Family Velloziaceae
Order Bromeliales
Family Bromeliaceae
Order Haemodorales
Family Haemodoraceae
Order Philydrales
Family Philyraceae
Order Pontederiales
Family Pontederiaceae
Order Typhales

Family Typhaceae
Superorder Zingiberanae
  Order Zingiberales
    Family Lowiaceae
    Family Musaceae
    Family Heliconiaceae
    Family Strelitziaceae
    Family Zingiberaceae
    Family Costaceae
    Family Cannaceae
    Family Marantaceae
Superorder Commelinanae
  Order Commelinales
    Family Mayacaceae
    Family Commelinaceae
    Family Xyridaceae
    Family Rapateaceae
    Family Eriocaulaceae
  Order Hydatellales
    Family Hydatellaceae
  Order Cyperales

Family Juncaceae
Family Thurniaceae
Family Cyperaceae
Order Poales
  Family Flagellariaceae
  Family Joinvilleaceae
  Family Restionaceae
  Family Centrolepidaceae
  Family Poaceae
Superorder Arecanae
  Order Hanguanales
    Family Hanguanaceae
  Order Arecales
    Family Arecaceae
Superorder Cyclanthanae
  Order Cyclanthales
    Family Cyclanthaceae
Superorder Pandananae
  Order Pandanales
    Family Pandanaceae

This is perhaps the most updated of Angiosperm classifications, incorporating more biochemical evidence but still developed and presented in traditional terms.

References

Abbott, L.A., Bisby, F.A. and Rogers, D.J. (1985) *Taxonomic Analysis in Biology: Computers, Models and Databases*, Columbia University Press, New York.

Abele, L.G. (1991) Comparison of morphological and molecular phylogeny of the Decapoda. *Mem. Queensl. Mus.*, **31**, 101–8.

Abele, L.G. and Felgenhauer, B.E. (1982) Decapoda. Amphionidacea, in *Synopsis and Classification of Living Organisms*, (ed S.P. Parker), McGraw-Hill, New York, **2**, pp. 295–326.

Ackery, P.R. (1988) Host plants and classification: A review of nymphalid butterflies. *Biol. J. Linn. Soc.*, **33**, 95–203.

Ackery, P.R. and Vane-Wright, R.I. (1984) *Milkweed Butterflies: Their Cladistics and Biology*, British Museum (Natural History), London.

Adams, E.N. (1972) Consensus techniques and the comparison of taxonomic trees. *Syst. Zool.*, **21**, 390–7.

Adanson, M. (1763) *Familles des Plantes*, Vincent, Paris.

Adkins, R.M. and Honeycutt, R.L. (1991) Molecular phylogeny of the superorder Archonta. *Proc. Natl. Acad. Sci. USA*, **88**, 10317–21.

Agassiz, L. (1857) Essay on classification, in *Contributions to the Natural History of the United States*, Vol. 1, Little, Brown & Co., Boston, MA.

Ahmad, I. and Schaefer, C.W.(1987) Food plants and feeding biology of the Pyrrhocoroidea (Hemiptera). *Phytophaga*, **1**, 75–92.

Albert, V.A. and Mishler, B.D. (1992) On the rationale and utility of weighting nucleotide sequence data. *Cladistics*, **8**, 73–83.

Aldrich, J.R. (1988) Chemical ecology of the Heteroptera. *Annu. Rev. Entomol.*, **33**, 211–38.

Allard, M.W., Myiamoto, M.M. and Honeycutt, R.L. (1991) Tests for rodent polyphyly. *Nature*, **353**, 610–11.

Allen, G.M. (1916) Bats of the genus *Corynorhinus*. *Cambridge, Mass., Bull. Mus. Comp. Zool. Harvard Coll.*, **60**, 333–56.

Allen, J.E. (1877) The influence of physical conditions in the genesis of species. *Radical Rev.*, **1**, 108–40.

Allmon, W.D. (1992) Genera in paleontology. *Historical Biol.*, **6**, 149–58.

Altschul, S.F. and Lipman, D.J. (1990) Equal animals. *Nature*, **348**, 493–4.

Amadon, D. (1966) The superspecies concept. *Syst. Zool.*, **15**, 245–9.

Amadon, D. and Short, L.R. (1976) Treatment of subspecies approaching species status. *Syst. Zool.*, **25**, 161–7.

Anderson. D.T. (1981) Origins and relationships among the animal phyla. *Proc. Linn. Soc. N. S.W.*, **106**, 151–66.

Andràssy, I. (1976) *Evolution as a Basis for the Systematization of Nematodes*, Pitman, London.

Andrews, P. (1987) Aspects of hominoid phylogeny, in *Molecules and Morphology in Evolution: Conflict or Compromise?* (ed C. Patterson), Cambridge University Press, Cambridge, pp. 23–53.

Anonymous (1981) Origine dei grandi phyla dei Metazoi. *Atti dei Convegni Lincei*, **49**, Accademia Naz. dei Lincei, Rome.

Anonymous (1986) The genera of the Cactaceae: towards a new consensus. *Bradleya*, **4**, 65–78.

Anonymous (1988) Discovery of a new species of lemur. *Nature*, **333**, 206.

Anonymous (1989) Index Kewensis on computer. *Taxon*, **38**, 529–30.

Anthony, H.E. (1923) Mammals from Mexico and South America. *Am. Mus. Novit.*, **54**, 1–10.

Applegate, M., Abele, L.G. and Spears, T. (1991) A phylogenetic study of the Cirripedia based on 18S ribosomal DNA sequences. *Am. Zool.*, **31**, 101A.

Archie, J.W. (1989) Homoplasy excess ratios: New indices for measuring levels of homoplasy in phylogenetic systematics and a critique of the consistency index. *Syst. Zool.*, **38**, 253–69.

Archie, J.W. (1991) Tests to distinguish between phylogenetic information and random noise in nucleotide sequence data, in *The Unity of Evolutionary Biology. Proceedings of the Fourth International Congress of Systematic and Evolutionary Biology*, Vol. 2, (ed E.C. Dudley), Dioscorides Press, Portland, OR, pp. 923–34.

Aritzia, E.V., Andersen, R.A. and Sogin, M.L. (1991) A new phylogeny for chromophyte algae using 16S-like rRNA sequences from *Mallomonas papillosa* (Synurphyceae) and *Tribonema aequale* (Xanthophyceae). *J. Phycol.*, **27**, 428–36.

Arnett, R.H. Jr. and Samuelson, G.A. (1986) *The insect and spider collections of the world*, E.J. Brill/Fauna and Flora Publications, Gainesville, FL.

Arthur, W. (1984) *Mechanisms of Morphological Evolution*, Wiley, Chichester.

Arthur, W. (1988) *A Theory of the Evolution of Development*, Wiley, Chichester.

Ashlock, P.D. (1971) Monopoly and associated terms. *Syst. Zool.*, **20**, 63–9.

Ashlock, P.D. (1972) Monophyly again. *Syst. Zool.*, **21**, 430–8.

Ashlock, P.D. (1974) The uses of cladistics. *Annu. Rev. Syst. Ecol.*, **5**, 81–99.

Ashlock, P.D. (1979) An evolutionary systematist's view of classification. *Syst. Zool.*, **28**, 441–50.

Ashlock, P.D. (1984) Monophyly: its meaning and importance, in *Cladistics: Perspectives on the Reconstruction of Evolutionary History*, (eds T. Duncan and T.F. Stuessy), Columbia University Press, New York, pp. 39–46.

Aspöck, H. (1986) The Raphidioptera of the world: A review of present knowledge, in *Recent Research in Neuropterology*, (eds J. Gepp, H. Aspöck and H. Hölzel), (publisher not given), Graz, pp. 15–29.

Aspöck, H., Aspöck, V. and Hölzel, H. (1980) *Die Neuropteren Europas*, Goecke and Evers, Krefeld, 2 vols.

Assali, N.-E., Martin, W.F., Sommerville, C.C. *et al.* (1991) Evolution of the Rubisco operon from prokaryotes to algae: Structure and analysis of the ercS gene of the brown alga *Pylaiella littoralis. Plant Mol. Biol.*, **17**, 853–63.

Atran, S. (1986) *Fondements de L'Histoire Naturelle*, Editions Complexe, Paris.

Atran, S. (1987) The early history of the species concept: An anthropological reading, in *Histoire du Concept d'Espèce dans les Sciences de la Vie*, (eds S. Atran, R.W. Burkhardt, Jr., P. Corsi *et al.*), Fondation Singer-Polignac, Paris, pp. 1–36.

Atran, S., Burkhardt, R.W., Jr., Corsi, P. *et al.* (1987) *Histoire du Concept d'Espèce dans les Sciences de la Vie*, Fondation Singer-Polignac, Paris.

Aubert, J. and Solignac, M. (1990) Experimental evidence for a mitochondrial DNA introgression between *Drosophila* species. *Evolution*, **44**, 1272–82.

Avinoff, V. (1913) On some new forms of the genus *Parnassius* Lat. *Hor. Soc. Ent. Ross.*, **40**, 161–7.

Avise, J.C. (1983) Protein variation and phylogenetic reconstruction, in *Protein Polymorphism: Adaptive and Taxonomic Significance*, (eds G.S. Oxford and D. Rollinson), Academic Press, New York, pp. 103–10.

Avise, J.C. (1991) Ten unorthodox perspectives on evolution prompted by comparative population genetic findings on mitochondrial DNA. *Annu. Rev. Genet.*, **25**, 45–69.

Avise, J.C., Arnold, J., Ball, R.M. *et al.* (1987) Intraspecific phylogeography: The mitochondrial DNA bridge between population genetics and systematics. *Annu. Rev. Ecol. Syst.*, **18**, 489–522.

Avise, J.C., Bermingham, E., Kessler, L.G. *et al.* (1984) Characterization of mitochondrial DNA variability in a hybrid swarm between subspecies of bluegill sunfish (*Lepomis macrochirus*). *Evolution*, **38**, 931–41.

Avise, J.C., Trexler, J.C. Travis, J. *et al.* (1991) *Poecilia mexicana* is the recent female parent of the unisexual fish *P. formosa*. *Evolution*, **45**, 1530–3.

Ax, P. (1956) Die Gnathostomulida, eine rätselhafte Wurmgruppe aus dem Meeressand. *Akad. Wiss. Lit. Mainz, Math. Naturw. Kl.*, **1956**, 531–62.

Ax, P. (1961) Verwandtschaftsbeziehungen und Phylogenie der Turbellarien. *Ergebn. Biol.*, **24**, 1–68.

Ax, P. (1963) Relationships and phylogeny of the Turbellaria, in *The Lower Metazoa. Comparative Biology and Phylogeny*, (ed E.C. Dougherthy), University of California Press, Berkley, CA, pp. 191–224.

Ax, P. (1984) *Das phylogenetische System*, Gustav Fischer, Stuttgart.

Ax, P. (1985a) The position of the Gnathostomulida and Platyhelminthes in the phylogenetic system of the Bilateria, in *The Origins and Relationships of Lower Invertebrates. The Systematics Association Special Volume No. 28*, (eds S. Conway Morris, J.D. George, R. Gibson *et al.*), Clarendon Press, Oxford, pp. 168–80.

Ax, P. (1985b) Stem species and the stem lineage concept. *Cladistics* , **1**, 279–87.

Ax, P. (1987) *The Phylogenetic System. The Systematization of Organisms on the Basis of their Phylogenesis*, Wiley, Chichester.

Ax, P. (1988) *Systematik in der Biologie*, Gustav Fischer, Stuttgart.

Ax, P. (1989a) Basic phylogenetic systematization of the Metazoa, in *The Hierarchy of Life*, (eds B. Fernholm, K. Bremer and H. Jörnvall), Elsevier, Amsterdam, pp. 229–45.

Ax, P. (1989b) The integration of fossils in the phylogenetic system of organisms. *Abh. Naturwiss. Ver. Hamburg*, **28**, 27–42.

Ax, P. (1989c) Homologie in der Biologie – ein Relationsbegriff im Vergleich von Arten. *Zool. Beitr. (Berlin)* , **32**, 487–96.

Ayala, F.J. (1982) Gradualism versus punctualism in speciation: Reproductive isolation, morphology, genetics, in *Mechanisms of Speciation*, (ed C. Barigozzi), Alan R. Liss, New York, pp. 51–66.

Ayala, F.J. and Kiger, J.A. (1980) *Modern Genetics*, Benjamin Cummings, Menlo Park.

Ayala, F.J. and Valentine, J.W. (1979) *Evolving: The Theory and Process of Organic Evolution*, Benjamin Cummings, Menlo Park.

Azariah, J. (1982) Cephalocordata, in *Synopsis and Classification of Living Organisms*, (ed S.P. Parker), McGraw-Hill, New York, Vol. 2, pp. 829–30.

Baccetti, B. (1989) The spermatozoon of Strepsiptera (Insecta) and its value in the systematic position of the group. *J.Submicrosc. Cytol. Pathol.*, **21**, 397–8.

Baer, K.E. von (1828) *Entwicklungsgeschichte der Thiere: Beobachtung und Reflexion*, Gebrüder Bornträger, Königsberg.

Bailey, W.J., Slightom, J.L. and Goodman, M. (1992) Rejection of the 'Flying Primate' hypothesis by phylogenetic evidence from the ε-globin gene. *Science*, **256**, 86–9.

Baker, A.N., Rowe, F.W.E. and Clark, H.E.S. (1986) A new class of Echinodermata from New Zealand. *Nature*, **321**, 862–4.

Baker, H.G. (1959) Reproductive methods as factors in speciation in flowering plants. *Cold Spring Harbor Symp. Quant. Biol.*, **24**, 177–91.

Baker, R.J., Novacek, M.J. and Simmons, N.B. (1991) On the monophyly of bats. *Syst. Zool.*, **40**, 216–31.

Balows, A., Trüper, H.G., Dworkin, M. *et al.* (eds) (1992) *The Prokaryotes*, 2nd edn, 4 vols, Springer, Berlin.

Bardack, D. (1991) First fossil hagfish (Myxinoidea): A record from the Pennysylvanian of Illinois. *Science*, **254**, 701–3.

Barel, C.D.N. (ed) (1986) *The Decline of Lake Victoria's Cichlid Species Flock*, Research-Group Ecological Morphology, Zoologisch Laboratorium, Leiden.

Barigozzi, C. (ed) (1982) *Mechanisms of Speciation*. Alan R. Liss, New York.

Barnes, R.D. (1985) Current perspectives on the origins and relationships of lower invertebrates, in *The Origins and Relationships of Lower Invertebrates. The Systematics Association Special Volume No. 28*, (eds S. Conway Morris *et al.*), Clarendon Press, Oxford, pp. 360–7.

Baroin, A., Perasso, R., Qu, L.H. *et al.* (1988) Partial phylogeny of the unicellular eukaryotes based on rapid sequencing of a portion of 28S ribosomal RNA. *Proc. Natl. Acad. Sci. USA*, **85**, 3474–8.

Barr, A.R. (1980) Cytoplasmatic incompatibility in natural populations of a mosquito, *Culex pipiens* L. *Nature*, **283**, 71–2.

Barr, D.J.S. (1983) The zoosporic grouping of fungi – Entity or nonentity, in *Zoosporic Plant Pathogens – A Modern Perspective*, (ed S.T. Buczacki), Academic Press, New York, pp. 43–83.

Barrett, M., Donoghue, M.J. and Sober, E. (1991) Against consensus. *Syst. Zool.*, **40**, 486–93.

Barton, N.H. (1989) Founder effect speciation, in *Speciation and its Consequences*, (eds D. Otte and J.A. Endler), Sinauer Associates, Sunderland, MA, pp. 229–56.

Barton, N.H. and Charlesworth, B. (1984) Genetic revolutions, founder effects, and speciation. *Annu. Rev. Ecol. Syst.*, **15**, 133–64.

Barton, N.H. and Hewitt, G.M. (1981a) Hybrid zones and speciation, in *Evolution and Speciation, Essays in Honor of M.J.D. White*, (eds W.R. Atchley and D.S. Woodruff), Cambridge University Press, Cambridge, pp. 109–45.

Barton, N.H. and Hewitt, G.M. (1981b) The genetic basis of hybrid inviability between two chromosomal races of the grasshopper *Podisma pedestris*. *Heredity*, **47**, 367–83.

Barton, N.H. and Hewitt, G.M. (1989) Adaptation, speciation and hybrid zones. *Nature*, **341**, 497–503.

Baum, D.A. and Larson, A. (1991) Adaptation reviewed: A phylogenetic methodology for studying character macroevolution. *Syst. Zool.*, **40**, 1–18.

Baverstock, P.R. and Adams, M. (1987) Comparative rates of molecular chromosomal and morphological evolution in some Australian vertebrates, in *Rates of Evolution*, (eds K.S.W. Campbell and M.F. Day), Allen and Unwin, London, pp. 175–88.

Beckenbach, A.T., Thomas, W.K. and Sohrabi, H. (1990) Intraspecific sequence variation in the mitochondrial genome of rainbow trout (*Onchorhynchus mykiss*). *Genome*, **33**, 13–15.

De Beer, G.R. (1958) *Embryos and Ancestors*, 3rd edn. Clarendon Press, Oxford.

Beier, M. (ed) (1962–) *Orthopterorum Catalogus*, W. Junk, The Hague.

Belk, D. (1982) Branchiopoda, in *Synopsis and Classification of Living Organisms*, (ed S.P. Parker), McGraw-Hill, New York, **2**, pp. 174–80.

Bell, G. (1989) A comparative method. *Am. Nat.*, **133**, 553–77.

Bell, M.A. (1987) Interacting evolutionary constraints in pelvic reduction of three-spine sticklebacks, *Gasterosteus aculeatus* (Pisces, Gasterosteidae). *Biol. J. Linn. Soc.*, **31**, 347–82.

Belosludov, B.A. and Bashanov, A.V. (1938) A new genus and species of rodent from the Central Kazakstan. *Uchen. Zap. Kazak. Univ. Alma-Ata, Biol.*, **1**, 81–6 (in Russian).

Benado, M., Aguilera, M., Rejè, O.A. and Ayala, F. (1979) Biochemical genetics of chromosome forms of Venezuelan spiny rats of the *Proechimys guairae* and *Proechimys trinitatus* superspecies. *Genetica*, **50**, 89–97.

Benedix, J.H. Jr. and Howard, D.J. (1991) Calling song displacement in a zone of overlap and hybridization. *Evolution*, **45**, 1751–9.

Benito, J. (1982) Hemichordata, in *Synopsis and Classification of Living Organisms*, (ed S.P. Parker), McGraw-Hill, New York, **2**, pp. 819–21.

Bentham, G. and Hooker, J.D. (1862–1883) *Genera Plantarum*, London, 3 vols.

Benton, M.J. (ed) (1988a) *The Phylogeny and Classification of the Tetrapods. The Systematics Association Special Volume No. 35*, Clarendon Press, Oxford.

Benton, M.J. (1988b) The relationships of the major group (sic) of mammals: New approaches. *Trends Ecol. Evolut.*, **3**, 40–5.

Bergquist, P.R. (1985) Poriferan relationships, in *The Origins and Relationships of Lower Invertebrates. The Systematics Association Special Volume No. 28*, (eds S. Conway Morris *et al.*), Clarendon Press, Oxford, pp. 14–27.

Bergström, J. (1991) Metazoan evolution around the Precambrian–Cambrian transition, in *The Early Evolution of Metazoa and the Significance of Problematic Taxa*, (eds A.M. Simonetta and S. Conway Morris), Cambridge University Press, Cambridge, pp. 25–34.

Berlin, B. (1992) *Ethnobiological Classification. Principles of Categorization of Plants and Animals in Traditional Societies*, Princeton University Press, Princeton.

Berlin, B., Breedlove, D. and Raven, P. (1973) Folk taxonomies and biological classifications. *Science*, **154**, 273–5.

Bernardi, G. (1980) Les catégories taxonomiques de la systématique évolutive, in *Les Problèmes de l'Espèce dans le Régne Animal*, (eds Ch. Bocquet, J. Génermont and M. Lamotte), *Mém. Soc. Zool. France*, **40**, pp. 373–425.

Bernhauer, M. and Scheerpeltz, O. (1926) Staphylinidae. III. *Coleopterorum Catalogus*, **82**, 499–988.

Berzin, A.A. and Vladimirov, V.L. (1983) A new species of killer whale (Cetacea, Delphinidae) from the Antarctic waters. *Zool. Zhurnal*, **62**, 287–95 (in Russian; English summary).

Bickel, D.J. (1982) Diptera in *Synopsis and Classification of Living Organisms*, (ed S.P. Parker), McGraw-Hill, New York, **2**, 563–99.

Bisby, F.A. (1984) Information services in taxonomy, in *Databases in Systematics*, (eds R. Allkin and F.A. Bisby), Academic Press, London, pp. 17–33.

Bisby, F.A. (1988) Communications in taxonomy, in *Prospects in Systematics*, (ed D.L. Hawksworth), Clarendon Press, Oxford, pp. 277–91.

Bishop, M.J. and Friday, A.E. (1988) Estimating the interrelationships of tetrapod groups on the basis of molecular sequence data, in *The Phylogeny and Classification of the Tetrapods. Systematics Association Special Volume No. 35A*, (ed M.J. Benton), Clarendon Press, Oxford, pp. 33–58.

Blackwelder, R.E. (1952) The generic names of the beetle family Staphylinidae, with an essay on genotypy. *Bull. U.S. Nat. Mus.*, **200**, 483 pp.

Blackwelder, R.E. (1967) *Taxonomy, A Text and Reference Book*, Wiley, New York.

Blair, W.F. (1943) Criteria for species and their subdivisions from the point of view of genetics. *Ann. New York Acad. Sci.*, **44**, 178–88.

Blake, D.B. (1987) A classification and phylogeny of post–Palaeozoic sea stars (Asteroidea, Echinodermata). *J. Nat. Hist.*, **21**, 481-528.

Blanchard, R. (1894) Hirudinées de l'Italie continentale et insulaire. *Boll. Mus. Zool. Anat. Comp. Torino*, **9(192)**, 1–84.

Blouin, M.S., Dame J.B., Tarrant, C.A. and Courtney, C.H. (1922) Unusual population genetics of a parasitic nematode, mtDNA variation within and among populations. *Evolution*, **46**, 470–6.

Boaden, P.J.S. (1985) Why is a gastrotrich? in *The Origins and Relationships of Lower Invertebrates. The Systematics Association Special Volume No. 28*, (eds S. Conway Morris *et al.*), Clarendon Press, Oxford, pp. 248–60.

Bock, W.J. (1977) Foundations and methods of evolutionary classification, in *Major Patterns in Vertebrate Evolution*, (eds M.K. Hecht, P.C. Goody and B. Hecht), Plenum Press, New York, pp. 851–95.

Bock, W.J. (1982a) Aves, in *Synopsis and Classification of Living Organisms*, (ed S.P. Parker), McGraw-Hill, New York, **2**, 967–1015.

Bock, W.J. (1982b) Biological classification, in *Synopsis and Classification of Living Organisms*, (ed S.P. Parker), McGraw-Hill, New York, **2**, 1067–71.

Bock, W.J. (1986) Species concepts, speciation, and macroevolution, in *Modern Aspects of Species*, (eds K. Iwatsuki, P.H. Raven and W.J. Bock), University of Tokyo Press, Tokyo, pp. 31–57.

Bocquet, C., Génermont, J. and Lamotte, M. (eds) (1976, 1977, 1980) *Les problèmes de l'espèce dans le régne animal. Mem. Soc. Zool. France*. 3 vols.

Böcher, T.W. (1967) Continuous variation and taxonomy. *Taxon*, **16**, 255–8.

Böhme, W. (1978) Das Kühneltsche Prinzip der regionalen Stenözie und seine Bedeutung für das Subspezies-Problem, ein theoretischer Ansatz. *Z. Zool. Syst. Evol.-forsch.*, **16**, 256–66.

Boero, F. and Bouillon, J. (1987) Inconsistent evolution and paedomorphosis among the hydroids and medusae of the Athecatae/Anthomedusae and the Thecatae/Leptomedusae (Cnidaria, Hydrozoa), in *Modern Trends in the Systematics, Ecology and Evolution of Hydroids and Hydromedusae*, (eds J. Bouillon *et al.*), Oxford University Press, Oxford, pp. 229–50.

Bohart, R.M. and Menke, A.S. (1976) *Sphecid Wasps of the World, a Generic Revision*, University of California Press, Berkeley, CA.

Bohlken, H. (1961) Haustiere und zoologische Systematik. *Z. Tierzüchtung Züchtungsbiol.*, **76**, 107–13.

Bohm, B.A. and Chan, J. (1992) Flavonoids and affinities of Greyiaceae with a discussion of the occurrence of B-ring deoxyflavonoids in dicotyledonous families. *Syst. Bot.*, **17**, 272–81.

Bold, H.C., Alexopoulos, C.J. and Delevoryas, T. (1980) *Morphology of Plants and Fungi*, Harper and Row, New York.

Bonik, K. (1982) Gibt es Arten bei Diatomeen? Eine evolutionsbiologische Deutung am Beispiel der Gattung, *Nitzschia. Senckenbergiana Biol.*, **62** (1981), 413–34.

Bordeu, T. de (1751) *Recherches Anatomiques sur la Position des Glandes et sur leur Action*, G.F. Quillau, Paris.

Boss, K.J. (1982) Mollusca, in *Synopsis and Classification of Living Organisms*, (ed S.P. Parker), McGraw-Hill, New York, **1**, 945-1166.

Boucek, Z. (1974) A revision of the Leucospidae (Hymenoptera Chalcidoidea) of the world. *Bull. Br. Mus. (Nat. Hist.) Entomol., Suppl.*, **23**, 1–241.

Boucomont, A. and Gillet, J.J.E. (1927) Scarabaeidae, Coprinae II. Termitotrogidae. *Coleopterorum Catalogus*, **90**, 103–264.

Boudreaux, H.B. (1979) *Arthropod Phylogeny with Special Reference to Insects*, Wiley, New York.

Bourne, W.R.P., Nelson, G.J., Eldredge, N. and Harlow, G.E. (1988) Cuts at British Museum (NH). *Nature*, **333**, 292.

Bousfield, E.L. (1982) Thermosbaenacea. Spelaeogriphacea, in *Synopsis and Classification of Living Organisms*, (ed S.P. Parker), McGraw-Hill, New York, **2**, 241–2.

Bovee, E.C. (1971) The Trypanosomatidae: Possibly descendant of euglenids? *J. Protozool.*, **18**, suppl. 14.

Bowman, T.E. (1982a) Mystacocarida, in *Synopsis and Classification of Living Organisms*, (ed S.P. Parker), McGraw-Hill, New York, **2**, 202.

Bowman, T.E. (1982b) Cephalocarida, in *Synopsis and Classification of Living Organisms*, (ed S.P. Parker), McGraw-Hill, New York, **2**, 174.

Bowman, T.E., Garner., S.P., Hessler, R.R. *et al.* (1985) Mictacea, a new order of Crustacea Peracarida. *J. Crust. Biol.*, **5**, 74–8.

Bowman, T.E. and Iliffe, T.M. (1985) *Mictocaris halope*, a new unusual peracaridan crustacean from marine caves in Bermuda. *J. Crust. Biol.* , **5**, 58–73.

Boxshall, G.A. (1983) A comparative functional analysis of the major maxillopodan groups, in *Crustacean Phylogeny (Crustacean Issues, 1)*, (ed F.R. Schram), A.A. Balkema, Rotterdam, pp. 121–43.

Brandenburg, W.A. (1991) The need for stabilized plant names in agriculture and horticulture, in *Improving the Stability of Names, Needs and Options. Regnum Vegetabile No. 123*, (ed D.L. Hawksworth), Koeltz, Königstein, pp. 23–31.

Brandl, R., Mann, W. and Sprinzl, M. (1992) Mitochondrial tRNA and the phylogenetic position of Nematoda. *Biochem. Syst. Ecol.*, **20**, 325–30.

Bremer, B. (1991) Restriction data from chloroplast DNA for phylogenetic reconstruction, is there only one accurate way of scoring? *Plant Syst. Evol.*, **175**, 39–54.

Bremer, K. (1985) Summary of green plant phylogeny and classification. *Cladistics*, **1**, 369–85.

Bremer, K. (1987) Tribal interrelationships of the Asteraceae. *Cladistics*, **3**, 210–53.

Bremer, K. and Wanntorp, H.E. (1981) A cladistic classification of green plants. *Nord. J. Bot.*, **1**, 1–3.

Bretfeld, G. (1986) Phylogenetic systematics of the higher taxa of Symphypleona Börner, 1901 (Insecta, Entognatha, Collembola), in *2nd International Seminar on Apterygota*, (ed R. Dallai), University of Siena, Siena, pp. 307–11.

Brickell, C.D., Richens, R.H., Kelly, A.F. *et al.* (eds) (1980) *International Code of Nomenclature of Cultivated Plants 1980*, Bohn, Scheltema and Holkema, Deventer.

Briggs, D. and Walters, S.M. (1984) *Plant Variation and Evolution*, 2nd edn, Cambridge University Press, Cambridge.

Briggs, D.E.G. (1992) Conodonts: A major extinct group added to the vertebrates. *Science*, **256**, 1285–6.

Briggs, D.E.G. and Fortey, R.A. (1989) The early radiation and relationships of the major arthropod groups. *Science*, **246**, 241–3.

Briggs, D.E.G., Fortey, R.A. and Wills, M.A. (1992) Morphological disparity in the Cambrian. *Science*, **256**, 1670–3.

Bright, P.M. and Leeds, H.A. (1938) *A Monograph of the British Aberrations of the Chalk-hill Blue Butterfly* Lysandra coridon *(Poda, 1761)*, Bournemouth.

Brignoli, P.M. (1983) *A Catalogue of the Araneae Described Between 1940 and 1981*, Manchester University Press, Manchester.

Brinkhurst, R.O. (1980) Postscript, in *Aquatic Oligochaete Biology*, (eds R.O. Brinkhurst and D.G. Cook), Plenum, New York, pp. 507–19.

Brinkhurst, R.O. and Jamieson, B.G.M. (1971) *Aquatic Oligochaeta of the World*, University of Toronto Press, Toronto.

Brinkhurst, R.O. and Nemec, F.L. (1987) A comparison of phenetic and phylogenetic methods applied to the systematics of Oligochaeta. *Hydrobiologia*, **155**, 65–74.

Brinkhurst, R.O. and Wetzel, M.J. (1984) *Aquatic Oligochaeta of the World: Supplement. A Catalogue of New Freshwater Species, Descriptions, and Revisions*, Canad. Techn. Rep. Hydrogr. and Ocean Sciences.

Britschgi, T.B. and Giovannoni, S.J. (1991) Phylogenetic analysis of a natural marine bakterioplankton population by rRNA gene cloning and sequencing. *Appl. Environ. Microbiol.*, **57**, 1707–13.

Brochmann, C., Soltis, P.S. and Soltis, D.E. (1992a) Recurrent formation and polyphyly of Nordic polyploids in *Draba* (Brassicaceae). *Am. J. Bot.*, **79**, 673–88.

Brochmann, C., Soltis, P.S. and Soltis, D.E. (1992b) Multiple origins of the octoploid Scandinavian endemic *Draba cacuminum*, electrophoretic and morphological evidence. *Nord. J. Bot.*, **12**, 257–72.

Brookfield, J.F.Y. (1986) Genetic changes in speciation. *Trends Ecol. Evol.*, **1**, 117–18.

Brooks, D.R. (1982) Higher level classification of parasitic Platyhelminthes and fundamentals of cestode classification, in *Parasites – Their World and Ours*, (eds D.F. Mettrick and S.S. Desser), Elsevier Biomedical Press, Amsterdam, pp. 189–93.

Brooks, D.R. (1989a) A summary of the database pertaining to the phylogeny of the major groups of parasitic platyhelminths, with a revised classification. *Can. J. Zool.*, **67**, 714–20.

Brooks, D.R. (1989b) The phylogeny of the Cercomeria (Platyhelminthes, Rhabdocoela) and general evolutionary principles. *J. Parasitol.*, **75**, 606–16.

Brooks, D.R., Bandoni, S.M., Macdonald, Ch.A. and O'Grady, R.T. (1989) Aspects of the phylogeny of the Trematoda Rudolphi, 1808 (Platyhelminthes, Cercomeria). *Can. J. Zool.*, **67**, 2609–24.

Brooks, D.R. and McLennan, D.A. (1991) *Phylogeny, Ecology, and Behavior*, The University of Chicago Press, Chicago.

Broom, R. (1896) On a small fossil Marsupial with large premolars. *Proc. Linn. Soc. N.S. Wales*, **20**, 553–67.

Brown, W.L. Jr. (1982a) Zoraptera, in *Synopsis and Classification of Living Organisms*, (ed S.P. Parker), McGraw-Hill, New York, **2**, 393.

Brown, W.L. Jr. (1982b) Mecoptera in *Synopsis and Classification of Living Organisms*, (ed S.P. Parker), McGraw-Hill, New York, **2**, 555–7.

Brown, W.L. Jr. (1982c) Strepsiptera, in *Synopsis and Classification of Living Organisms*, (ed S.P. Parker), McGraw-Hill, New York, **2**, 553–5.

Brown, W.L. Jr. (1982d) Hymenoptera, in *Synopsis and Classification of Living Organisms*, (ed S.P. Parker), McGraw-Hill, New York, **2**, 652–80.

Brown, W.M. and Wright, J.W. (1979) Mitochondrial DNA analyses and the origin and relative age of parthenogenetic lizards (genus *Cnemidophorus*). *Science*, **203**, 1247–9.

Brundin, L. (1972) Evolution, causal biology, and classification. *Zool. Scr.*, **1**, 107–20.

Brundin, L. (1976) A Neocomian Chironomid and Podonomidae-Aphroteniinae (Diptera) in the light of phylogenetics and biogeography. *Zool. Scr.*, **5**, 139–60.

Brusca, R.C. and Wilson, G.D.F. (1991) A phylogenetic analysis of the Isopoda with some classificatory recommendations. *Mem. Queensl. Mus.*, **31**, 143–204.

Bryant, H.N. (1989) An evaluation of cladistic and character analyses as hypothetico-deductive procedures and the consequences for character weighting. *Syst. Zool.*, **38**, 214–27.

Bütschli, O. (1884) Bemerkungen zur Gastraeatheorie. *Morph. Jb.*, **9**, 415–27.

Bullini, L. (1985) Speciation by hybridization in animals. *Boll. Zool.*, **52**, 121–37.

Burger-Wiersma, T., Veenhuis, M., Korthals, H.J. *et al.* (1986) A new prokaryote containing chlorophylls *a* and *b*. *Nature*, **320**, 262–4.

Burlando, B. (1990) The fractal dimension of taxonomic systems. *J. Theor. Biol.*, **146**, 99–146.

Burtt, B.L. (1970) Infraspecific categories in flowering plants. *Biol. J. Linn. Soc.*, **2**, 233–8.

Bush, G.L. (1969) Sympatric host race formation and speciation in frugivorous flies of genus *Rhagoletis* (Diptera, Tephritidae). *Evolution*, **23**, 237–51.

Bush, G.L. (1975a) Modes of animal speciation. *Annu. Rev. Ecol. Syst.*, **6**, 339–64.

Bush, G.L. (1975b) Sympatric speciation in phytophagous parasitic insects, in *Evolutionary Strategies of Parasitic Insects and Mites*, (ed P.W. Price), Plenum, New York, pp. 187–206.

Busslinger, M., Rusconi, S. and Birnstiel, M.L. (1982) An unusual evolutionary behaviour of a sea urchin histone gene cluster. *EMBO J.*, **1**, 27–33.

Butlin, R. (1987a) Speciation by reinforcement. *Trends Ecol. Evol.*, **2**, 8–13.

Butlin, R. (1987b) A new approach to sympatric speciation. *Trends Ecol. Evol.*, **2**, 310–11.

Butlin, R.K. (1987c) Species, speciation and reinforcement. *Am. Nat.*, **130**, 461–4.

Caesalpinus, A. (1583) *De Plantis Libri XVI*, Apud Georgium Marescottum, Florentiae.

Cain, A.J. (1954) *Animal Species and their Evolution*, Hutchinson University Library, London.

Cain, A.J. (1958) Logic and memory in Linnaeus' system of taxonomy. *Proc. Linn. Soc. Lond.*, **169**, 144–63.

Caisse, M. and Antonovics, J. (1978) Evolution in closely adjacent plant populations, IX. Evolution of reproductive isolation in clinal populations. *Heredity*, **40**, 371–84.

Calman, R.T. (1909) Crustacea, in *A Treatise on Zoology*, (ed R. Lankester), Adam and Charles Black, London, **7**, 346 pp.

Camin, J.H. and Sokal, R.R. (1965) A method for deducing branching sequences in phylogeny. *Evolution*, **19**, 311–26.

Camp, W.H. (1951) Biosystematy. *Brittonia*, **7**, 113–27.

Cantino, P.D. (1982) Affinities of the Lamiales: A cladistic analysis. *Syst. Bot.*, **7**, 237–48.

Capanna, E. (1982) Robertsonian numerical variation in animal speciation, *Mus musculus*, an emblematic model, in *Mechanisms of Speciation*, (ed C. Barigozzi), Alan R. Liss, New York, pp. 155–77.

Carpenter, J.M. (1989) Testing scenarios: Wasp social behavior. *Cladistics*, **5**, 131–44.

Carpenter, J.M. (1992) Distances, assumptions and social wasps. *Cladistics*, **8**, 155–60.

Carr, S.M., Ballinger, S.W., Derr, J.N. *et al.* (1986) Mitochondrial DNA analysis of hybridization between sympatric whitetailed and mule deer in West Texas. *Proc. Natl. Acad. Sci. USA*, **83**, 9576–80.

Carroll, R.L. (1987) *Vertebrate Palaeontology and Evolution*, W.H. Freeman, New York.

Carson H.L. (1971) Speciation and the founder principle. *Stadler Genet. Symp.*, **3**, 51–70.

Carson, H.L. (1973) Reorganization of the gene pool during speciation, in *Genetic Structure of Populations*, (ed N.E. Morton), University of Hawaii Press, Honolulu, pp. 274–80.

Carson, H.L. (1975) The genetics of speciation at the diploid level. *Am. Nat.*, **109**, 83–92.

Carson, H.L. (1982) Speciation as a major reorganization of polygenic balances, in *Mechanisms of Speciation*, (ed C. Barigozzi), Alan R. Liss, New York, pp. 411–33.

Carson, H.L. (1985) Unification of speciation theory in plants and animals. *Syst. Bot.*, **10**, 380–90.

Carson, H.L. (1989) Genetic imbalance, realigned selection, and the origin of species, in *Genetics, Speciation, and the Founder Principle*, (eds L.V. Giddings, K.Y. Kaneshiro and W.W. Anderson), Oxford University Press, New York, pp. 345–62.

Cartmill, M. (1981) Hypothesis testing and phylogenetic reconstruction. *Z. Zool. Syst. Evol.-forsch.*, **19**, 73–96.

Carvalho, J.C.M. (1957–1960) *A Catalogue of the Miridae of the World*, Arq. Mus. Nac. (Brazil), Rio de Janeiro, **I–V**.

Catzeflis, F.M., Aguilar, J.-P. and Jaeger, J.-J. (1992) Muroid rodents, phylogeny and evolution. *Trends Ecol. Evol.*, **7**, 122–6.

Cavalier-Smith, T. (1978) The evolutionary origin an phylogeny of microtubles, mitotic spindles and eukaryote flagella. *BioSystems*, **10**, 93–114.

Cavalier-Smith, T. (1981) Eukaryotic kingdoms, seven or nine? *BioSystems*, **14**, 461–81.

Cavalier-Smith, T. (1987) The simultaneous symbiotic origin of mitochondria, chloroplasts and microbodies. *Ann. N. Y. Acad. Sci.*, **503**, 55–71.

Cavalier-Smith, T. (1989) Archaebacteria and Archezoa. *Nature*, **339**, 100–1.

Cavalier-Smith, T. (1991) Intron phylogeny, a new hypothesis. *Trends Genet.*, **7**, 145–8.

Cavender, J.A. (1978) Taxonomy with confidence. *Math. Biosci.*, **40**, 271–80.

Cesalpino, A. (1583) *De Plantis*, Florentiae.

Chandler, C.R. and Gromko, M.H. (1989) On the relationship between species concepts and speciation processes. *Syst. Zool.*, **38**, 116–25.

Chang, M.M. and Yu, X. B. (1983) Structure and phylogenetic significance of *Diabolichthys speratus* gen. et sp. nov., a new dipnoan-like form from the Lower Devonian of Eastern Yunnan, China. *Proc. Linn. Soc. N.S.W.* **107**, 171–84.

Charig, A.J. (1982) Systematics in biology: A fundamental comparison of some major schools of thought, in *Problems of Phylogenetic Reconstruction. The Systematics Association Special Volume No. 21*, (eds K.A. Joysey and A.E. Friday), Academic Press, London, pp. 363–440.

Chase, M.W. and Palmer, J.D. (1988) Chloroplast DNA variation, geographical distribution, and morphological parallelism in subtribe Oncidiinae (Orchidaceae). *Am. J. Bot.*, **75**, 163–4.

Chase, M.W. and Palmer, J.D. (1992) Floral morphology and chromosome number in subtribe Oncidiinae (Orchidaceae): Evolutionary insights from a phylogenetic analysis of chloroplast DNA restriction site variation, in *Molecular Systematics of Plants*, (eds P.S. Soltis, D.E. Soltis and J.J. Doyle), Chapman & Hall, New York, pp. 324–39.

Chatterjee, S. (1991) Cranial anatomy and relationships of a new Triassic bird from Texas. *Phil. Trans. R. Soc. London*, B, **332**, 277–342.

Chauvet, M. (1991) The needs for stability in names for germoplasm conservation; in *Improving the Stability of Names, Needs and Options. Regnum Vegetabile No. 123*, (ed D.L. Hawksworth), Koeltz, Königstein, pp. 33–8.

Chen, J.-Y. and Erdtmann, B.-D. (1991) Lower Cambrian fossil Lagerstätte from Chengjiang, Yunnan, China. Insights for reconstructing early metazoan life, in *The Early Evolution of Metazoa and the Significance of Problematic Taxa*, (eds A.M. Simonetta and S. Conway Morris), Cambridge University Press, Cambridge, pp. 57–76.

Cheverud, J.M., Dow, M.M. and Leutenegger, W. (1985) The quantitative assessment of phylogenetic constraints in comparative dimorphism in body weight among primates. *Evolution*, **39**, 1335–51.

Christen, N.R., Ratto, A., Baroin, A. *et al.* (1991a) An analysis of the origin of metazoans, using comparisons of partial sequences of the 28S RNA, reveals an early emergence of triploblasts. *EMBO J.*, **10**, 499–503.

Christen, N.R., Ratto, A., Baroin, A. *et al.* (1991b) Origin of metazoans. A phylogeny deduced from sequences of the 28S ribosomal RNA, in *The Early Evolution of Metazoa and the Significance of Problematic Taxa*, (eds A.M. Simonetta and S. Conway Morris), Cambridge University Press, Cambridge, pp. 1–9.

Christensen, T. (1962) Alger, in *Botanik 2, Systematisk Botanik, pt. 1*, (eds T.W. Böcher, M. Lange and T. Sørensen), Munksgaard, Copenhagen.

Claridge, M.F. (1985a) Acoustic signals in the Homoptera: Behaviour, taxonomy and evolution. *Ann. Rev. Entomol.* **30**, 297–317.

Claridge, M.F. (1985b) Acoustic behaviour of leafhoppers and planthoppers, Species problems and speciation, in *The Leafhoppers and Planthoppers*, (eds L.R. Nault and J.G. Rodriguez), Wiley, New York, pp. 103–25.

Claridge, M.F. (1988) Species concepts and speciation in parasites, in *Prospects in Systematics*, (ed D.L. Hawksworth), Clarendon Press, Oxford, pp. 92–111.

Claridge, M.F., den Hollander, J. and Morgan, J.C. (1985) Variation in courtship signals and hybridization between geographically definable populations of the rice brown planthopper, *Nilaparvata lugens* (Stål). *Biol. J. Linn. Soc.*, **24**, 35–49.

Claridge, M.F. and Nixon, G.A. (1986) *Oncopsis flavicollis* (L.) associated with tree birches (*Betula*). A complex of biological species or a host plant utilization polymorphism? *Biol. J. Linn. Soc.*, **27**, 381–97.

Clausen, J., Keck, D.D. and Hiesey, W.M. (1945) *Experimental Studies on the Nature of Species. II.* Plant Evolution through Amphiploidy and Autoploidy, with Examples from the Madiineae. Carnegie Inst. Washington Publ., **564**.

Clay, T. (1970) The Amblycera (Phthiraptera, Insecta). *Bull. Br. Mus. Nat. Hist. (Entomol.)*, **25**, 73–98.

Clayton, W.D. (1972) Some aspects of the genus concept. *Kew Bull.*, **27**, 281–7.

Clayton, W.D. (1974) The logarithmic distribution of Angiosperm families. *Kew Bull.*, **29**, 271–9.

Cleevely, R.J. (1983) *World Palaeontological Collections, an Index of Collectors and Collections*, British Museum (Natural History), London.

Clément, P. (1985) The relationships of rotifers, in *The Origins and Relationships of Lower Invertebrates. The Systematics Association Special Volume No. 28*, (eds S. Conway Morris *et al.*), Clarendon Press, Oxford, pp. 224–47.

Clifford, H.T., Rogers, R.W. and Dettmann, M.E. (1990) Where now for taxonomy? *Nature*, **346**, 602.

Coddington, J.A. (1986) The monophyletic origin of the orb web, in *Spiders, Webs, Behavior and Evolution*, (ed W.A. Shear), Stanford University Press, Stanford, pp. 319–63.

Coddington, J.A. (1988) Cladistic tests of adaptational hypotheses. *Cladistics*, **4**, 1–20.

Coddington, J.A. (1990) Bridges between evolutionary pattern and process. *Cladistics*, **6**, 379–86.

Coddington, J.A. and Levi, H.W. (1991) Systematics and the evolution of spiders (Araneae). *Annu. Rev. Ecol. Syst.*, **22**, 565–92.

Coluzzi, M. (1982) Spatial distribution of chromosomal inversions and speciation in anopheline mosquitoes, in *Mechanisms of Speciation*, (ed C. Barigozzi), Alan R. Liss, New York, pp. 143–53.

Connor, E.F. (1986) The role of Pleistocene forest refugia in the evolution and biogeography of tropical biotas. *Trends Ecol. Evol.*, **1**, 165–8.

Conway Morris, S. (1977) Fossil priapulid worms. *Spec. Pap. Palaeontol.*, **20**, 1–95.

Conway Morris, S. (1989) Burgess Shale faunas and the Cambrian explosion. *Science*, **246**, 339–46.

Conway Morris, S., George, J.D., Gibson, R. and Platt, H.M. (eds) (1985) *The Origins and Relationships of Lower Invertebrates. Systematics Association Special Volume No. 28*, Clarendon Press, Oxford.

Coomans, A. (1981) Aspects of the phylogeny of nematodes, in *Origine dei Grandi Phyla dei Metazoi*, Atti dei Convegni Lincei, Rome, **49**, pp. 161–74.

Cooper, A., Mourer-Chauviré, C., Chambers, G.K. *et al.* (1992) Independent origins of New Zealand moas and kiwis. *Proc. Natl. Acad. Sci. USA*, **89**, 8741–4.

Copeland, H.F. (1938) The kingdoms of organisms. *Quart. Rev. Biol.*, **13**, 383–420.

Corbet, G.B. and Clutton-Brock, J. (1984) Taxonomy and nomenclature, in *Evolution of Domesticated Animals*, (ed I.L. Mason), Longman, London, pp. 434–8.

Corbet, G.B. and Hill, J.E. (1986) *A World List of Mammalian Species*, British Museum (Natural History), London.

Corliss, J.O. (1974) Time for evolutionary biologists to take more interest in protozoan phylogenetics? *Taxon*, **23**, 497–522.

Corliss, J.O. (1979) *The Ciliated Protozoa, Characterization, Classification, and Guide to the Literature*, 2nd edn, Pergamon Press, London.

Corliss, J.O. (1983) Consequences of creating new kingdoms of organisms. *Bioscience*, **33**, 314–18.

Corliss, J.O. (1984) The Kingdom Protista and its 45 phyla. *BioSystems*, **17**, 87–126.

Corliss, J.O. (1986a) The kingdoms of organisms – from a microscopist's point of view. *Trans. Am. Microsc. Soc.*, **105**, 1–10.

Corliss, J.O. (1986b) Progress in protistology during the first decade following reemergence of the field as a respectable interdisciplinary area in modern biological research. *Prog. Protistol.*, **1**, 11–63.

Corliss, J.O. (1986c) The development of ciliate systematics from the era of Jozsef Gelei until the present time. *Symp. Biol. Hung.*, **33**, 67–86.

Corliss, J.O. (1987) Protistan phylogeny and eukaryogenesis. *Int. Rev. Cytol.*, **100**, 319–70.

Corliss, J.O. (1990) Toward a nomenclatural protist perspective, in *The Handbook of Protoctists*, (eds L. Margulis *et al.*), Jones and Bartlett, Boston, pp. XXV–XXX.

Corliss, J.O. (unpublished) Some proposed 'working' classification systems for Protists involving 'Protozoa', 'Algae', and 'Lower Fungi'.

Corner, E.J.H. (1981) Angiosperm classification and phylogeny: A criticism. *Bot. J. Linn. Soc.*, **82**, 81–7.

Coyne, J.A. (1992) Genetics and speciation. *Nature*, **355**, 511–15.

Coyne, J.A. and Kreitman, M. (1986) Evolutionary genetics of two sibling species, *Drosophila simulans* and *D. sechellia. Evolution*, **40**, 673–91.

Coyne, J.A. and Orr, H.A. (1989) Patterns of speciation in *Drosophila. Evolution*, **43**, 362–81.

Cracraft, J. (1979) Phylogenetic analysis, evolutionary models and paleontology, in

Phylogenetic Analysis and Paleontology, (eds J. Cracraft and N. Eldredge), Columbia University Press, New York, pp. 7–39.

Cracraft, J. (1981) Toward a phylogenetic classification of the recent birds of the world. *Auk*, **98**, 691–714.

Cracraft, J. (1983) Species concepts and speciation analysis. *Curr. Ornithol.* **1**, 159–87.

Cracraft, J. (1985) Early evolution of birds, in *A Dictionary of Birds*, (eds B. Campbell and E. Lack), Calton, T. and A.D. Poyser for The British Ornithologists' Union, pp. 163–5.

Cracraft, J. (1986) The origin and early diversification of birds. *Paleobiology*, **12**, 383–99.

Cracraft, J. (1987a) Species concepts and the ontology of evolution. *Biol. Philos.*, **2**, 329–46.

Cracraft, J. (1987b) DNA hybridization and avian phylogenetics, in *Evolutionary Biology*, (eds M.K. Hecht, B. Wallace and G.T. Prance), Plenum, New York, **21**, pp. 47–96.

Cracraft, J. (1988) The major clades of birds, in *The Phylogeny and Classification of the Tetrapods. Systematics Association Special Volume No. 35A*, (ed M.J. Benton), Clarendon Press, Oxford, **1**, pp. 339–61.

Cracraft, J. (1922) The species of the birds-of-paradise (Paradisaeidae): Applying the phylogenetic species concept to a complex pattern of diversification. *Cladistics*, **8**, 1–43.

Cracraft, J. and Mindell, D.P. (1989) The early history of modern birds: A comparison of molecular and morphological evidence, in *The Hierarchy of Life*, (eds B. Fernholm, K. Bremer and H. Jörnvall), Elsevier, Amsterdam, pp. 389–403.

Crampton, G.M. (1929) The terminal abdominal structures of female insects compared throughout the orders from the standpoint of phylogeny. *J. New York Entomol. Soc.*, **37**, 453–96.

Crane, P.R. (1985a) Phylogenetic analysis of seed plants and the origin of angiosperms. *Ann. Missouri Bot. Garden*, **72**, 716–93.

Crane, P.R. (1985b) Phylogenetic relationships in seed plants. *Cladistics*, **1**, 329–48.

Craske, A.J. and Jefferies, R.P.S. (1989) A new mitrate from the Upper Ordovician of Norway, and a new approach to subdividing a plesion. *Palaeontology*, **32**, 69–99.

Crease, T.J., Lynch, M. and Spitze, K. (1990) Hierarchical analysis of population genetic variation in mitochondrial and nuclear genes of *Daphnia pulex*. *Mol. Biol. Evol.*, **7**, 444–58.

Creighton, G.K. and Strauss, R.E. (1986) Comparative patterns of growth and development in Cricetine rodents and the evolution of ontogeny. *Evolution*, **40**, 94–106.

Crezée, M. (1982) Turbellaria, in *Synopsis and Classification of Living Organisms*, (ed S.P. Parker), McGraw-Hill, New York, **1**, pp. 718–40.

Crisp, D. and Fogg, G.E. (1988) Taxonomic instability continues to irritate. *Nature*, **335**, 120–1.

Cronk, Q.C.B. (1989) Measurement of biological and historical influences in plant classifications. *Taxon*, **38**, 357–70.

Cronquist, A. (1969) On the relationship between taxonomy and evolution. *Taxon*, **18**, 177–87.

Cronquist, A. (1981) *An Integrated System of Classification of Flowering Plants*, Columbia University Press, New York.

Cronquist, A. (1982) Magnoliophyta, in *Synopsis and Classification of Living Organisms*, (ed S.P. Parker), McGraw-Hill, New York, **1**, pp. 357–487.

Cronquist, A. (1987) A botanical critique of cladism. *Bot. Rev.*, **53**, 1–52.

Cronquist, A. (1988) *The Evolution and Classification of Flowering Plants*, 2nd edn, The New York Botanical Garden, New York.

Crosskey, R.W., Cogan, B.R., Freeman, P. *et al.* (eds) (1980) *Catalogue of the Diptera of the Afrotropical Regional*, British Museum (Natural History), London.

Crowson, R.A. (1955) *The Natural Classification of the Families of Coleoptera*, N. Lloyd, London.

Crowson, R.A. (1960) The phylogeny of Coleoptera. *Annu. Rev. Entomol.*, **5**, 111–34.

Crowson, R.A. (1970) *Classification and Biology*, Heinemann, London.

Crowson, R.A. (1981) *The Biology of the Coleoptera*, Academic Press, New York.

Crum, H.A. (1985) Traditional make-do taxonomy. *The Bryologist*, , **88**, 21–2.

Culberson, W.L., Culberson, C.F. and Johnson, A. (1977) Correlations between secondary-product chemistry and ecogeography in the *Ramalina siliquosa* complex. *Plant Syst. Evol.*, **127**, 191–200.

Cunningham, C.W., Blackstone, N.W. and Buss, L.W. (1992) Evolution of king crabs from hermit crab ancestors. *Nature*, **355**, 539–42.

Cunningham Vaught, K. (1989) *A Classification of the Living Mollusca*, American Malacologists Inc, Florida.

Cutler, E.B. (1982) Pogonophora, in *Synopsis and Classification of Living Organisms*, (ed S.P. Parker), McGraw-Hill, New York, Vol. 2, pp. 63–4.

Cuvier, G. (1799–1805) *Leçons d'Anatomie Comparée*, 5 vols, Baudoin, Paris.

Cuvier, G. (1812) Sur un nouveau rapprochment à établir entre les classes qui composent le règne animal. *Ann. Mus. Hist. Nat. Paris*, **19**, 173–84.

Dahl, E. (1983) Malacostracan phylogeny and evolution, in *Crustacean Phylogeny (Crustacean Issues, 1)*, (ed F.R. Schram), A.A. Balkema, Rotterdam, pp. 189–212.

Dahlgren, G. (1989a) An updated angiosperm classification. *Bot. J. Linn. Soc.*, **100**, 197–203.

Dahlgren, G. (1989b) The last Dahlgrenogram. System of classification of the Dicotyledons, in *Plant Taxonomy, Phytogeography and Related Subjects. The Davis and Hedge Festschrift*, (ed K. Tan), Edinburgh University Press, Edinburgh, pp. 249–60.

Dahlgren, R. (1975) A system of classification of the angiosperms to be used to demonstrate the distribution of characters. *Bot. Not.*, **128**, 119–47.

Dahlgren, R.M.T. (1980) A revised system of classification of the angiosperms. *Bot. J. Linn. Soc.*, **80**, 91–124.

Dahlgren, R. (1983) General aspects of angiosperm evolution and macrosystematics. *Nordic J. Bot.*, **3**, 119–49.

Dahlgren, R. and Bremer, K. (1985) Major clades of angiosperms. *Cladistics*, **1**, 349–68.

Dahlgren, R., Clifford, H.T. and Yeo, P.F. (1985) *The Families of the Monocotyledons*, Springer, Berlin.

Dance, S.P. (1986) *A History of Shell Collecting*, E.J. Brill/Dr. W. Backhuys, Leiden.

Danser, B.H. (1929) Ueber die Begriffe Komparium, Kommiskuum und Konvivium und die Entstehungsweise der Konvivien. *Genetica*, **11**, 399–450.

Danser, B.H. (1950) A theory of systematics. *Bibliotheca Biotheoretica*, **4**, 117–80.

Darwin, C. (1859) *On the Origin of Species by Means of Natural Selection, or the Preservation of Favored Races in the Struggle for Life*, John Murray, London.

Daugherty, C.H., Cree, A., Hay, J.M. *et al.* (1990) Neglected taxonomy and continuing extinctions of tuatara (*Sphenodon*). *Nature*, **347**, 177–9.

Davidson, E.H., Hough-Evans, B.R. and Britten, R.J. (1982) Molecular biology of the sea urchin embryo. *Science*, **217**, 17–26.

Davies, D.A.L. and Tobin, P. (1984–5) The dragonflies of the world: a systematic list of the extant species of Odonata. *Soc. Internat. Odonatologica, Rapid Communications*, **3**, 127 pp.; **5**, 151 pp.

Davis, P.H. and Heywood, V.H. (1963) *Principles of Angiosperm Taxonomy*, Oliver and Boyd, Edinburgh.

Davis, J.I. and Gilmartin, A.J. (1985) Morphological variation and speciation. *Syst. Bot.*, **10**, 417–25.

Dawley, R.M. and Bogart, J.P. (eds), *Evolution and Ecology of Unisexual Vertebrates*, New York State Museum, Albany.

Dayhoff, M.O. (ed) (1969) *Atlas of Protein Sequences and Structure*, National Biomedical Research Foundation, Silverspring.

DeBry, R.W. and Slade, N.A. (1985) Cladistic analysis of restriction endonuclease cleavage maps within a maximum-likelihood framework. *Syst. Zool.*, **34**, 21–34.

De Candolle, A.P. (1817–1821) *Regni Vegetabilis Systema Naturale, sive Ordines, Genera et Species Plantarum Secundum Methodi Naturalis Normas Digestarum et Descriptarum*, Paris, 2 vols.

De Candolle, A.P. (1824–1874) *Prodromus Systematis Naturalis Regni Vegetabilis, sive Enumeratio Contracta Ordinum, Generum, Specierumque Plantarum Hucusque Cognitarum Juxta Methodi Naturalis Normas Digesta*, Paris, 17 vols.

Dehalu, M. and Leclercq, J. (1951) Applications des séries logarithmiques de Fisher-Williams à la classification des Hyménoptères Crabroniens. *Ann. Soc. R. Zool. Belg.*, **82**, 67–82.

Delfinado, M.D. and Hardy, D.E. (eds) (1973–1977) *A Catalog of the Diptera of the Oriental Region*, 3 vols, University of Hawaii Press, Honolulu.

Delle Cave, L. and Simonetta, A.M. (1991) Early Palaeozoic arthropods and problems of arthropod phylogeny; with some notes on taxa of doubtful affinities, in *The Early Evolution of Metazoa and the Significance of Problematic Taxa*, (eds A.M. Simonetta and S. Conway Morris), Cambridge University Press, Cambridge, pp. 189-244.

DeLong, E.F. (1992) Archaea in coastal marine enviornments. *Proc. Natl. Acad. Sci. USA*, **89**, 5685–9.

De Pamphilis, C.W. and Palmer, J.D. (1990) Loss of photosynthetic and chlororespiratory genes from the plastid genome of a parasitic flowering plant. *Nature*, **348**, 337–9.

De Pinna, M.C.C. (1992) Concepts and tests of homology in the cladistic paradigm. *Cladistics*, **7**, 367–94.

De Salle, R., Freedman, T., Praeger, E.M. *et al.* (1987) Tempo and mode of sequence evolution in mitochondrial DNA of Hawaiian *Drosophila*. *J. Mol. Evol.*, **26**, 157–64.

De Salle, R., Gatesy, J., Wheeler, W. *et al.* (1992) DNA sequences from a fossil termite in Oligo-Miocene amber and their phylogenetic implications. *Science*, **257**, 1933–6.

Dettner, K. (1987) Chemosystematics and evolution of beetle chemical defenses. *Ann. Rev. Entomol.*, **32**, 17–48.

Dextre Clarke, S.G. (1988) The use and future of bibliographic database systems, in *Prospects in Systematics, The Systematics Association Special Volume No. 36*, (ed D.L. Hawksworth), Clarendon Press, Oxford, pp. 305–14.

Dial, K.P. and Marzluff, J.M. (1989) Nonrandom diversification within taxonomic assemblages. *Syst. Zool.*, **38**, 26–37.

Diamond, J. (1985) How many unknown species are yet to be discovered? *Nature*, **315**, 538–9.

Diamond, J.M. (1990) Old dead rats are valuable. *Nature*, **347**, 334–5.

Didier, R. and Séguy, E. (1953) *Catalogue Illustré des Lucanides du Globe*, Lechevalier, Paris, Encyclopédie Entomol., **27**.

Dobzhansky, T. (1937) *Genetics and the Origin of Species*, Columbia University Press, New York.

Dobzhansky, T. (1951) *Genetics and the Origin of Species*, Third Edition, Columbia University Press, New York.

Dobzhansky, T. (1970) *Genetics of the Evolutionary Process*, Columbia University Press, New York.

Doebley, J. (1992) Molecular systematics and crop evolution, in *Molecular Systematics of Plants*, (eds P.S. Soltis, D.E. Soltis and J.J. Doyle), Chapman & Hall, New York, pp. 202–22.

Dohle, W. (1980) Sind die Myriapoden eine monophyletische Gruppe? Eine Diskussion der Verwandtschaftsbeziehungen der Antennaten. *Abh. Naturwiss. Ver. Hamburg.*, **23**, 45–104.

Dohle, W. (1985) Phylogenetic pathways in the Chilopoda. *Bijdr. Dierkunde*, **55**, 55–66.

Dohle, W. (1988) *Myriapoda and the Ancestry of Insects. The Charles H. Brookes Memorial Lecture*, The Manchester Polytechnic in collaboration with the British Myriapod Group, Manchester, 28 pp.

Dohle, W. (1990) Some observations on morphology and affinities of *Craterostigmus tasmanianus* (Chilopoda), in *Proceedings of the 7th International Congress of Myriapodology*, (ed A. Minelli), E.J. Brill, Leiden, pp. 69–79.

Doll, R. (1982) Grundriss der Evolution der Gattung *Taraxacum* Zinn. *Feddes Rep.*, **93**, 481–624.

Dolling, W.R. (1991) *The Hemiptera*, Oxford University Press (Natural History Museum Publications), Oxford.

Donoghue, M.J. (1985) A critique of the biological species concept and recommendations for a phylogenetic alternative. *Bryologist*, **88**, 172–81.

Donoghue, M.J. (1990) Sociology, selection and success, a critique of David Hull's analysis of science and systematics. *Biol. Philoso.*, **5**, 459–72.

Donoghue, M.J. and Cantino, P.D. (1988) Paraphyly, ancestors and the goals of taxonomy: A botanical defense of cladism. *Bot. Rev.*, **54**, 107–28.

Donoghue, M.J. and Doyle, J.A. (1989) Phylogenetic analysis of angiosperms and the relationships of Hamamelidae, in *Evolution, Systematics, and Fossil History of the Hamamelidae. Vol. 1: Introduction and 'Lower' Hamamelidae*, (eds P.R. Crane and S. Blackmore), Clarendon Press, Oxford, pp. 17–45.

Donoghue, M.J., Doyle, J.A., Gauthier, J. *et al.* (1989) The importance of fossils in phylogeny reconstruction. *Annu. Rev. Ecol. Syst.*, **20**, 431–60.

Donoghue, M.J. and Sanderson, M.J. (1992) The suitability of molecular and morphological evidence in reconstructing plant phylogeny, in *Molecular Plant Systematics*, (eds P.S. Soltis, D.E. Soltis and J.J. Doyle), Chapman & Hall, New York, pp. 340–68.

Dorado, O., Rieseberg, L.H. and Arias, D.M. (1992) Chloroplast DNA introgression in Southern California sunflowers. *Evolution*, **46**, 566–72.

Dougherty, E.C., Gordon, H.T. and Allen, M.B. (1957) The absence of α-ε-diamino-pimelic acid from the primitive red alga, *Porphyridium cruentum*. *Exp. Cell Res.*, **13**, 171–3.

Douglas, S.E., Murphy, C.A., Spencer, D.F. *et al.* (1991) Cryptomonad algae are evolutionary chimaeras of two phylogenetically distinct unicellular eukaryotes. *Nature*, **350**, 148–51.

Dover, G. (1982a) Molecular drive: A cohesive mode of species evolution. *Nature*, **299**, 111–17.

Dover, G. (1982b) A role for the genome in the origin of species, in *Mechanisms of Speciation*, (ed C. Barigozzi), Alan R. Liss, New York, pp. 435–59.

Dover, G.A. (1988a) rDNA world falling to pieces. *Nature* **336**, 623–4.

Dover, G.A. (1988b) The new genetics, in *Prospects in Systematics. The Systematic Association Special Volume No. 36*, (ed D.L. Hawksworth), Clarendon Press, Oxford, pp. 151–68.

Dover, G., Brown, S., Coen, E. *et al.* (1982) The dynamics of genome evolution and species differentiation, in *Genome Evolution*, (eds G.A. Dover and R.B. Flavell), Academic Press, London, pp. 343–74.

Dowling, T.E., Moritz, C. and Palmer, J.D. (1990) Nucleic acids II: Restriction site analysis, in *Molecular Systematics*, (eds D.M. Hillis and C. Moritz), Sinauer Associates, Sunderland, MA, pp. 250–317.

Downie, S.R. and Palmer, J.D. (1992) Use of chloroplast rearrangements in reconstructing plant phylogeny, in *Molecular Systematics of Plants*, (eds P.S. Soltis, D.E. Soltis and J.J. Doyle), Chapman & Hall, New York, pp. 14–35.

Doyle, J.A. (1978) Origin of angiosperms. *Annu. Rev. Ecol. Syst.*, **9**, 365–92.

Doyle, J.A. and Donoghue, M.J. (1987a) The origin of angiosperms: A cladistic approach, in *The Origins of Angiosperms and their Biological Consequences*, (eds E.M. Friis, W.G. Chaloner and P.R. Crane), Cambridge University Press, Cambridge, pp. 17–49.

Doyle, J.A. and Donoghue, M.J. (1987b) Seed plant phylogeny and the origin of angiosperms: An experimental cladistic approach. *Bot. Rev.*, **52**, 321–431.

Doyle, J.J. (1992) Gene trees and species trees: Molecular systematics as one-character taxonomy. *Syst. Bot.*, **17**, 144–63.

Doyle, J.J., Lavin, M. and Bruneau, A. (1992) Contributions of molecular data to papilionoid legume systematics, in *Molecular Systematics of Plants*, (eds P.S. Soltis, D.E. Soltis and J.J. Doyle), Chapman & Hall, New York, pp. 223–51.

Drake, C.J. and Ruhoff, F.A. (1965) Lacebugs of the world: A catalog (Hemiptera: Tingidae). *U.S. Mus. Bull.*, **243**, 1–634.

Dressler, R.L. (1981) *The Orchids. Natural History and Classification*, Harvard University Press, Cambridge, MA.

Drouet, F. (1978) Revision of the Nostocaceae with constricted trichomes. *Beihefte zur Nova Hedwigia*, **57**, 369 pp.

Drouet, F. (1981) Revision of the Stigonemataceae with a summary of the classification of the blue-green algae. *Beihefte zur Nova Hedwigia*, **66**, 221 pp.

Dubois, A. (1977) Les problèmes de l'espèce chez les Amphibiens Anoures, in *Les Problèmes de l'Espèce dans le Règne Animal*, (eds Ch. Bocquet, J. Génermont and M. Lamotte), *Mém. Soc. Zool. France*, **39**, 161–284.

Dubois, A. (1982) Les notions de genre, sous-genre et group d'espèce en zoologie à la lumière de la systématique évolutive. *Monit. Zool. Ital.*, **16**, 9–63.

Dubois, A. (1986) A propos de l'emploi controversé du terme 'monophylétique': nouvelles propositions. *Bull. Soc. Linn. Lyon*, **55**, 248–54.

Dubois, A. (1988) Le genre en zoologie: Essai de systématique théorique. *Mém. Mus. Natl. Hist. Nat.*, Paris, **139**, 1–130.

Duellman, W.E. (1982) Reptilia, in *Synopsis and Classification of Living Organisms*, (ed S.P. Parker), McGraw-Hill, New York, **2**, 955–66.

Duffels, J.P. and Van der Laan, P.A. (1985) *Catalogue of the Cicadoidea* (Homoptera, Auchenorhyncha) *1956–1980*. W. Yunk, Dordrecht.

Dujardin, F. (1965) Description de sous-espèces et formes nouvelles de *Zygaena* F. d'Europe occidentale, méridionale et d'Afrique du Nord. *Entomops*, **1**, 16–22; **2**, 33–64.

Duméril, A. (1806) *Zoologie Analitique*, Allais Paris.

Duncan, T. and Stuessy, T.F. (eds) (1984) *Cladistics: Perspectives on the Reconstruction of Evolutionary History*. Columbia University Press, New York.

Dupuis, C. (1979) Permanence et actualité de la Systématique: La Systématique: La 'Systématique phylogénétique' de W. Hennig (Historique, discussion, choix de références). *Cah. Nat.*, **34**, 1–69.

Dupuis, C. (1984) Willi Hennig's impact on taxonomic thought. *Annu. Rev. Ecol. Syst.*, 15, 1–24.

Dwight, J. (1909) The burden of nomenclature. *Science*, **30**, 526–7.

Dzik, J. (1991) Is fossil evidence consistent with traditional views of the early metazoan phylogeny? in *The Early Evolution of Metazoa and the Significance of Problematic Taxa*, (eds A.M. Simonetta and S. Conway Morris), Cambridge University Press, Cambridge, pp. 47–56.

Eastop, V.F. and Lambers, D.H.R. (1976) *Survey of the World's Aphids*, W. Junk, The Hague.

Eberhard, W.G. (1990) Function and phylogeny of spider webs. *Annu. Rev. Ecol. Syst.*, 21, 341–72.

Echelle, A.A. and Kornfield, I. (eds) (1984) *Evolution of Fish Species Flocks*, University of Maine Press, Orono, Maine.

Edmonds, S.J. (1982) Echiura, in *Synopsis and Classifications of Living Organisms*, (ed S.P. Parker), McGraw-Hill, New York, **1**, 65–6.

Edmunds, G.F. (1982) Ephemeroptera, in *Synopsis and Classification of Living Organisms*, (ed S.P. Parker), McGraw-Hill, New York, **2**, 330–8.

Edwards, A.W.F. and Cavalli Sforza, L.L. (1964) Reconstruction of evolutionary trees, in *Phenetic and Phylogenetic classification*, (eds V.H. Heywood and J. McNeill), Systematic Association, London, pp. 67–76.

Edwards, P. (1976) A classification of plants into higher taxa based on cytological and biochemical criteria. *Taxon*, **25**, 529–42.

Ehlers, U. (1985) *Das phylogenetische System der Plathelminthes*, Gustav Fischer, Stuttgart.

Ehrendorfer, F. (1970) Evolutionary patterns and strategies in seed plants. *Taxon*, **19**, 185–95.

Eichler, W. (1989) Phthiraptera, in *Exkursionsfauna für die Gebiete de DDR und der BRD. 8. Auflage*, (ed E. Stresemann), Volk and Wissen Volkseigener Verlag, Berlin, 2/1, pp. 126–31.

Eigen, M., Winkler-Oswatisch, R. and Dress, A. (1988) Statistical geometry in sequence space: a method of quantitative comparative sequence analysis. *Proc. Natl. Acad. Sci. USA*, **85**, 5913–17.

Eisner, T., Alsop, D., Hicks, K. and Meinwald, J. (1978) Defensive secretions of millipeds, in *Arthropod Venoms (Handbook of Experimental Pharmacology. 48)*, (ed S. Bettini), Springer, Berlin, pp. 41–72.

Eldredge, N. (1979) Cladism and common sense, in *Phylogenetic Analysis and Paleontology*, (eds J. Cracraft and N. Eldredge), Columbia University Press, New York, pp. 165–98.

Eldredge, N. (1989) *Macroevolutionary Dynamics. Species, Niches, and Adaptive Peaks*, McGraw-Hill, New York.

Eldredge, N. and Cracraft, J. (1980) *Phylogenetic Patterns and the Evolutionary Process. Methods and Theory in Comparative Biology*, Columbia University Press, New York.

Elkington, T.T. (1986) Patterns of variation in wild plants in relation to selection pressures and infraspecific categories, in *Infraspecific Classification of Wild and Cultivated Plants. The Systematics Association Special Volume No. 29*, (ed B.T. Styles), Clarendon Press, Oxford, pp. 33–52.

Ellgren, H. (1991) DNA typing of museum birds. *Nature*, **354**, 113.

Emerson, M.J. and Schram, F.R. (1990) A new view of arthropod evolution. *Am. Zool.*, **30**, 19A.

Emig, Ch.C. (1982) Phoronida, in *Synopsis and Classification of Living Organisms*, (ed S.P. Parker), McGraw-Hill, New York, **2**, 741.

Endler, J.A. (1977) *Geographic Variation, Speciation and Clines. Monographs in Population Biology. 10*, Princeton University Press, Princeton, NY.

Endler, J.A. (1986) *Natural Selection in the Wild. Monographs in Population Biology.* **21**, Princeton University Press, Princeton, NY.

Endler, J.A. (1989) Conceptual and other problems in speciation, in *Speciation and its Consequences*, (eds D. Otte and J.A. Endler), Sinauer Associates, Sunderland, MA, pp. 625–48.

Engel, J.J. (1982) Hepaticopsida, in *Synopsis and Classification of Living Organisms*, (ed S.P. Parker), McGraw-Hill, New York, **1**, 271–304.

Enghoff, H. (1982) The millipede genus *Cylindroiulus* on Madera – an insular species swarm (Diplopoda, Julida: Julidae). *Entomol. Scand. Suppl.*, **18**, 1–142.

Enghoff, H. (1983) Adaptive radiation of the millipede genus *Cylindroiulus* on Madera: Habitat, body size and morphology (Diplopoda, Julida: Julidae). *Rev. Ecol. Biol. Sol*, **20**, 403–15.

Enghoff, H. (1984) Phylogeny of millipedes – a cladistic analysis. *Z. Zool. Syst. Evol.-forsch.*, **22**, 8–26.

Enghoff, H. (1990) The ground-plan of the chilognathan millipedes (external morphology), in *Proceedings of the 7th International Congress of Myriapodology*, (ed A. Minelli), E.J. Brill, Leiden, pp. 1–21.

Enghoff, H. (1992) Macaronesian millipedes (Diplopoda) with emphasis on endemic species swarms on Madeira and the Canary Islands. *Biol. J. Linn. Soc.*, **46**, 153–61.

Eriksson, O. (1981) The families of bitunicate Ascomycetes. *Opera Bot.*, **60**, 1–220.

Eriksson, O.E. (1989) Eclectic mycology: A response to Reynolds. *Taxon*, **38**, 64–7.

Eriksson, O. and Hawksworth, D.L. (1986) Outline of the ascomycetes – 1986. *Systema Ascomycetum*, **5**, 185–324.

Erséus, C. (1987) Phylogenetic analysis of the aquatic Oligochaeta under the principle of parsimony. *Hydrobiologia*, **15**, 75–89.

Erwin, T.L. (1982) Tropical forests: Their richness in Coleoptera and other arthropod species. *Coleopterists Bull.*, **36**, 74–5.

Erwin, T.L. (1988) The tropical forest canopy. The heart of biotic diversity, in *Biodiversity*, (ed E.O. Wilson), National Academy Press, Washington, pp. 123–9.

Eschmeyer, W.N. (1990) *Catalog of the Genera of Recent Fishes*, California Academy of Sciences, San Francisco.

Esslinger, T.L. (1977) A chemosystematic revision of the brown Parmeliae. *J. Hattori Bot. Lab.*, **42**, 1–211.

Estabrook, G.F. (1972) Cladistic methodology: A discussion of the theoretical basis for the induction of evolutionary history. *Annu. Rev. Ecol. Syst.*, **3**, 427–56.

Estabrook, G.F. (1983) The causes of character incompatibility, in *Numerical Taxonomy. NATO ASI Series Vol. G1*, (ed J. Felsenstein), Springer, Berlin, pp. 279–95.

Estabrook, G.F., Johnson, C.S., Jr. and McMorris, F.R. (1975) An idealized concept of the true cladistic character. *Math. Biosci.*, **23**, 263–72.

Estabrook, G.F., Johnson, C.S., Jr. and McMorris, F.R. (1976a) A mathematical foundation for the analysis of character compatibility. *Math. Biosci.*, **29**, 181–7.

Estabrook, G.F., Johnson, C.S., Jr. and McMorris, F.R. (1976b) An algebraic analysis of cladistic characters. *Discrete Math.*, **16**, 141–7.

Estes, R. (1988) Recent perspectives on phylogenetic relationships of reptiles, in

Symposium on the Evolution of Terrestrial Vertebrates, University of Naples, Naples, abstract 21.

Evenhuis, N.L. (1989) *Catalog of the Diptera of the Australasian and Oceanic Regions*.

Fahrenholz, H. (1913) Ectoparasiten und Abstammungslehre. *Zool. Anz.*, **41**, 371–4.

Farr, E.R, Leussink, J.A. and Stafleu, F.A. (eds) (1979) *Index Nominum Genericorum (Plantarum)*, International Association for Plant Taxonomy, The Hague.

Farr, E.R., Leussink, J.A. and Zijlstra, G. (eds) (1986) *Supplement to Index Nominum Genericorum (Plantarum)*, International Association for Plant Taxonomy, The Hague.

Farris, J.S. (1969) A successive approximations approach to character weighting. *Syst. Zool.*, **18**, 374–85.

Farris, J.S. (1970) Methods for computing Wagner trees. *Syst. Zool.*, **19**, 83–92.

Farris, J.S. (1973) A probabilistic model for inferring evolutionary trees. *Syst. Zool.*, **22**, 250–6.

Farris, J.S. (1974) Formal definitions of paraphyly and polyphyly. *Syst. Zool.*, **23**, 548–54.

Farris, J.S. (1977) Phylogenetic analysis under Dollo's Law. *Syst. Zool.*, **26**, 77–88.

Farris, J.S. (1983) The logical basis of phylogenetic analysis, in *Advances in Cladistics, Volume 2: Proceedings of the Second Meeting of the Willi Hennig Society*, (eds N.I. Platnick and V.A. Funk), Columbia University Press, New York, pp. 7–36.

Farris, J.S. (1989) The retention index and the rescaled consistency index. *Cladistics*, **5**, 417–19.

Farris, J.S. (1990a) Haeckel, History, and Hull. *Syst. Zool.*, **39**, 81–8.

Farris, J.S. (1990b) Phenetics in camouflage. *Cladistics*, **6**, 91–100.

Farris, J.S. (1991) Hennig defined paraphyly. *Cladistics*, **7**, 297–304.

Farris, J.S., Kluge, A.G. and Eckhardt, M.J. (1970) A numerical approach to phylogenetic systematics. *Syst. Zool.*, **19**, 172–91.

Fauchald, K. (1977) The polychaete worms: Definitions and keys to the orders, families and genera. *Nat. Hist. Mus. Los Angeles Co. Sci. Ser.*, **28**, 1–190.

Fauchald, K. (1990) The uses of types: An object lesson from the study of palolo worm. *Am. Zool.*, **30**, 60A.

Feder, J.L. and Bush, G.L. (1991) Genetic variation among apple and hawthorn host races of *Rhagoletis pomonella* across an ecological transition zone in the Mid-Western United States. *Ent. Exp. Appl.*, **59**, 249–65.

Feder, J.L., Chilcote, C.A. and Bush, G.L. (1988) Genetic differentiation between sympatric host races of the apple maggotfly *Rhagoletis pomonella*. *Nature*, **336**, 61–4.

Fell, F.J. (1982) Echinodermata (p.p.), in *Synopsis and Classification of Living Organisms*, (ed S.P. Parker), McGraw-Hill, New York, **2**, 785–813.

Felsenstein, J. (1973) Maximum likelihood and minimum-step methods for estimating evolutionary trees from data on discrete characters. *Syst. Zool.*, **22**, 240–9.

Felsenstein, J. (1978a) Cases in which parsimony or compatibility methods will be positively misleading. *Syst. Zool.*, **27**, 401–10.

Felsenstein, J. (1978b) The number of evolutionary trees. *Syst. Zool.*, **27**, 27–33.

Felsenstein, J. (1979) Alternative methods of phylogenetic inference and their interrelationships. *Syst. Zool.*, **28**, 49–62.

Felsenstein, J. (1981a) A likelihood approach to character weighting and what it tells us about parsimony and compatibility. *Biol. J. Linn. Soc.*, **16**, 183–96.

Felsenstein, J. (1981b) Evolutionary trees from gene frequencies and quantitative characters: Finding maximum likelihood estimates. *Evolution*, **35**, 1229–42.

Felsenstein, J. (1985a) Confidence limits on phylogenies: An approach using the bootstrap. *Evolution*, **39**, 783–91.

Felsenstein, J. (1985b) Phylogenies and the comparative method. *Am. Nat.*, **125**, 1–15.

Felsenstein, J. (1988a) The detection of phylogeny, in *Prospects in Systematics*, (ed D.L. Hawksworth), Clarendon Press, Oxford, pp. 112–27.

Felsenstein, J. (1988b) Phylogenies from molecular sequences: Influence and reliability. *Annu. Rev. Genet.*, **22**, 521–65.

Ferguson, A. (1988) Isozyme studies and their interpretation, in *Prospects in Systematics*, (ed D.L. Hawksworth), Clarendon Press, Oxford, pp. 184–201.

Fiedler, K. (1991) Fische, in *A. Kaestner's Lehrbuch der Speziellen Zoologie. Band II. Wirbeltiere*, 2, (ed D. Starck), Gustav Fischer, Jena.

Field, K.G., Olsen, G.J., Lane, D.J. *et al.* (1988) Molecular phylogeny of the animal kingdom. *Science*, **239**, 748–53.

Field, K.G., Turbeville, J.M., Raff, R.A. and Best, B. (1990) Evolutionary relationships of the Phylum Cnidaria inferred from 18S rRNA sequence data. *ICSEB IV, Fourth International Congress of Systematic and Evolutionary Biology*, College Park, Abstracts.

Figueroa, F., Günther, E. and Klein, J. (1988) MHC polymorphism pre-dating speciation. *Nature*, **335**, 265–7.

Fioroni, P. (1987) *Allgemeine und vergleichende Embryologie der Tiere. Ein Lehrbuch*, Springer, Berlin.

Fischer, F.C.J. (1960–73) *Trichopterorum Catalogues*, Nederl. Entomol. Vereen., Amsterdam, **I–XV**.

Fitch, W.M. (1970) Distinguishing homologous from analogous proteins. *Syst. Zool.*, **19**, 99–113.

Fitch, W.M. (1971) Toward defining the course of evolution: Minimal change for a specific tree topology. *Syst. Zool.*, **20**, 406–16.

Fitch, W.M. and Margoliash, E. (1967) Construction of phylogenetic trees. *Science*, **155**, 279–84.

Fitzpatrick, J.W. (1988) Why so many passerine birds? A response to Raikov. *Syst. Zool.*, **37**, 71–6.

Flynn, J.J. (1988) Ancestry of sea mammals. *Nature*, **334**, 383–4.

Flynn, J.J., Neff, N.A. and Jedford, R.H. (1988) Phylogeny of the Carnivora, in *The Phylogeny and Classification of the Tetrapods. Systematics Association Special Volume No. 35B*, (ed M.J. Benton), Clarendon Press, Oxford, pp. 73–115.

Ford, E.B. (1964) *Ecological Genetics*, 2nd edn, Methuen, London.

Forey, P. (1987) Relationships of lungfishes. *J. Morphol., Suppl.*, **1**, 75–91.

Forey, P. (1991) Blood lines of the coelacanth. *Nature*, **351**, 347–8.

Fortey, R.A. and Jefferies, R.P.S. (1982) Fossils and phylogeny – a compromise approach, in *Problems of Phylogenetic Reconstruction. Systematics Association Special Volume No. 21*, (eds K.A. Joysey and A.E. Friday), Academic Press, London, pp. 197–234.

Fortey, R.A. and Whittington, H.B. (1989) The Trilobita as a natural group. *Historical Biol.*, **2**, 125–38.

Foster, B.A. and Buckeridge, J.S. (1987) Barnacle palaeontology, in *Barnacle Biology (Crustacean Issues, 5)*, (ed A.J. Southward), A.A. Balkema, Rotterdam, pp. 43–61.

Foster, M.W. (1982) Brachiopoda, in *Synopsis and Classification of Living Organisms*, (ed S.P. Parker), McGraw-Hill, New York, Vol. 2, 773–80.

Fott, B. (1974) The phylogeny of eucaryotic algae. *Taxon*, **23**, 449–61.

Fox, G.E., Pechman, K.R. and Woese, C.R. (1977) Comparative cataloging of 16S rRNA: Molecular approach to prokaryotic systematics. *Int. J. Syst. Bacteriol.*, **27**, 44–57.

Fox, R.C., Youzwyshyn, G.P. and Krause, D.W. (1992) Post-Jurassic mammal-like reptile from the Palaeocene. *Nature*, **358**, 233–5.

Francke, O.F. (1982) Scorpiones, in *Synopsis and Classification of Living Organisms*, (ed S.P. Parker), McGraw-Hill, New York, **2**, 73–5.

Frank, J.H. and Curtis, G.A. (1979) Trend lines and the number of species of Staphylinidae. *Coleopterists Bull.*, **33**, 133–49.

Franki, R.I.B., Fauquet, C.M., Knudson, D.L. and Brown, F. (eds) (1991) *Classification and Nomenclature of Viruses. Arch. Virol. Suppl.* **2**, 1–450.

Friedmann, E.I. (1982) Cyanophycota, in *Synopsis and Classification of Living Organisms*, (ed S.P. Parker), McGraw-Hill, New York, **1**, 45–52.

Frost, D.R. (ed) (1985) *Amphibian Species of the World*, Allen Press, Lawrence, Kansas, and The Association of Systematics Collections.

Frost, D.R. and Wright, J.W. (1988) The taxonomy of uniparental species, with special reference to parthenogenetic *Cnemidophorus* (Squamata, Teiidae). *Syst. Zool.*, **37**, 200–9.

Frost, J.S. and Platz, J.E. (1983) Comparative assessment of modes of reproductive isolation among four species of leopard frogs (*Rana pipiens* complex). *Evolution*, **37**, 66–78.

Fryer, G. (1982) Branchiura, in *Synopsis and Classification of Living Organisms*, (ed S.P. Parker), McGraw-Hill, New York, **2**, 218–20.

Fryer, G. (1987) A new classification of the branchiopod Crustacea. *Zool. J. Linn. Soc.*, **91**, 357–83.

Fuhrman, J.A., McCallum, K. and Davis, A.A. (1992) Novel major archaebacterial group from marine plankton. *Nature*, **356**, 148–9.

Funk, V.A. and Brooks, D.R. (1990) *Phylogenetic Systematics as the Basis of Comparative Biology*, Smithsonian Institution Press, Washington.

Futuyma, D.J. (1989) Speciational trends and the role of species in macroevolution. *Am. Nat.*, **134**, 318–21.

Futuyma, D.J. and Mayer, G.C. (1980) Non-allopatric speciation in animals. *Syst. Zool.*, **29**, 254–71.

Futuyma, D.J. and Philippi, T.E. (1987) Genetic variation and covariation responses to host plants by *Alsophila pometaria* (Lepidoptera: Geometridae). *Evolution*, **41**, 269–79.

Futuyma, D.J. and Slatkin, M. (eds) (1983) *Coevolution*, Sinauer Associates, Sunderland, MA.

Gaffney, E.S. and Meylan, P.A. (1988) A phylogeny of turtles, in *The Phylogeny and Classification of the Tetrapods. Systematics Association Special Volume No. 35A*, (ed M.J. Benton), Clarendon Press, Oxford, **1**, 157–219.

Gajadhar, A.A., Marquardt, W.C., Hall, R. *et al.* (1991) Evolutionary relationships among apicomplexans, dinoflagellates, and ciliates: ribosomal RNA sequences of *Sarcocystis muris, Theileria annulata* and *Crypthecodinium cohnii. Mol. Biochem. Parasitol.*, **45**, 147–54.

Gale, A.S. (1987) Phylogeny and classification of the Asteroidea (Echinodermata). *Zool. J. Linn. Soc.*, **89**, 107–32.

Gardiner, B. (1982) Tetrapod classification. *Zool. J. Linn. Soc.*, **74**, 207–32.

Gaston, K.J. and May, R.M. (1992) Taxonomy of taxonomists. *Nature*, **356**, 281–2.

Gates, R.R. (1951) The taxonomic units in relation to cytogenetics and gene ecology. *Am. Nat.*, **85**, 31–50.

Gauld, I.D. and Bolton, B. (eds) (1988) *The Hymenoptera*, British Museum (Natural History), London and Oxford University Press, Oxford.

Gauld, I.D. and Mound, L.A. (1982) Homoplasy and the delineation of holophyletic genera in some insect groups. *Syst. Entomol.*, **7**, 73–86.

Gauthier, J., Cannatella, D., de Queiroz, K. *et al.* (1989) Tetrapod phylogeny, in *The*

Hierarchy of Life, (eds B. Fernholm, K. Bremer and H. Jörnvall), Elsevier, Amsterdam, pp. 337–53.

Gauthier, J., Kluge, A.G. and Rowe, T. (1988a) Amniote phylogeny and the importance of fossils. *Cladistics*, **4**, 105–209.

Gauthier, J.A., Kluge, A.G. and Rowe, T. (1988b) The early evolution of the Amniota, in *The Phylogeny and Classification of the Tetrapods. Systematics Association Special Volume No. 35A*, (ed M.J. Benton), Clarendon Press, Oxford, pp. 103–155.

Gee, H. (1988) Friends and relations. *Nature*, **334**, 13–14.

Génermont, J. (1980) Les animaux à reproduction uniparentale, in *Les Problèmes de l'Espèce dans le Régne Animal* (eds Ch. Bocquet, J. Génermont and M. Lamotte), *Mém. Soc. Zool. de France*, **40**, 287–320.

Gentry, A.H. and Dodson, C.H. (1987) Diversity and biogeography of neotropical vascular epiphytes. *Ann. Missouri Bot. Gard.*, **74**, 205–33.

Geoffroy Saint-Hilaire, E. (1830) *Principles de Philosophie Zoologique Discutés en Mars 1830, au Sein de l'Académie Royale des Sciences*, Pichon et Didier and Rousseau, Paris.

George, J.D. and Hartmann-Schröder, G. (1985) *Polychaetes: British Amphinomida, Spintherida and Eunicida. Synopses of the British Fauna (New Series) No. 32*, E.J. Brill/Dr. W. Backhuys, London.

Ghirardelli, E. (1968) Some aspects of the biology of the chaetognaths. *Adv. Mar. Biol.*, **6**, 271–375.

Ghiselin, M.T. (1966a) On psychologism in the logic of taxonomic controversies. *Syst. Zool.*, **15**, 207–15.

Ghiselin, M.T. (1966b) Reproductive function and the phylogeny of opistobranch gastropods. *Malacologia*, **3**, 327–78.

Ghiselin, M.T. (1974) A radical solution to the species problem. *Syst. Zool.*, **23**, 536–44.

Ghiselin, M.T. (1981) Categories, life, and thinking. *Behavior Brain Sci.*, **4**, 269–313.

Ghiselin, M.T. (1984a) 'Definition', 'Character', and other equivocal terms. *Syst. Zool.*, **33**, 104–10.

Ghiselin, M.T. (1984b) Narrow approaches to phylogeny: A review of nine books on cladism, in *Oxford Surveys in Evolutionary Biology*, (eds R. Dawkins and M. Ridley), Oxford University Press, Oxford, Vol. 1, pp. 209–222.

Ghiselin, M.T. (1987) Species concepts, individuality, and objectivity. *Biol. Philos.*, **2**, 127–43.

Ghiselin, M.T. (1988a) The origin of molluscs in the light of molecular evidence, in *Oxford Surveys in Evolutionary Biology*, (eds P.H. Harvey and L. Partridge), Oxford University Press, Oxford, Vol. 5, pp. 66–95.

Ghiselin, M.T. (1988b) The individuality thesis, essences, and laws of nature. *Biol. Philos.*, **3**, 467–71.

Ghiselin, M.T. and Lowenstein, J.M. (1988) Phylogeny of echinoderms: Comparison of trees based on radioimmunoassay and 18S ribosomal RNA sequence data. *Am. Zool.*, **28**, 9A.

Giannasi, D.E. (1979) Systematic aspects of flavonoid biosynthesis and evolution. *Bot. Rev.*, **44**, 399–429.

Giannasi, D.E. (1988) Flavonoids and evolution in dicotyledons, in *The Flavonoids, Advances in Research Since 1980*, (ed J.B. Harborne), Chapman & Hall, London, pp. 479–504.

Giannasi, D.E. and Crawford, D.J. (1986) Biochemical systematics. II. A reprise, in *Evolutionary Biology*, (eds M.K. Hecht, B. Wallace and G.T. Prance) Plenum Press, New York, Vol. 20, pp. 25–248.

Gibbons, N.E. and Murray, R.G.E. (1978) Proposals concerning the higher taxa of bacteria. *Int. J. Syst. Bact.*, **28**, 1–6.

Gibbs, R.D. (1974) *Chemotaxonomy of Flowering Plants*, McGill-Queen's University Press, Montreal.

Gibson, G.A.P. (1985) Some pro- and mesothoracic structures important for phylogenetic analysis of Hymenoptera, with a review of terms used for the structures. *Can. Entomol.*, **117**, 1395–443.

Giddings, L.V., Kaneshiro, K.Y. and Anderson, W.W. (eds) (1989) *Genetics, Speciation, and the Founder Principle*, Oxford University Press, Oxford.

Gilmartin, A.J. (1981) Morphological variation within five angiosperm families: Asclepiadaceae, Bromeliaceae, Melastomataceae, Piperaceae, and Rubiaceae. *Syst. Bot.*, **6**, 331–45.

Gilmour, J.S.L. (1940) Taxonomy and phylosophy, in Huxley, J. (ed), *The New Systematics*, (ed J. Huxley), Oxford University Press, Oxford, pp. 461–74.

Gilmour, J.S.L. and Heslop-Harrison, J. (1954) The deme terminology and the units of micro-evolutionary change. *Genetica*, **27**, 147–61.

Giovannoni, S.J., Britschgi, T.B., Moyer, C.L. and Field, K.G. (1990) Genetic diversity in Sargasso Sea bacterioplankton. *Nature*, **345**, 60–3.

Gisin, H. (1967) La systématique idéale. *Z. Zool. Syst. Evol.-forsch.*, **5**, 111–28.

Götting, K.-J. (1980) Origin and relationships of the Mollusca. *Z. Zool. Syst. Evol.-forsch.*, **18**, 24–7.

Goldberg, A. (1986) Classification, evolution, and phylogeny of the families of dicotyledons. *Smithsonian Contrib. Bot.*, **58**, 314 pp.

Goldberg, A. (1989) Classification, evolution, and phylogeny of the families of monocotyledons. *Smithsonian Contrib. Bot.*, **71**, 74 pp.

Goldman, N. and Barton, N.H. (1992) Genetics and geography. *Nature*, **357**, 440–1.

Golenberg, E.M., Giannasi, D.E., Clegg, M.T. *et al.* (1990) Chloroplast sequence from a Miocene *Magnolia* species. *Nature*, **344**, 656–8.

Goloboff, P.A. (1991a) Homoplasy and the choice among cladograms. *Cladistics*, **7**, 215–32.

Goloboff, P.A. (1991b) Random data, homoplasy and information. *Cladistics*, **7**, 395–406.

Goodbody, I. (1982) Tunicata, in *Synopsis and Classification of Living Organisms*, (ed S.P. Parker), McGraw-Hill, New York, Vol. 2, pp. 823–9.

Goodman, M., Miyamoto, M.M. and Czelusniak, J. (1987) Pattern and process in vertebrate phylogeny revealed by coevolution of molecules and morphologies, in *Molecules and Morphology in Evolution: Conflict or Compromise?* (ed C. Patterson), Cambridge University Press, Cambridge, pp. 141–76.

Goodman, M., Weiss, M.L. and Czelusniak, J. (1982) Molecular evolution above the species level: Branching patterns, rates, and mechanisms. *Syst. Zool.*, **31**, pp. 376–99.

Goodrich, E.S. (1916) On the classification of the Reptilia. *Proc. R. Soc. London*, **89**, 261–76.

Goodwin, B.C. (1984) Changing from an evolutionary to a generative paradigm in biology, in *Evolutionary Theory: Paths into the Future*, (ed J.W. Pollard), Wiley, Chichester, pp. 99–120.

Goodwin, B.C. and Trainor, L. (1983) The ontogeny and phylogeny of the penta-dactyl limb, in *Development and Evolution*, (eds B. Goodwin, N. Holder and C. Wylie), Cambridge University Press, Cambridge, pp. 75–98.

Gordh, G. and Moczar, L. (1990) *A Catalogue of the World Bethylidae (Hymenoptera Aculeata)*, Am. Entomol. Inst. Mem. No. 46.

Gorr, T., Kleinschmidt, T. and Fricke, H. (1991) Close tetrapod relationships of the coelacanth *Latimeria* indicated by haemoglobin sequences. *Nature*, **351**, 394–7.

Gosliner, T.M. and Ghiselin, M.T. (1984) Parallel evolution in opisthobranch gastropods and its implications for phylogenetic methodology. *Syst. Zool.*, **33**, 255–74.

Gottlieb, L.D. (1984) Genetics and morphological evolution in plants. *Am. Nat.*, **123**, 681–709.

Gould, S.J. (1977) *Ontogeny and Phylogeny*, The Belknap Press of Harvard University Press, Cambridge, MA.

Gould, S.J. (1989) *Wonderful Life. The Burgess Shale and the Nature of History*, W.W. Norton, New York.

Grant, P.R. and Grant, B.R. (1992) Hybridization of bird species. *Science*, **256**, 193–7.

Grant, V. (1958) The regulation of recombination in plants. *Cold Spring Harbor Symp. Quant. Biol.*, **23**, 337–63.

Grant, V. (1963) *The Origin of Adaptations*, Columbia University Press, New York.

Grant, V. (1971) *Plant Speciation*, Columbia University Press, New York.

Grant, V. (1981) *Plant Speciation*, 2nd edn, Columbia University Press, New York.

Grant, V. (1985) *The Evolutionary Process. A Critical Review of Evolutionary Theory*, Columbia University Press, New York.

Grant, V. (1992) Comments on the ecological species concept. *Taxon*, **41**, 300–12.

Grassle, J.F. (1989) Species diversity in deep-sea communities. *Trends Ecol. Evol.*, **4**, 12–15.

Grassle, J.F. and Maciolek, N.J. (1992) Deep-sea species richness: Regional and local diversity estimates from quantitative bottom samples. *Am. Nat.*, **139**, 313–41.

Graur, D., Hide, W.A. and Li, W.-H. (1991) Is the guinea-pig a rodent? *Nature*, **351**, 649–52.

Green, D.M. (1991) Chaos, fractals and nonlinear dynamics in evolution and phylogeny. *Trends Ecol. Evol.*, **6**, 333-7.

Greene, S.W. and Harrington, A.J. (1988, 1989) The conspectus of bryological taxonomic literature. Parts 1 and 2. *Bryophytorum Bibliotheca*, **35**, 1–272; **37**, 1–322.

Greenwood, P.H. (1974) The cichlid fishes of Lake Victoria, East Africa: the biology and evolution of a species flock. *Bull. Br. Mus. Nat. Hist. (Zool.), Suppl.*, **6**, 1–134.

Greene, P.H., Miles, S. and Patterson, C. (eds) (1973) *Interrelationships of Fishes*, Academic Press, London.

Greenwood, P.H., Rosen, D.E., Weitzman, S.H. *et al.* (1966) Phyletic studies of teleostean fishes, with a provisional classification of living forms. *Bull. Am. Mus. Nat. His.*, **131**, 339–456.

Grell, K.G. (1971) *Trichoplax adhaerens* F.E. Schulze und die Entstehung der Metazoen. *Naturwiss. Rundschau*, **24**, 160–1.

Grell, K.G. (1974) Vom Einzeller zum Vielzeller, Hundert Jahre Gastraea-Theorie. *Biol. im unseren Zeit*, **4**, 65–71.

Grell, K.G. (1982) Placozoa, in *Synopsis and Classification of Living Organisms*, (ed S.P. Parker), McGraw-Hill, New York, Vol. 1, 639 pp.

Greuter, W., Burdet, H.M., Chaloner, W.G. *et al.* (eds) (1988) *International Code of Botanical Nomenclature Adopted by the Fourteenth International Botanical Congress, Berlin, July–August 1987 (Regnum Vegetabile, Vol. 118)*, Koeltz, Königstein.

Greuter, W., McNeill, J. and Nicolson, D.H. (1989) Report on Botanical Nomenclature – Berlin 1987. *Englera*, **9**, 1–288.

Griffiths, G.C.D. (1974) On the foundations of biological systematics. *Acta Biotheor.*, **23**, 85–131.

Griffiths, G.C.D. (1976) The future of Linnean nomenclature. *Syst. Zool.*, **25**, 168–73.

Grimaldi, D.A. (1990) A phylogenetic, revised classification of genera in the Drosophilidae (Diptera). *Bull. Am. Mus. Nat. Hist.*, **197**, 1–139.

Grolle, R. (1983) Nomina generica Hepaticarum; references, types and synonymies. *Acta Bot. Fennica*, **121**, 1–62.

Groves, C.P. and Lay, D.M. (1985) A new species of the genus *Gazella* (Mammalia, Artiodactyla: Bovidae) from the Arabian Peninsula. *Mammalia*, **49**, 27–36.

Gruner, H.-E. and Holthuis, L.B. (eds) (1967) *Crustaceorum Catalogus*, W. Junk, Den Haag.

Grygier, M.J. (1983) Ascothoracica and the unity of Maxillopoda, in *Crustacean Phylogeny (Crustacean Issues, 1)*, (ed F.R. Schram), A.A. Balkema, Rotterdam, pp. 73–104.

Gunderson, J.H., Elwood, H., Ingold, A. *et al.* (1987) Phylogenetic relationships between chlorophytes, chrysophytes and oomycetes. *Proc. Natl. Acad. Sci. USA*, **84**, 5823–7.

Gunn, C.R., Wiersema, J.H. and Kirkbride, J.H. Jr (1991) Agricultural perspective on stabilizing scientific names of spermatophytes, in *Improving the Stability of Names: Needs and Options. Regnum Vegetabile No. 123*, (ed D.L. Hawksworth), Koeltz, Königstein, pp. 17–21.

Gutmann, W.F. (1966) Coelomgliederung, Myomerie und die Frage der Vertebraten-Antezedenten. *Z. Zool. Syst. Evol.-forsch.*, **4**, 13–57.

Gutmann, W.F. (1967) Nachtrag zur 'Wurmtheorie' der Vertebraten-Evolution. *Z. Zool. Syst. Evol.-forsch.*, **5**, 314–32.

Guyer, C. and Slowinski, J.B. (1991) Comparisons of observed phylogenetic topologies with null expectations among three monophyletic lineages. *Evolution*, **45**, 340–50.

Gwynn, D.T. and Morris, G.K. (1986) Heterospecific recognition and behavioral isolation in acoustic Orthoptera (Insecta). *Evol. Theory*, **8**, 33–8.

Gyllenhaal, C., Soejarto, D.D., Farnsworth, N.R. and Huft, M.J. (1990) The value of herbaria. *Nature*, **347**, 704.

Hadlington, S. (1988) Natural History Museum in a decline? *Nature*, **333**, 289.

Haeckel, E. (1866) *Generelle Morphologie der Organismem*, G. Reimer, Berlin.

Haeckel, E. (1874) *Anthropogenie oder Entwicklungsgeschichte des Menschen (Keimes- und Stammesgeschichte)*, Leipzig.

Haeckel, E. (1888) Report on the Siphonophorae. *Rep. Sci. Results Voy. Challenger*, **28**.

Haeckel, E. (1894) *Systematische Phylogenie. 1. Systematische Phylogenie der Protisten und Pflanzen*, Reimer, Berlin.

Haffer, J. (1969) Speciation in Amazonian forest birds. *Science*, **165**, 131–7.

Haffer, J. (1974) Avian speciation in tropical South America, with a systematic study of the toucans (Rhamphastidiae) and jacamars (Galbulidae). *Publ. Nuttall Ornith. Club*, **14**, 1–370.

Hahn, G. (1989) Reconstruction of the phylogenetic relationships among the higher taxa of trilobites. *Abh. Naturwiss. Ver. Hamburg*, **28**, 187–99.

Hall, A.V. (1991) An unifying theory for methods of systematic analysis. *Biol. J. Linn. Soc.*, **42**, 425–56.

Hallam, A. (1988) The contribution of paleontology to systematics and evolution, in *Prospects in Systematics*, (ed D.L. Hawksworth), Clarendon Press, Oxford, pp. 128–47.

Hamby, R.K. and Zimmer, E.A. (1992) Ribosomal RNA as a phylogenetic tool in plant systematics, in *Molecular Systematics of Plants*, (eds P.S. Soltis, D.E. Soltis and J.J. Doyle), Chapman & Hall, New York, pp. 50–91.

Hamilton, C.W. and Reichards, S.H. (1992) Current practice in the use of subspecies, variety and forma in the classification of wild plants. *Taxon*, **41**, 485–98.

Hamilton, K.G.A. (1988) Cretaceous Homoptera from Brazil: a test of cladistics. *Proc. XVIII Int. Congr. Entomology, Vancouver*, 7.

van der Hammen, L. (1977) A new classification of Chelicerata. *Zool. Meded., Leiden*, **51**, 307–19.

Hanelt, P. (1986) Formal and informal classifications of the infraspecific variability of cultivated plants – advantages and limitations, in *Infraspecific Classification of Wild and Cultivated Plants. The Systematics Association Special Volume No. 29*, (ed B.T. Styles), Clarendon Press, Oxford, pp. 139–56.

Harbison, G.R. (1985) On the classification and evolution of the Ctenophora, in *The Origins and Relationships of Lower Invertebrates. The Systematics Association Special Volume no. 28*, (eds S. Conway Morris *et al.*), Clarendon Press, Oxford, pp. 78–100.

Harbison, G.R. and Madin, L.P. (1982) Ctenophora, in *Synopsis and Classification of Living Organisms*, (ed S.P. Parker), McGraw-Hill, New York, Vol. 1, pp. 707–15.

Harlan, J.R. and de Wet, J.M.J. (1971) Toward a rational classification of cultivated plants. *Taxon*, **20**, 509–17.

Harlan, J.R. and de Wet, J.M.J. (1986) Problems in merging populations and counterfeit hybrids, in *Infraspecific Classification of Wild and Cultivated Plants. The Systematics Association Special Volume No. 29*, (ed B.T. Styles), Clarendon Press, Oxford, pp. 71–6.

Harper, C.W. Jr (1976) Phylogenetic inference in paleontology. *J. Paleontol.*, **50**, 180–93.

Harris, S.A. and Ingram, R. (1991) Chloroplast DNA and biosystematics: The effects of intraspecific diversity and plastid transmission. *Taxon*, **40**, 393–412.

Harrison, R.G. (1986) Pattern and process in a narrow hybrid zone. *Heredity*, **56**, 337–49.

Harrison, R.G. (1989) Animal mitochondrial DNA as a genetic marker in population and evolutionary biology. *Trends Ecol. Evol.*, **4**, 6–11.

Harrison, R.G., Rand, D.M. and Wheeler, W.C. (1987) Mitochondrial DNA variation in field crickets across a narrow hybrid zone. *Mol. Biol. Evol.*, **4**, 144–58.

Hartmann, O. (1959–1965) Catalogue of the Polychaetous Annelids of the World. *Allen Hancock Found. Publ. Occas. Pap.*, **23**.

Harvey, M.S. (1990) *Catalogue of the Pseudoscorpionida*, Manchester University Press, Manchester.

Hartwich, G. (1984) 10. Stamm Nemathelminthes, 11. Stamm Priapulida, in *Kaestner's Lehrbuch der speziellen Zoologie*, (ed H.-E. Gruner), Gustav Fischer, Stuttgart, **1(2)**, 463–548.

Harvey, M.S. (1990) *Catalogue of the Pseudoscorpionida*, Manchester University Press, Manchester.

Harvey, P.H. and Pagel, M.D. (1991) *The Comparative Method in Evolutionary Biology*, Oxford University Press, Oxford.

Harvey, P.H. and Purvis, A. (1991) Comparative methods for explaining adaptations. *Nature*, **351**, 619–24.

Hasegawa, M. Cao, Y., Adachi, J. *et al.* (1992) Rodent polyphyly? *Nature*, **355**, 595.

Hasegawa, M. and Kishino, H. (1989) Heterogeneity of tempo and mode of mitochondrial DNA evolution among mammalian orders. *Japan. J. Genet.*, **64**, 243–58.

Hasegawa, M., Kishino, H. and Yano, T. (1987) Man's place in Hominoidea as inferred from molecular clocks of DNA. *J. Mol. Evol.*, **26**, 132–47.

Haskell, P.T. and Morgan, P.J. (1988) User needs in systematics and obstacles to their fulfillment, in *Prospects in Systematics*, (ed D.L. Hawksworth), Clarendon Press, Oxford, pp. 399–413.

Haszprunar, G. (1985) The Heterobranchia – a new concept of the phylogeny of the higher Gastropoda. *Z. Zool. Syst. Evol.-forsch.*, **23**, 15–37.

Haszprunar, G. (1986) Die Klado-evolutionäre Klassifikation-Versuch einer Synthese. *Z. Zool. Syst. Evol.-forsch.*, **24**, 89–109.

Haszprunar, G. (1992a) The first molluscs – small animals. *Boll. Zool.*, **59**, 1–16.

Haszprunar, G. (1992b) The types of homology and their significance for evolutionary biology and phylogenetics. *J. Evol. Biol.*, **5**, 13–24.

Haszprunar, G., Rieger, R.M. and Schuchert, P. (1991) Extant 'Problematica' within or near the Metazoa, in *The Early Evolution of Metazoa and the Significance of Problematic Taxa*, (eds A.M. Simonetta and S. Conway Morris), Cambridge University Press, Cambridge, pp. 99–105.

Hauser, D.L. and Presch, W. (1991) The effect of ordered characters on phylogenetic reconstruction. *Cladistics*, **7**, 243–65.

Hawkes, J.G. (1986) Infraspecific classification – the problems, in *Infraspecific Classification of Wild and Cultivated Plants. The Systematics Association Special Volume No. 29*, (ed B.T. Styles), Clarendon Press, Oxford, pp. 1–7.

Hawksworth, D.L. (1988) Improved stability for biological nomenclature. *Nature*, **334**, 301.

Hawksworth, D.L. (1989) Taxonomic stability. *Nature*, **337**, 416.

Hawksworth, D.L. (ed) (1991a) *Improving the Stability of Names: Needs and Options. Regnum Vegetabile Volume 123*, Koeltz, Königstein.

Hawksworth, D.L. (1991b) The fungal dimension of biodiversity: Magnitude, significance, and conservation. *Mycol. Res.*, **95**, 641–55.

Hawksworth, D.L. and Bisby, F.A. (1988) Systematics: the keystone of biology, in *Prospects in Systematics*, (ed D.L. Hawksworth), Clarendon Press, Oxford, pp. 3–30.

Hawksworth, D.L., Sutton, B.C. and Ainsworth, G.C. (1983) *Ainsworth and Bisby's Dictionary of the Fungi*, Commonwealth Mycological Institute, Kew, Surrey.

Healy, J.M. (1992) Sperm morphology as an indicator of major divisions among basal heterobranch gastropods, in *Abstracts of the 11th International Malacological Congress, Siena 1992*, (eds F. Giusti and G. Manganelli), University of Siena, Siena, pp. 56–9.

Heath, I.B. (1988) Gut fungi. *Trends Ecol. Evol.*, **3**, 167–71.

Hecht, M.K. and Edwards, J.L. (1977) The methodology of phylogenetic inference above the special level, in *Major Patterns in Vertebrate Evolution*, (eds M.K. Hecht, P.C. Goody and B.M. Hecht), Plenum Press, New York, pp. 3–51.

Hedges, S.B., Bogart, J.P. and Maxson, L.R. (1992) Ancestry of unisexual salamanders. *Nature*, **356**, 708–10.

Hedgpeth, J.W. (1961) Taxonomy: Man's oldest profession. *11th Annual University of the Pacific Faculty Lecture*, pp. 1–19.

Hedgpeth, J.W. (1982) Pycnogonida, in *Synopsis and Classification of Living Organisms*, (ed S.P. Parker), McGraw-Hill, New York, Vol. 2, pp. 169–73.

Hegnauer, R. (1962–73) *Chemotaxonomie der Pflanzen. 6 vols.*, Birkhäuser, Basel.

Hendriks, L., Huysmans, E., Vandenberghe, A. and De Wachter, R. (1986) Primary structure of the 5S ribosomal RNAs of 11 arthropods and applicability of 5S RNA to the study of metazoan evolution. *J. Mol. Evol.*, **24**, 103–9.

Hendy, M.D. and Penny, D. (1982) Branch and bound algorithms to determine minimal evolutionary trees. *Math. Biosci.*, **59**, 277–90.

Hennig, W. (1950) *Grundzüge einer Theorie der phylogenetischen Systematik*, Deutscher Zentralverlag, Berlin.

Hennig, W. (1965) Phylogenetic Systematics. *Annu. Rev. Entomol.*, **10**, 97–116.

Hennig, W. (1966) *Phylogenetic Systematics*, University of Illinois Press, Urbana, IL.

Hennig, W. (1969) *Die Stammesgeschichte der Insekten*, Kramer, Frankfurt am Main.

Hennig, W. (1970) Insektenfossilien aus der unteren Kreide. II. Empididae (Diptera, Brachycera). *Stuttg. Beitr. Naturk.*, **214**, 1–12.

Hennig, W. (1972) Insektenfossilien aus der unteren Kreide. IV. Psychodidae (Phlebotominae) mit einer kritischen Übersicht über das phylogenetische System der Familie und die bisher beschriebenen Fossilien (Diptera). *Stuttg. Beitr. Naturk.*, **241**, 1–69.

Hennig, W. (1974) Kritische Bemerkungen zur Frage 'Cladistic analysis or cladistic classification?' *Z. Zool. Syst. Evol.-forsch.*, **12**, 279–94.

Hennig, W. (1975) 'Cladistic analysis or cladistic classification?': A reply of Ernst Mayr. *Syst. Zool.*, **25**, 244–56.

Hennig, W. (1982) *Phylogenetische Systematik*, Verlag Paul Parey, Berlin.

Hennig, W. (1985) Stammesgeschichte der Chordaten. *Ztschr. Zool. Syst. Evol.-forsch.*, *Beihefte: Fortschr. Zool. Syst. Evol.-forsch.*, **2**, 1–208.

Heppell, D. (1981) The evolution of the Code of Zoological Nomenclature, in *History in the Service of Systematics. Society for the Bibliography of Natural History Special Publication Number 1*, (eds A. Wheeler and J.H. Price), The Society for the Bibliography of Natural History, London, pp. 135–41.

Herre, W. and Röhrs, M. (1990) *Haustiere – Zoologisch gesehen. 2. Auflage*, Gustav Fischer, Stuttgart.

Hessler, R.R. (1983) A defense of the caridoid facies: wherein the early evolution of the Eumalacostraca is discussed, in *Crustacean Phylogeny (Crustacean Issues, 1)*, (ed F.R. Schram), A.A. Balkema, Rotterdam, pp. 145–64.

Hewitt, G.M. (1988) Hybrid zones – natural laboratories for evolutionary studies. *Trends Ecol. Evol.*, **3**, 158–67.

Heywood, P. and Rothschild, L.J. (1987) Reconciliation of evolution and nomenclature among the higher taxa of protists. *Biol. J. Linn. Soc.*, **30**, 91–8.

Heywood, V.H. (1973) Ecological data in practical taxonomy, in *Taxonomy and Ecology. Systematics Association Special Volume No. 5*, (ed V.H. Heywood), Academic Press, London, pp. 329–47.

Heywood, V.H. (1977) Principles and concepts in the classification of higher taxa. *Pl. Syst. Evol., Suppl.*, **1**, 1–12.

Heywood, V.H. (1988) The structure of systematics, in *Prospects in Systematics*, (ed D.L. Hawksworth), Clarendon Press, Oxford, pp. 44–56.

Heywood, V.H. (1991) Needs for stability of nomenclature in conservation, in *Improving the Stability of Names: Needs and Options. Regnum Vegetabile No. 123*, (ed D.L. Hawksworth), Koeltz, Königstein, pp. 53–8.

Heywood, V.H. and Clark, R.B. (eds) (1982) *Taxonomy in Europe. European Science Foundation, ESR Review 17*, North Holland Publishers, Amsterdam.

Heywood, V.H., Harborne, J.B. and Turner, B.L. (eds) (1977) *The Biology and Chemistry of the Compositae*, Academic Press, New York.

Higgins, R.P. (1982) Kinorhyncha, in *Synopsis and Classification of Living Organisms*, (ed S.P. Parker), McGraw-Hill, New York, Vol. 1, pp. 873–7.

Hill, C.R. and Crane, P.R. (1982) Evolutionary cladistics and the origin of angiosperms, in *Problems of Phylogenetic Reconstruction. The Systematics Association Special Volume No. 21*, (eds K.A. Joysey and A.E. Friday), Academic Press, London, pp. 269–361.

Hill, J.E. (1974) A new family, genus and species of bat (Mammalia, Chiroptera) from Thailand. *Bull. Br. Mus. Nat. Hist. (Zool.)*, **27**, 303–36.

Hill, L.R., Skerman, V.B.D. and Sneath, P.H.A. (eds) (1984) Corrigenda to the approved lists of bacterial names. *Int. J. Syst. Bacteriol.*, **34**, 508–11.

Hillis, D.M. (1981) Premating isolating mechanisms among three species of the *Rana pipiens* complex in Texas and Southern Oklahoma. *Copeia*, **1982**, 168–74.

Hillis, D.M. (1987) Molecular versus morphological approaches to systematics. *Annu. Rev. Ecol. Syst.*, **18**, 23–42.

Hillis, D.M. and Moritz, C. (eds) (1990) *Molecular Systematics*, Sinauer Associates, Sunderland, MA.

Hodges, R.W. (1976) Presidential address 1976 – What insects can we identify. *J. Lepid. Soc.*, **30**, 245–51.

Hodkinson, I.D. and Casson, D. (1991) A lesser predilection for bugs: Hemiptera (Insecta) diversity in tropical rain forests. *Biol. J. Linn. Soc.*, **43**, 101–9.

Hoeh, W.R., Blakley, K.H. and Brown, W.M. (1991) Heteroplasmy suggests limited biparental inheritance of *Mytilus* mitochondrial DNA. *Science*, **251**, 1488–90.

Hoffman, R.L. (1979) *Classification of the Diplopoda*, Muséum d'Histoire Naturelle, Geneva.

Hoffman, R.L. (1982) Diplopoda, in *Synopsis and Classification of Living Organisms*, (ed S.P. Parker), McGraw-Hill, New York, Vol. 2, 689–724.

Hogg, J. (1860) On the distinctions of a plant and an animal and on a fourth kingdom of nature. *Edinburgh New Phil. J.*, **12**, 216–25.

Holdich, D.M. and Jones, J.A. (1983) *Tanaids. Synopses of the British Fauna (New Series) No. 27*, Cambridge University Press, Cambridge.

Holloway, J.D. (1987) Macrolepidoptera diversity in the Indo-Australian tropics, geographic, biotopic and taxonomic variations. *Biol. J. Linn. Soc.*, **30**, 325–41.

Holman, E. (1987) Recognizability of sexual and asexual species of rotifers. *Syst. Zool.*, **36**, 381–6.

Holmes, E.B. (1980) Reconsideration of some systematic concepts and terms. *Evol. Theory*, **5**, 35–87.

Holmes, E.B. (1985) Are lungfishes the sister group of tetrapods? *Biol. J. Linn. Soc.*, **25**, 379–97.

Holmes, E.C. (1991) Different rates of substitution may produce different phylogenies of the eutherian mammals. *J. Mol. Evol.*, **33**, 209–15.

Holmgren, N. (1933) On the origin of the tetrapod limb. *Acta Zool. (Stockholm)*, **14**, 185–295.

Holmgren, N. (1939) Contribution to the question of the origin of the tetrapod limb. *Acta Zool. (Stockholm)*, **20**, 89–124.

Holmgren, N. (1949) Contributions to the question of the origin of tetrapods. *Acta Zool. (Stockholm)*, **30**, 459–84.

Honacki, J.H., Kinman, K.E. and Koeppl, J.W. (1982) *Mammal Species of the World*, Allen Press and The Association of Systematics Collections, Lawrence, Kansas.

d'Hondt, J.-L. (1988) Remarques sur quelques difficultés et ambiguités de la taxonomie et de la systématique zoologiques. *Bull. Soc. Zool. France*, **113**, 5–19.

Hooker, J.D. (1872–1897) *The Flora of British India*, London, 7 vols.

Hopkins, G.H.E. and Clay, T. (1952) *A Check List of the Genera and Species of Mallophaga*, British Museum (Natural History), London.

Hori, H. and Osawa, S. (1987) Origin and evolution of organisms as deduced from 5S ribosomal RNA sequencies. *Mol. Biol. Evol.*, **4**, 445–72.

Horn, W., Kahle, I., Friese, G. and Gaedike, R. (1990) *Collectiones entomologicae. Ein Kompendium über den Verbleib entomologischer Sammlungen der Welt bis 1960*,

Akademie der Landwirtschaftswissenschaften der Deutschen Demokratischen Republik, Berlin, 2 vols.

Hou, X.-G. and Bergström, J. (1991) The arthropods of the Lower Cambrian Chengjiang fauna, with relationships and evolutionary significance, in *The Early Evolution of Metazoa and the Significance of Problematic Taxa*, (eds A.M. Simonetta and S. Conway Morris), Cambridge University Press, Cambridge, pp. 179–87.

House, M.R. (ed) (1979) *The Origin of Major Invertebrate Groups. Systematics Association Special Volume No. 12*, Academic Press, London.

Howard, R. and Moore, A. (1980) *A Complete Checklist of the Birds of the World*, Oxford University Press, Oxford.

Huber, H. (1982) Die zweikeimblättrigen Gehölze im System der Angiospermen. *Mitt. Bot. Staatssamml. München*, **18**, 59–78.

Hughes, A.L. (1992) Avian species described on the basis of DNA only. *Trends Ecol. Evol.*, **7**, 2–3.

Hull, D.L. (1976) Are species really individuals? *Syst. Zool.*, **25**, 174–91.

Hull, D.L. (1978) A matter of individuality. *Philos. Sci.*, **45**, 335–60.

Hull, D.L. (1980) Individuality and selection. *Annu. Rev. Ecol. Syst.*, **11**, 311–32.

Hull, D. (1981) Metaphysics and common usage. *Behav. Brain Sci.*, **4**, 290–1.

Hull, D.L. (1988) *Science as a Process: An Evolutionary Account of the Social and Conceptual Development of Science*, University of Chicago Press, Chicago.

Hull, F.M. (1962) Robber Flies of the World: The Genera of the Asilidae, Parts 1 and 2. *Bull. U.S. Nat. Mus.*, **224**.

Hull, F.M. (1973) Bee Flies of the World. The Genera of the Family Bombyliidae. *Bull. U.S. Nat. Mus.*, **286**.

Humphries, C.J. (ed) (1989) *Ontogeny and Systematics*, British Museum (Natural History), London.

Humphries, C.J. (1991) The implication of pragmatism for systematics. In *Improving the Stability of Names: Needs and Options. Regnum Vegetabile No. 123*, (ed D.L. Hawksworth), Koeltz, Königstein, pp. 313–22.

Humphries, C.J. and Chappill, J.A. (1988) Systematics as science: A response to Cronquist. *Bot. Rev.*, **54**, 129–44.

Hunt, P.F. (1986) The nomenclature and registration of orchid hybrids at specific and generic levels, in *Infraspecific Classification of Wild and Cultivated Plants. The Systematics Association Special Volume No. 29*, (ed B.T. Styles), Clarendon Press, Oxford, pp. 367–74.

Huxley, J. (ed) (1940) *The New Systematics*, Oxford University Press, Oxford.

Huxley, J.S. (1958) Evolutionary process and taxonomy with special reference to grades. *Uppsala Univ. Arsskr.*, **1958**, 21–39.

Illies, J. (1966) Katalog der rezenten Plecoptera. *Das Tierreich*, **82**, 631 pp.

International Commission on Zoological Nomenclature (1985) *International Code of Zoological Nomenclature, Adopted by the XX General Assembly of the International Union of Biological Sciences*. International Trust for Zoological Nomenclature in Association with British Museum (Natural History), London.

International Commission on Zoological Nomenclature (1990) General Session of the Commission, University of Maryland, 4 July 1990. *Bull. Zool. Nomencl.*, **47**, 246–9.

Ish-Horowicz, D. (1982) Transposable elements, hybrid incompatibility and speciation. *Nature*, **299**, 676–7.

Ivanov, A.V. (1970) Verwandtschaft und Evolution der Pogonophoren. *Z. Zool. Syst. Evol.-forsch.*, **8**, 109–19.

Iwatsuki, K., Raven, P.H. and Bock, W.J. (eds) *Modern Aspects of Species*, Tokyo University Press, Tokyo.

Jacobs, M. (1969) Large families – not alone! *Taxon*, **18**, 253–62.

Jacot-Guillarmod, C.F. (1970–79) Catalogue of the Thysanoptera of the world. *Ann. Cape Prov. Mus. (Nat. Hist.)*, **7**, 1–1724.

Jaenicke, J. (1981) Criteria for ascertaining the existence of host races. *Am. Nat.*, **117**, 830–4.

Jamieson, B.G.M. (1988) On the phylogeny and higher classification of the Oligochaeta. *Cladistics*, **4**, 367–410.

Jamieson, B.G.M. (1991) Ultrastructure and phylogeny of crustacean spermatozoa. *Mem. Queensl. Mus.*, **31**, 109–42.

Janis, Ch.M. (1988) New ideas in ungulate phylogeny and evolution. *Trends Ecol. Evol.*, **3**, 291–7.

Jankowski, A.W. (1980) Conspectus of a new system of the phylum Ciliophora. *Trudy Zool. Inst. Leningrad*, **94**, 103–21 (in Russian).

Jansen, R.K., Holsinger, K.E., Michaels, H.J. and Palmer, J.D. (1990) Phylogenetic analysis of chloroplast DNA restriction site data at higher taxonomic levels: An example from the Asteraceae. *Evolution*, **44**, 2089–105.

Jansen, R.K., Michaels, H.J., Wallace, R.S. *et al.* (1992) Chloroplast DNA variation in the Asteraceae: Phylogenetic and evolutionary implications, in *Molecular Systematics of Plants*, (eds P.S. Soltis, D.E. Soltis and J.J. Doyle), Chapman & Hall, New York, pp. 252–94.

Jansen, R.K. and Palmer, J.D. (1988) Phylogenetic implications of chloroplast DNA restriction site variation in the Mutisieae (Asteraceae). *Am. J. Bot.*, **75**, 753–66.

Janvier, Ph. (1984) Cladistics: Theory, purpose and evolutionary implications, in *Evolutionary Theory: Paths into the Future*, (ed J.W. Pollard), Wiley, Chichester, pp. 39–75.

Janzen, D.H. (1977) What are dandelions and aphids? *Am. Nat.*, **111**, 586–9.

Janzen, D.H. (1979) Reply. *Am. Nat.*, **114**, 156–7.

Jardine, N. (1967) The concept of homology in biology. *Br. J. Philos. Sci.*, **18**, 125–39.

Jarvik, E. (1980) *Basic Structure and Evolution of Vertebrates*, Academic Press, London, 2 vols.

Jarvik, E. (1986) The origin of the Amphibia, in *Studies in Herpetology*, (ed Z. Rocek), Charles University, Prague, pp. 1–24.

Jeanmonod, D. (1984) Speciation: Various aspects and recent models. *Candollea*, **39**, 151–94.

Jeekel, C.A.W. (1971) *Nomenclator Generum et Familiarum Diplopodorum. A List of the Genus and Family-group Names in the Class Diplopoda from the 10th Edition of Linnaeus, 1758 to the End of 1957*, Nederl. Entomol. Vereen. Monogr. No. 5, Amsterdam.

Jefferies, R.P.S. (1968) The subphylum Calcichordata (Jefferies 1968) – primitive fossil chordates with echinoderm affinities. *Bull. Br. Mus. Nat. Hist. (Geol.)*, **16**, 243–339.

Jefferies, R.P.S. (1979) The origin of chordates – a methodological essay, in *The Origin of Major Invertebrate Groups. Systematics Association Volume No. 12*, (ed M.R. House), Academic Press, London, pp. 443–77.

Jefferies, R.P.S. (1986) *The Ancestry of the Vertebrates*, British Museum (Nat. Hist.), London.

Jeffrey, C. (1968) Systematic categories for cultivated plants. *Taxon*, **17**, 104–14.

Jeffrey, C. (1971) Thallophytes and kingdoms – a critique. *Kew Bull.*, **25**, 291–9.

Jeffrey, C. (1986) Some differences between the botanical and zoological codes, in *Biological Nomenclature Today. A Review of the Present State and Current Issues of Biological Nomenclature of Animals, Plants, Bacteria and Viruses. IUBS Monograph Series No. 2*, (eds W.D.L. Ride and T. Younès), IRL Press, Eynsham, Oxford, pp. 62–5.

Jeffrey, C. (1989) *Biological Nomenclature*, 3rd edn, Edward Arnold, London.

Jermy, A.C. (1990) Selaginellaceae, in *The Families and Genera of Vascular Plants. I. Pteridophytes and Gymnosperms*, (ed K. Kubitzki), Springer-Verlag, Berlin, pp. 39–45.

Jirásek, V. (1961) Evolution of proposals of taxonomic categories for the classification of cultivated plants. *Taxon*, **10**, 34–45.

Johnson, N.F. (1992) *Catalog of World Species of Proctotrupoidea, Exclusive of Platygastridae (Hymenoptera)*, Am. Entomol. Inst. Mem. No. 51.

Johnston, D.E. (1982) Acari, in *Synopsis and Classification of Living Organisms*, (ed S.P. Parker), McGraw-Hill, New York, Vol. 2, 111.

Jones, M.L. (1981) *Riftia pachyptila*, new genus, new species, the vestimentiferan worm from the Galapagos Rift geothermal vents (Pogonophora). *Proc. Biol. Soc. Washington*, **93**, 1295–313.

Jones, M.L. (1985) Vestimentiferan pogonophores: their biology and affinities, in *The Origins and Relationships of Lower Invertebrates. The Systematics Association Special Volume No. 28*, (eds S. Conway Morris *et al.*), Clarendon Press, Oxford, pp. 327–42.

Jones, T.R. and Collins, J.P. (1992) Analysis of a hybrid zone between subspecies of the tiger salamander (*Ambystoma tigrinum*) in central New Mexico, USA. *J. Evol. Biol.*, **5**, 375–402.

de Jussieu, A. (1848) Taxonomie, in *Dictionnaire Universel d'Histoire Naturelle, No. 12*, (ed Ch. D'Orbigny), Renard, Paris, pp. 368–434.

de Jussieu, A.L. (1789) *Genera Plantarum secundum Ordines Naturales Disposita*, Paris.

Kabata, Z. (1979) *Parasitic Copepoda of British Fishes*, Ray Society, London.

Kato, M. (1988) The phylogenetic relationship of Ophioglossaceae. *Taxon*, **37**, 381–6.

Kelly, M.C. (1991) Literature databases, in *The Unity of Evolutionary Biology. Proceedings of the Fourth International Congress of Systematic and Evolutionary Biology*, (ed E.C. Dudley), Dioscorides Press, Portland, OR, Vol. 2, pp. 955–65.

Kemp, T.S. (1988a) Interrelationships of the Synapsida, in *The Phylogeny and Classification of the Tetrapods. Systematics Association Special Volume No. 35B*, (ed M.J. Benton), Clarendon Press, Oxford, Vol. 2, pp. 1–22.

Kemp, T.S. (1988b) A note on the Mesozoic mammals and the origin of therians, in *The Phylogeny and Classification of the Tetrapods. Systematic Association Special Volume No. 35B*, (ed. M.J. Benton), Clarendon Press, Oxford, Vol. 2, pp. 23–9.

Kemp, T.S. (1988c) Haemothermia or Archosauria? The interrelationships of mammals, birds and crocodiles. *Zool. J. Linn. Soc.*, **92**, 67–104.

Kevan, D.K.McE. (1973) The place of classical taxonomy in modern systematic entomology, with particular reference to orthopteroid insects. *Can. Entomol.*, **105**, 1211–22.

Kevan, D.K.McE. (1982) Orthoptera. Phasmatoptera, in *Synopsis and Classification of Living Organisms*, (ed S.P. Parker), McGraw-Hill, New York, Vol. 2, pp. 352–83.

Kilian, F. (1984) Stamm Mesozoa, in *Kaestner's Lehrbuch der speziellen Zoologie*, (ed H.E. Gruner), Gustav Fischer, Stuttgart, Vol. 1, pp. 336–40.

Kim, K.C. and Knutson, L. (1986) *Foundations for a National Biological Survey*, Association of Systematics Collections, Lawrence, Kansas.

Kim, K.C. and Ludwig, H.W. (1982) Parallel evolution, cladistics, and classification of parasitic Psocodea. *Ann. Entomol. Soc. America*, **75**, 537–48.

Kimsey, L.S. and Bohart, R.M. (1990) *The Chrysidid Wasps of the World*, Oxford University Press, Oxford.

Kimura, M. (1983) *The Neutral Theory of Molecular Evolution*, Cambridge University Press, Cambridge.

King, B.L. (1977a) Flavonoid analysis of hybridization in *Rhododendron* section *Pentanthera* (Ericaceae). *Syst. Bot.*, **2**, 14–27.

King, B.L. (1977b) The flavonoids of the deciduous *Rhododendron* of North America. *Am. J. Bot.*, **64**, 350–60.

King, R.M. and Robinson, H. (1970) *Eupatorium*, a composite genus of Arcto-Tertiary distribution. *Taxon*, **19**, 769–74.

King, R.M. and Robinson, H. (1987) *The Genera of the Eupatorieae (Asteraceae)*, Botanical Garden, St. Louis, Missouri.

Kingsbury, D.W. (1986) Nomenclature of plant viruses, in *Biological Nomenclature Today. A Review of the Present State and Current Issues of Biological Nomenclature of Animals, Plants, Bacteria and Viruses. IUBS Monograph Series No. 2*, (eds D.L. Ride and T. Younès), IRL Press, Eynsham, Oxford, pp. 49–53.

Kirkpatrick, P.A. and Pugh, P.R. (1984) *Siphonophores and Velellids. Synopses of the British Fauna (New Series), No. 29*, E.J. Brill/Dr. W. Backhuys, London.

Kitzmiller, J.B. and Laven, H. (1958) Current concepts of evolutionary mechanisms in mosquitoes. *Cold Spring Harbor Symp. Quant. Biol.*, **24**, 173–5.

Klein, R.M. and Cronquist, A. (1967) A consideration of the evolutionary and taxonomic significance of some biochemical, micromorphological, and physiological characters in the Thallophytes. *Quart. Rev. Biol.*, **42**, 105–296.

Kluge, A.G. (1984) The relevance of parsimony to phylogenetic inference, in *Cladistics: Perspectives on the Reconstruction of Evolutionary History*, (eds T. Duncan and T.F. Stuessy), Columbia University Press, New York, pp. 24–38.

Kluge, A.G. (1988) The characteristics of ontogeny, in *Ontogeny and Systematics*, (ed C.J. Humphries), British Museum (Natural History), London, pp. 57–81.

Kluge, A.G. (1989) A concern for evidence and a phylogenetic hypothesis of relationships among *Epichrates* (Boidae, Serpentes). *Syst. Zool.*, **38**, 7–25.

Kluge, A.G. (1990) On the special treatment of fossils and taxonomic burden: A response to Loconte. *Cladistics*, **6**, 191–3.

Kluge, A.G. and Farris, J.S. (1969) Quantitative phyletics and the evolution of anurans. *Syst. Zool.*, **18**, 1–32.

Kluge, A.G. and Strauss, R.E. (1985) Ontogeny and systematics. *Annu. Rev. Ecol. Syst.*, **16**, 247–68.

Knight, D. (1981) *Ordering the World. A History of Classifying Man*, Burnett Books, London.

Knobloch, I.W. (1972) Intergeneric hybridization in flowering plants. *Taxon*, **21**, 97–103.

Knutson, L. and Murphy, W.L. (1988) *Systematics: Relevance, Resources, Services, and Management: A Bibliography. Association of Systematics Collections Special Publications No. 1*, The Association of Systematics Collections, Washington.

Kochmer, J.P. and Wagner, R.H. (1988) Why are there so many kinds of passerine birds? Because they are small. A reply to Raikov. *Syst. Zool.*, **37**, 68–9.

Kohn, A.J. (1991) Diversification patterns in the most diverse marine snail genus, in *The Unity of Evolutionary Biology. Proceedings of the Fourth International Congress of Systematic and Evolutionary Biology*, (ed E.C. Dudley), Dioscorides Press, Portland, OR, pp. 253–4.

Kondo, R., Satta, Y., Matsuura, E.T. *et al.* (1990) Incomplete maternal transmission of mitochondrial DNA in *Drosophila*. *Genetics*, **126**, 657–63.

Königsmann, E. (1960) Zur Phylogenie der Parametabola. *Beitr. Entomol.*, **10**, 705–44.

Kormilev, N.A. and Froeschner, R.C. (1987) Flat bugs of the world, a synonymic list. *Entomography*, **5**, 246 pp.

Kramer, K.U. (1990) Notes on the higher level classification of the recent ferns, in *The Families and Genera of Vascular Plants: I. Pteridophytes and Gymnosperms*, (ed K. Kubitzki), Springer-Verlag, Berlin, pp. 49–52.

Kraus, F. and Miyamoto, M.M. (1990) Mitochondrial genotype of a unisexual salamander of hybrid origin is unrelated to either of its nuclear haplotypes. *Proc. Natl. Acad. Sci. USA*, **87**, 2235–8.

Kretzoi, M. (1955) *Dolomys* and *Ondatra*. *Acta Geol. Acad. Sci. Hung*, **3**, 347–55.

Kristensen, N.P. (1975) The phylogeny of hexapod 'orders'. A critical review of recent accounts. *Z. Zool. Syst. Evol.-forsch.*, **13**, 1–44.

Kristensen, N.P. (1976) Remarks on the family-level phylogeny of butterflies. *Z. Zool. Syst. Evol.-forsch.*, **14**, 25–33.

Kristensen, N.P. (1981) Phylogeny of insect orders. *Annu. Rev. Entom.*, **26**, 135–57.

Kristensen, N.P. (1991) Phylogeny of extant hexapods, in *The Insects of Australia*, 2nd edn, (ed I.D. Nauman), Melbourne University Press, Melbourne, Vol. 1, pp. 125–40.

Kristensen, R.M. (1983) Loricifera, a new phylum with Aschelminthes characters from the meiobenthos. *Z. Zool. Syst. Evol.-forsch.*, **21**, 163–80.

Kronick, D.A. (1985) *The Literature of the Life Sciences*, ISI Press, Philadelphia.

Kubai, D.F. (1978) Mitosis and fungal phylogeny, in *Nuclear Division in the Fungi*, (ed I.B. Heath), Academic Press, New York, pp. 177–229.

Kugler, Ch. (1982) Apterygota, in *Synopsis and Classification of Living Organisms*, (ed S.P. Parker), McGraw-Hill, New York, Vol. 2, pp. 328–30.

Kukalova-Peck, J. (1987) New Carboniferous Diplura, Monura, and Thysanura, the hexapod ground plan, and the role of thoracic side lobes in the origin of wings (Insecta). *Can. J. Zool.*, **65**, 2327–45.

Kukalova-Peck, J. (1991) Fossil history and the evolution of hexapod structures, in *The Insects of Australia*, 2nd edn, (ed I.D. Nauman), Melbourne University Press, Melbourne, Vol. 1, pp. 141–79.

Kukalova-Peck, J. (1992) The 'Uniramia' do not exist: the ground plan of the Pterygota as revealed by Permian Diaphanopterodea from Russia (Insecta: Paleodictyopteroidea). *Can. J. Zool.*, **70**, 236–55.

Kullmann, E. (1964) Neue Ergebnisse über den Netzbau und das Sexualverhalten einiger Spinnenarten (*Cresmatoneta mutinensis*, *Drapetisca socialis*, *Lithyphantes paykullianus*, *Cyrtophora citricola*). *Z. Zool. Syst. Evol.-forsch.*, **2**, 41–122.

Kumazaki, T., Hori, H. and Osawa, S. (1983a) The nucleotide sequences of 5S rRNAs from two annelid species, *Perinereis brevicirrus* and *Sabellastarte japonica* and an Echiura species, *Urechis unicinctus*. *Nucleic Acid Res.*, **11**, 3347–50.

Kumazaki, T., Hori, H. and Osawa, S. (1983b) The nucleotide sequences of 5S rRNA from two ribbonworms: *Emplectonema gracile* contains two 5S rRNA species differing considerably in their sequences. *Nucleic Acid Res.*, **11**, 7141–4.

Kuschel, G. (1988) Thoughts on past classifications of the weevils – how a new scheme may be attempted. *Proc. XVIII Int. Congr. Entomology, Vancouver*, 40.

La Greca, M. (1987) L'uso delle categorie sistematiche sottogenere e sottospecie in tassonomia, alla luce della ricerca biogeografica. *Boll. Ist. Entomol. 'Guido Grandi', Bologna*, **41**, 159–71.

Lake, J.A. (1987) Rate-independent technique for analysis of nucleic acid sequences: Evolutionary parsimony. *Mol. Biol. Evol.*, **4**, 167–91.

Lake, J.A. (1988) Origin of the eukaryotic nucleus determined by rate-invariant analysis of rRNA sequences. *Nature*, **331**, 184–6.

Lake, J.A. (1989) Origin of the eukaryotic nucleus: Eukaryotes and eocytes are genotypically related. *Can. J. Microbiol.*, **35**, 109–18.

Lake, J.A. (1990) Origin of the Metazoa. *Proc. Natl. Acad. Sci. USA*, **87**, 763–6.

Lake, J.A., Henderson, E., Oakes, M. and Clark, M.V. (1984) Eocytes: A new ribosome structure indicates a kingdom with a close relationship to eukaryotes. *Proc. Natl. Acad. Sci. USA*, **81**, 3786–90.

Lake, J.A., de la Cruz, V.F., Ferreira, P.C.G. *et al.* (1988) Evolution of parasitism: Kinetoplastid protozoan history reconstructed from mitochondrial rRNA gene sequences. *Proc. Natl. Acad. Sci. USA*, **85**, 4779–83.

Lamb, I.M. (1963) *Index Nominum Lichenum*, Ronald Press, New York.

Lamb, T. and Avise, J.C. (1986) Directional introgression of mitochondrial DNA in a hybrid population of tree frogs: The influence of mating behavior. *Proc. Natl. Acad. Sci. USA*, **83**, 2526–30.

Lambert, D.M. and Paterson, H.E. (1982) Morphological resemblance and its relationship to genetic distance measures. *Evol. Theory*, **5**, 291–300.

Lambert, D.M. and Paterson, H.E.H. (1984) On 'bridging the gap between race and species': The isolation concept and an alternative. *Proc. Linn. Soc. N.S.W.*, **107**(1983), 501–14.

Land, J. van der and Nørrevang, A. (1975) The systematic position of *Lamellibrachia* (Annelida, Vestimentifera). *Z. Zool. Syst. Evol.-forsch.*, **1**, 86–101.

Land, J. van der and Nørrevang, A. (1985) Affinities and intraphyletic relationships of the Priapulida, in *The Origins and Relationships of Lower Invertebrates. The Systematics Association Special Volume No. 28*, (eds S. Conway Morris *et al.*), Clarendon Press, Oxford, pp. 261–73.

Lande, R. and Kirkpatrick, M. (1988) Ecological speciation by sexual selection. *J. Theor. Biol.*, **133**, 85–98.

Larson, A. (1989) The relationship between speciation and morphological evolution, in *Speciation and its Consequences*, (eds D. Otte and J.A. Endler), Sinauer Associates, Sunderland, MA, pp. 579–98.

Laubitz, D.R. (1982) Euphausiacea, in *Synopsis and Classification of Living Organisms*, (ed S.P. Parker), McGraw-Hill, New York, Vol. 2, 295 pp.

Laubmann, A. (1921) Die quaternäre Nomenclatur und ihre Anwendung in der Ornithologie. *Club van Nederl. Vogelkund*, **11**, 40–51.

Lauder, G. (1981) Form and function: Structural analysis in evolutionary morphology. *Paleobiology*, **7**, 430–42.

Lauder, G.V. and Liem, K.F. (1983) The evolution and interrelationships of the Actinopterygian fishes. *Bull. Mus. Comp. Zool., Cambridge, Mass.*, **150**, 95–197.

Lauterbach, K.-E. (1980) Schlüsselereignisse in der Evolution des Grundplanes der Arachnata (Arthropoda). *Abh. Naturwiss. Ver. Hamburg*, **23**, 163–327.

Lauterbach, K.-E. (1983) Synapomorphien zwischen Trilobiten- und Cheliceraten- zweig der Arachnata. *Zool. Anz.*, **210**, 213–38.

Lauterbach, K.-E. (1989) Trilobites and Phylogenetic Systematics: A reply to G. Hahn. *Abh. Naturwiss. Ver. Hamburg*, **28**, 201–11.

Laven, H. (1958) Speciation by cytoplasmic isolation in the *Culex pipiens* complex. *Cold Spring Harbor Symp. Quant. Biol.*, **24**, 166–73.

La Vergata, A. (1987) Au nom de l'espèce. Classification et nomenclature au XIX siècle, in *Histoire du Concept d'Espèce dans les Sciences de la Vie*, (eds S. Atran *et al.*), Fondation Singer-Polignac, Paris, pp. 193–225.

Lawlor, D.A., Ward, F.E., Ennis, P.D. *et al.* (1988) HLA-A and B polymorphisms predate the divergence of humans and chimpanzees. *Nature*, **335**, 268–71.

Lawrence, J.F. (1982) Coleoptera, in *Synopsis and Classification of Living Organisms*, (ed S.P. Parker), McGraw-Hill, New York, Vol. 2, pp. 482–553.

Leedale, G.F. (1974) How many are the kingdoms of organisms? *Taxon*, 23, 261–70.

Leedale, G.F. and Hibberd, D.J. (1985) Order 8. Volvocida Francé, 1894, in *An Illustrated Guide to the Protozoa*, (eds J.J. Lee, S.H. Hutner and E.C. Bovee), Society of Protozoologists, Lawrence, Kansas, pp. 88–97.

Leenhouts, P.W. (1967) A conspectus of the genus Allophylus (Sapindaceae). *Blumea*, **15**, 301–14.

Lehtinen, P.T. (1979) Evolutionary trends as taxonomic characters. *Zool. Scripta*, **8**, 315.

Le Quesne, W.J. (1969) A method of selection of characters in numerical taxonomy. *Syst. Zool.*, **18**, 201–5.

Le Quesne, W.J. (1972) Further studies based on the uniquely derived character concept. *Syst. Zool.*, **21**, 281–8.

Le Quesne, W.J. (1982) Compatibility analysis and its applications. *Zool. J. Linn. Soc.*, **74**, 267–75.

Levi, H.W. (1982a) Merostomata, in *Synopsis and Classification of Living Organisms*, (ed S.P. Parker), McGraw-Hill, New York, Vol. 2, 72 pp.

Levi, H.W. (1982b) Uropygi. Schizomida. Amblypygi. Palpigradi. Araneae. Ricinulei, in *Synopsis and Classification of Living Organisms*, (ed S.P. Parker), McGraw-Hill, New York, Vol. 2, pp. 75–96.

Levin, S.A. (1989) Challenges in the developments of a theory of community and ecosystem structure and function, in *Perspectives in Ecological Theory*, (eds J. Roughgarden, R.M. May and S.A. Levin), Princeton University Press, Princeton, pp. 242–55.

Levine, N.D. (1962) Protozoology today. *J. Protozool.*, **9**, 1–6.

Levine, N.D. (1970) Taxonomy of Sporozoa. *J. Parasit.*, **56**, 208–9.

Levine, N.D. (1982) Apicomplexa, in *Synopsis and Classification of Living Organisms*, (ed S.P. Parker), McGraw-Hill, New York, Vol. 1, pp. 571–97.

Levinton, J. (1988) *Genetics, Paleontology, and Macroevolution*, Cambridge University Press, Cambridge.

Levy, M. and Levin, D.A. (1974) Novel flavonoids and reticulate evolution in the *Phlox pilosa - P. drummondii* complex. *Am. J. Bot.*, **61**, 156–67.

Levy, M. and Levin D.A. (1975) The novel flavonoid chemistry and phylogenetic origin of *Phlox floridana. Evolution*, **29**, 487–99.

Lewin, R.A. and Withers, N.W. (1975) Extraordinary pigment composition of a prokaryotic alga. *Nature*, **256**, 735–7.

Li, W.H., Gouy, M., Sharp, P.M. et al. (1990) Molecular phylogeny of Rodentia, Lagomorpha, Primates, Artiodactyla, and Carnivora and molecular clocks. *Proc. Natl. Acad. Sci. USA*, **87**, 6703–7.

Li, W.-H., Hide, W.A. and Graur, D. (1992) Origin of rodents and guinea pigs. *Nature*, **359**, 276–7.

Li, Z.-X. (1981) On a new species of musk deer from China. *Zool. Res.*, **2**, 157–61 (in Chinese; English summary).

Lidén, M. (1992) Species – where's the problem? *Taxon*, **43**, 315–17.

Linnaeus, C. (1753) *Species Plantarum*. Laur Salvius, Holmiae.

Linnaeus, C. (1758) *Systema Naturae*, Editio decima. Laur Salvius, Holmiae.

Lipscomb, D.L. (1985) The eukaryotic kingdoms. *Cladistics*, **1**, 127–57.

Lipscomb, D.L. (1992) Parsimony, homology and the analysis of multistate characters. Cladistics, **8**, 45–65.

Loconte, H. (1990) Cladistic classification of Amniota: A response to Gauthier *et al. Cladistics*, **6**, 187–90.

Loconte, H, and Stevenson, D.W. (1990) Cladistics of the Spermatophyta. *Brittonia*, **42**, 197–211.

Loconte, H. and Stevenson, D.W. (1991) Cladistics of the Magnoliidae. Cladistics, **7**, 267–96.

Lorenzen, S. (1981) Entwurf eines phylogenetischen Systems der freilebenden Nematoden. *Veröffentl. Inst. f. Meeresforsch. Bremerhaven, Suppl.*, **7**, 1–472.

Lorenzen, S. (1985) Phylogenetic aspects of pseudocoelomate evolution, in *The Origins and Relationships of Lower Invertebrates. The Systematics Association Special Volume no. 28*, (eds S. Conway Morris et al.), Clarendon Press, Oxford, pp. 210–23.

Löve, À. (1964) The biological species concept and its evolutionary structure. *Taxon*, **13**, 33–45.

Löve, À., Löve D. and Raymond, M. (1957) Cytotaxonomy of *Carex* section Capillares. *Can. J. Bot.*, **35**, 715–61.

Løvtrup, S. (1974) *Epigenetics, a treatise on theoretical biology*, Wiley-Interscience, New York.

Løvtrup, S. (1977) *The Phylogeny of Vertebrata*, Wiley, London.

Løvtrup, S. (1985) On the classification of the taxon Tetrapoda. *Syst. Zool.*, **34**, 463–70.

Lugo, A.E. (1988) Estimating reductions in the diversity of tropical forest species, in *Biodiversity*, (ed E.O. Wilson), National Academy Press, Washington DC, pp. 58–70.

Lyal, C.H.C. (1985) Phylogeny and classification of the Psocodea, with particular reference to the lice (Psocodea: Phthiraptera). *Syst. Entomol.*, **10**, 145–65.

Lynn, D.H. and Small, E.B. (1988) An update on the systematics of the phylum Ciliophora Doflein, 1901: The implications of kinetid diversity. *BioSystems*, **21**, 317–22.

Lynn, D.H. and Sogin, M.L. (1988) Assessment of phylogenetic relationships among ciliated protists using partial ribosomal RNA sequences derived from reverse transcripts. *BioSystems*, 21, 249–54.

Mabberley, D.J. (1987) *The Plant Book*, Cambridge University Press, Cambridge.

Mabee, P.M. (1989) An empirical rejection of the ontogenetic polarity criterion. *Cladistics*, **5**, 409–16.

Macnair, M.R. (1989) The potential for rapid speciation in plants. *Genome*, **31**, 203–10.

Macnair, M.R. and Christie, P. (1983) Reproductive isolation as a pleiotropic effect of copper tolerance in *Mimulus guttatus*? *Heredity*, **50**, 295–302.

Maddison, D.R. (1991) The discovery and importance of multiple islands of most-parsimonious trees. *Syst. Zool.*, **40**, 315–28.

Maddison, W.P. (1990) A method for testing the correlated evolution of two binary characters: Are gains or losses concentrated on certain branches of a phylogenetic tree? *Evolution*, **44**, 539–57.

Maddison, W.P. (1991) Squared-change parsimony reconstructions of ancestral states for continuous-valued characters on a phylogenetic tree. *Syst. Zool.*, **40**, 304–14.

Maddison, W.P., Donoghue, M.J. and Maddison, D.R. (1984) Outgroup analysis and parsimony. *Syst. Zool.*, **33**, 83–103.

Maddison, W.P. and Slatkin, M. (1991) Null models for the number of evolutionary steps in a character on a phylogenetic tree. *Evolution*, **45**, 1184–97.

Madin, L.P. and Harbison, G.R. (1978) *Thalassocalyce inconstans*, an enigmatic ctenophore representing a new family and order. *Bull. Mar. Sci.*, **28**, 680–7.

Madison, M. (1977) Vascular epiphytes. Their systematic occurrence and salient features. *Selbyana*, 2, 1–13.

Maggenti, A.R. (1982) Nemata, in *Synopsis and Classification of Living Organisms*, (ed S.P. Parker), McGraw-Hill, New York, Vol. 1, pp. 879–929.

Maisey, J.G. (1986) Heads and tails: A chordate phylogeny. *Cladistics*, 2, 201–56.

Maisey, J.G. (1988) Phylogeny of early vertebrate skeletal induction and ossification patterns, in *Evolutionary Biology*, (eds M.K. Hecht, B. Wallace and G.T. Prance), Plenum Press, New York, Vol. 22, pp. 1–36.

Malakhov, V.V. (1980) Cephalorhyncha, a new type of animal kingdom uniting Priapulida, Kinorhyncha, Gordiacea, and a system of Aschelminthes worms. *Zool. Zh.*, **59**, 485–99 (in Russian with English summary).

Mandelbrot, B. (1982) *The Fractal Geometry of Nature*, Freeman, San Francisco.

Manhart, J.R. and Palmer, J.D. (1990) The gain of two chloroplast tRNA introns marks the green algal ancestors of land plants. *Nature*, **345**, 268–70.

Manning, R.B. (1982) Hoplocarida, in *Synopsis and Classification of Living Organisms*, (ed S.P. Parker), McGraw-Hill, New York, Vol. 2, pp. 236–41.

Mannucci, M.P. and Minelli, A. (1987) *Viaggi e Scoperte*, Mondadori, Milan.

Mansfeld, R. (1953) Zur allgemeinen Systematik der Kulturpflanzen. I. *Kulturpflanze*, **1**, 138–55.

Manton, S.M. (1973) Arthropod phylogeny – a modern synthesis. *J. Zool.*, **171**, 111–30.

Manton, S.M. (1977) *The Arthropoda. Habits, Functional Morphology and Evolution*, Clarendon Press, Oxford.

Manton, S.M. and Anderson, D.T. (1979) Polyphyly and the evolution of arthropods, in *The Origin of Major Invertebrate Groups*, (ed M.R. House), Academic Press, London, pp. 269–321.

Marcotte, B.M. (1983) The imperatives of copepod diversity; perception, cognition, competition, and predation, in *Crustacean Phylogeny (Crustacean Issues, 1)*, (ed F.R. Schram), A.A. Balkema, Rotterdam, pp. 47–72.

Margulis, L. (1990) Introduction, in *Handbook of Protoctista. The Structure, Cultivation, Habitats and Life Histories of the Eukaryotic Microorganisms and their Descendants Exclusive of Animals, Plants and Fungi. A Guide to the Algae, Ciliates, Foraminifera, Sporozoa, Water Molds, Slime Molds and Other Protoctists*, (L. Margulis *et al.* eds), Jones and Bartlett Publishers, Boston, pp. xi–xxiii.

Margulis, L., Corliss, J.O., Melkonian, M. and Chapman, D.J. (1990) *Handbook of Protoctista. The Structure, Cultivation, Habitats and Life Histories of the Eukaryotic Microorganisms and their Descendants Exclusive of Animals, Plants and Fungi. A Guide to the Algae, Ciliates, Foraminifera, Sporozoa, Water Molds, Slime Molds and Other Protoctists*, Jones and Bartlett Publishers, Boston.

Margulis, L. and Sagan, D. (1985) Order amidst animalcules: The Protoctista kingdom and its undulipodiated cells. *BioSystems*, **18**, 141–7.

Margulis, L. and Schwartz, K.V. (1982) *Five Kingdoms*. W.H. Freeman, San Francisco.

Markowicz, Y. and Loiseaux-de Goër, S. (1991) Plastid genomes of the Rhodophyta and Chromophyta constitute a distinct lineage which differs from that of the Chlorophyta and have a composite phylogenetic origin, perhaps like that of the Euglenophya. *Curr. Genet.*, **20**, 427–30.

Marshall, N.B. (1979) *Developments in Deep-sea Biology*, Blandford Press, Poole.

Martin, A.P., Naylor, G.J.P. and Palumbi, S.R. (1992) Rates of mitochondrial DNA evolution in shark are slow compared with mammals. *Nature*, **357**, 153–5.

Martin, J.A. and Pashley, D.P. (1992) Molecular systematic analysis of butterfly family and some subfamily relationships (Lepidoptera: Papilionoidea). *Ann. Entomol. Soc. Am.*, **85**, 127–39.

Martinez Wells, M. and Henry, C.S. (1992) The role of courtship songs in reproductive isolation among populations of green lacewings of the genus *Chrysoperla* (Neuroptera: Chrysopidae). *Evolution*, **46**, 31–42.

Martino, V. and Martino, E. (1922) Note on a new snow-vole from Montenegro. *Ann. Mag. Nat. Hist.*, (9) **9**, 413.

Maslin, T.P. (1952) Morphological criteria of phylogenetic relationship. *Syst. Zool.*, **1**, 49–70.

Matile, L., Tassy, P. and Goujet, D. (1987) Introduction à la systématique zoologique (Concepts, Principes, Methodes). *Biosystema, Soc. Franc. Systém.*, **1**, 126 pp.

Matthews, R.E.F. (1981) The classification and nomenclature of viruses: Summary of results of meetings of the International Committee on Taxonomy of Viruses in Strasbourg, 1981. *Intervirology*, **16**, 53–60.

Mattox, K.R. and Stewart, K.D. (1984) Classification of the green algae: A concept based on comparative cytology, in *The Systematics of Green Algae*, (eds D.E.G. Irvine and D. John), Academic Press, London, pp. 29–72.

Maxson, L.R. and Maxson, R.D. (1990) Proteins II: Immunological techniques, in *Molecular Systematics*, (eds D.M. Hillis and C. Moritz), Sinauer Associates, Sunderland, MA, pp. 127–55.

Maxson, R.D. and Maxson, L.R. (1986) Micro-complement fixation: A quantitative estimator of protein evolution. *Mol. Biol. Evol.*, **3**, 375–88.

May, R.M. (1988) How many species are there on earth? *Science*, **241**, 1441–9.

May, R.M. (1990) Taxonomy as destiny. *Nature*, **347**, 129–30.

May, R.M. (1991) A fondness for fungi. *Nature*, **352**, 475–6.

May, R.M. (1992) Bottoms up for the oceans. *Nature*, **357**, 278–9.

Maynard Smith, J. (1966) Sympatric speciation. *Am. Nat.*, **100**, 637–50.

Maynard Smith, J. (1989) Trees, bundles or nets? *Trends Ecol. Evol.*, **4**, 302–4.

Maynard Smith, J. (1990) The evolution of prokaryotes: Does sex matter? *Annu. Rev. Ecol. Syst.*, **21**, 1–12.

Maynard Smith, J., Burian, R., Kauffman, S. *et al.* (1985) Developmental constraints and evolution. *Quart. Rev. Biol.*, **60**, 265–87.

Mayr, E. (1931) Notes on *Halcyon chloris* and some of its subspecies. *Am. Mus Novit.*, **469**, 10 pp.

Mayr, E. (1940) Speciation phenomena in birds. *Am. Nat.*, **74**, 249–78.

Mayr, E. (1942) *Systematics and the Origin of Species from the Viewpoint of a Zoologist*, Columbia University Press, New York.

Mayr, E. (1954) Change of genetic environment and evolution, in *Evolution as a Process*, (eds J. Huxley, A.C. Hardy and E.B. Ford), Allen and Unwin, London, pp. 157–80.

Mayr, E. (1957) Species concepts and definitions, in *The Species Problem*, (ed E. Mayr), Am. Assoc. Adv. Sci. Publ., Washington, no. **50**, pp. 1–22.

Mayr, E. (1963) *Animal Species and Evolution*, Harvard University Press, Cambridge, MA.

Mayr, E. (1969) *Principles of Systematic Zoology*, McGraw-Hill, New York.

Mayr, E. (1974) Cladistic analysis or cladistic classification? *Z. Zool. Syst. Evol.,-forsch.*, **12**, 94–128.

Mayr, E. (1980) Problems of the classification of birds: A progress report. *Acta, XVII Congressus Internationalis Ornitologicus*, Deutsche Ornithologen Gesellschaft, Berlin, pp. 95–112.

Mayr, E. (1982a) Processes of speciation in animals, in *Mechanisms of Speciation*, (ed C. Barigozzi), Alan R. Liss, New York, pp. 1–19.

Mayr, E. (1982b) *The Growth of Biological Thought. Diversity, Evolution, and Inheritance*, Harvard University Press, Cambridge, MA.

Mayr, E. (1984) Evolution of fish species flocks: A commentary, in *Evolution of Fish Species Flocks*, (eds A.A. Echelle and I. Kornfield), University of Maine Press, Orono, Maine, pp. 3–11.

Mayr, E. (1987) The ontological status of species: Scientific progress and philosophical terminology. *Biol. Philos.*, **2**, 145–66.

Mayr, E. (1988) Recent historical developments, in *Prospects in Systematics*, (ed D.L. Hawksworth), Clarendon Press, Oxford, pp. 31–43.

Mayr, E. (1990) A natural system of organisms. *Nature*, **348**, 491.

Mayr, E. (1991) More natural classification. *Nature*, **353**, 122.

Mayr, E. and Ashlock, P.D. (1991) *Principles of Systematic Zoology*, 2nd edn, McGraw-Hill, New York.

Mayr, E., Linsley, E.G. and Usinger, R. (1953) *Methods and Principles of Systematic Zoology*, McGraw-Hill, New York.

Mayr, E. and Provine, W.B. (1980) *The Evolutionary Synthesis*, Harvard University Press, Cambridge, MA.

Mazur, S. (1984) A world catalogue of Histeridae. *Polskie Pismo ent.*, **54**, 1–379.

McAlpine, J.F. (1989) Phylogeny and classification of the Muscomorpha, in *Manual of Nearctic Diptera. Volume 3. Research Branch Agriculture Canada Monograph No. 32*, (ed J.F. McAlpine), pp. 1397–518.

McEachran, J.D. (1982) Chondrichthyes, in *Synopsis and Classification of Living Organisms*, (ed S.P. Parker), McGraw-Hill, New York, 3, pp. 831–58.

McKinney, M.L. and McNamara, K.J. (1991) *Heterochrony. The Evolution of Ontogeny*, Plenum Press, New York.

McKitrick, M.C. and Zink, R.M. (1988) Species concepts in ornithology. *Condor*, **90**, 1–14.

McNamara, K.J. (1982) Heterochrony and phylogenetic trends. *Paleobiology*, **8**, 130–42.

McNamara, K. J. (1983) Progenesis in trilobites. *Spec. Pap. Paleontol.*, **30**, 59–68.

McNamara, K. J. (1986a) A guide to the nomenclature of heterochrony. *J. Paleontol.*, **60**, 4–13.

McNamara, K.J. (1986b) The role of heterochrony in the evolution of Cambrian trilobites. *Biol. Rev.*, **61**, 121–56.

McNamara, K.J. (1988) Patterns of heterochrony in the fossil record. *Trends Ecol. Evol.*, **3**, 176–80.

McPheron, B.A., Smith, D.C. and Berlocher, S.H. (1988) Genetic differences between host races of *Rhagoletis pomonella*. *Nature*, **336**, 64–6.

Melchior, H. (ed) (1964) *Engler's Syllabus der Pflanzenfamilien*, 12th edn, Berlin, 2 vols.

Melville, R.V. and Smith, J.D.D. (eds) (1987) *Official Lists and Indexes of Names and Works in Zoology*, The International Trust for Zoological Nomenclature c/o British Museum (Natural History), London.

Menken, S. (1980) Inheritance of allozymes in *Yponomeuta*. II. Interspecific crosses within the *padellus*-complex and reproductive isolation. *Proc. K. Ned. Akad. Wet.*, C, **83**, 425–31.

Metcalf, Z.P. et al. (1927–68) *General Catalogue of the Hemiptera and Homoptera*, North Carolina State College, Raleigh.

Meyen, S.B. (1984) Basic features of gymnosperm systematics and phylogeny as evidenced by fossil record. *Bot. Rev.*, **50**, 1–111.

Meyer, A., Kocher, T.D., Basasibwaki, P. and Wilson, A.C. (1990) Monophyletic origin of Lake Victoria cichlid fishes suggested by mitochondrial DNA sequences. *Nature*, **347**, 550–3.

Michaelsen, W. (1926) *Agriodrilus vermivorus* aus dem Baikal-See, ein Mittelglied zwischen typischen Oligochäten und Hirudineen. *Mitt. Zool. Staatsinst. Zool. Mus. Hamburg*, **42**, 1–20.

Mickevich, M.F. and Lipscomb, D.L. (1991) Parsimony and the choice between different transformations for the same character set. *Cladistics*, **7**, 111–39.

Miller, E.H. (ed) (1985) Museum Collections: Their Role and Future in Biological Research. *Br. Columbia Provincial Mus. Occas. Pap. Ser.*, No. **25**, 219 pp.

Miller, H.A. (1982) Bryophyte evolution and geography. *Biol. J. Linn. Soc.*, **18**, 145–96.

Milner, A.R. (1988) The relationships and origin of living amphibians, in *The*

Phylogeny and Classification of the Tetrapods. Systematics Association Special Volume No. 35A, (ed M.J. Benton), Clarendon Press, Oxford, Vol. 1, 59–102.

Mindell, D.P., Dick, C.W. and Baker, R.J. (1991) Phylogenetic relationships among megabats, microbats, and primates. *Proc. Nat. Acad. Sci. USA*, **88**, 10322–6.

Minelli, A. (1983) Considerazioni sulla filogenesi e la sistematica dei miriapodi. *Atti XIII Congr. Naz. Ital. Entomol., Sestriere-Torino*, pp. 551–60.

Minelli, A. (1987) On the phylogeny of segmentation within Articulata. *European Society for Evolutionary Biology, Founding Congress, Basel*, Abstract 72.

Minelli, A. (1988a) Problemi dimensionali e topologici della specie, in *Il Problema Biologico della Specie*, Mucchi, Modena, pp. 279–85.

Minelli, A. (1988b) Mechanisms of segmentation of insects and myriapods and their bearing on phylogeny. *Proc. XVIII Int. Congress Entomology, Vancouver*, Abstract 72.

Minelli, A. (1991a) Names for the system and names for the classification, in *Improving the Stability of Names: Needs and Options. Regnum Vegetabile No. 123*, (ed D.L. Hawksworth), Koeltz, Königstein, pp. 183–9.

Minelli, A. (1991b) *Introduzione alla Sistematica Biologica*, Muzzio, Padova.

Minelli, A. (1991c) Towards a new comparative morphology of myriapods. *Ber. Nat.-med. Verein Innsbruck, Suppl.*, **10**, 37–46.

Minelli, A. and Bortoletto, S. (1988) Myriapod metamerism and arthropod segmentation. *Biol. J. Linn. Soc.*, **33**, 323–43.

Minelli, A. and Peruffo, B. (1991) Developmental pathways, homology and homonomy in metameric animals. *J. Evol. Biol.*, **4**, 429–65.

Minelli, A., Fusco, G. and Sartori, S. (1990) Supraspecific taxa: Is their size distribution fractal? *IV Int. Congr. Syst. Evol. Biol., College Park, MD, 1–7 July 1990*, Abstract 249.

Minelli, A., Fusco, G. and Sartori, S. (1991) Self-similarity in biological classifications. *BioSystems*, **26**, 89–97.

Mishler, B.D. (1988) Relationships between ontogeny and phylogeny, with reference to Bryophytes, in *Ontogeny and Systematics*, (ed C.J. Humphries), British Museum (Natural History), London, pp. 117–36.

Mishler, B.D. and Brandon, R.N. (1987) Individuality, pluralism and the phylogenetic species concept. *Biol. Philos.*, **2**, 397–414.

Mishler, B.D. and Churchill, S.P. (1984) A cladistic approach to the phylogeny of the 'bryophytes'. *Brittonia*, 36, 406–24.

Mishler, B.D. and Churchill, S.P. (1985) Transition to a landflora: Phylogenetic relationships of the green algae and bryophytes. *Cladistics*, **1**, 305–28.

Mishler, R.D. and Donoghue, M.J. (1982) Species concepts: A case for pluralism. *Syst. Zool.*, **31**, 491–503.

Mitchell, A.W. (1986) *The Enchanted Canopy*, Collins, London.

Miyamoto, M.M. (1985) Consensus cladograms and general classifications. *Cladistics*, **1**, 186–9.

Miyamoto, M.M. and Goodman, M. (1986) Biomolecular systematics of eutherian mammals: Phylogenetic patterns and classification. *Syst. Zool.*, **35**, 230–40.

Möhn, E. (1984) *System und Phylogenie der Lebewesen. Band 1. Physikalische, chemische und biologische Evolution. Prokaryonta. Eukaryonta (bis Ctenophora)*, E. Schweizerbartsche Verlagsbuchhandlung, Stuttgart.

Møller Andersen, N. (1982) *The Semiaquatic Bugs (Hemiptera, Gerromorpha). Phylogeny, Adaptations, Biogeography and Classification. Entomonograph vol. 3*, Scandinavian Science Press, Copenhagen.

Monnerot, M., Solignac, M. and Wolstenholme, D.R. (1990) Distancy in divergence of the mitochondrial and nuclear genomes of *Drosophila teissieri* and *Drosophila yakuba. J. Mol. Evol.*, **30**, 500–8.

Monticelli, F.S. (1896) Adelotacta zoologica. 2. *Treptoplax reptans* Montic. *Mitt. Zool. Stn. Neapel*, **12**, 444–58.

Moore, R.C. (ed) (1993–) *Treatise on Invertebrate Paleontology*, University of Kansas Press, Lawrence, Kansas.

Morden, C.W. and Golden, S.S. (1989a) psbA genes indicate common ancestry of prochlorophytes and chloroplasts. *Nature*, **339**, 400.

Moritz, C. (1991) The origin and evolution of parthenogenesis in *Heteronotia binoei* (Gekkonidae): Evidence for recent and localized origins of widespread clones. *Genetics*, **129**, 211–9.

Moritz, C., Dowling, T.E. and Brown, W.M. (1987) Evolution of animal mitochondrial DNA: Relevance for population biology and systematics. *Annu. Rev. Ecol. Syst.*, **18**, 269–92.

Moritz, C. and Hillis, D.M. (1990) Molecular systematics: Context and controversies, in *Molecular Systematics*, (eds D.M. Hillis and C. Moritz), Sinauer Associates, Sunderland, MA, pp. 1–10.

Morony, J.J. Jr, Bock, W.J. and Farrand, J. Jr (1975) *Reference List of the Birds of the World*, American Museum of Natural History, New York, pp. 1–207.

Morris, P. and Cobabe, E. (1991) Cuvier meets Watson and Crick: The utility of molecules as classical homologies. *Biol. J. Linn. Soc.*, **44**, 307–24.

Mound, L.A. and Halsey, S.H. (1978) White Fly of the World: A Systematic Catalogue of the Aleyrodidae (Homoptera) with Host Plant and Natural Enemy Data. *Br. Mus. (Nat. Hist.) Publ.*, no. **787**.

Mound, L.A., Heming, B.S. and Palmer, J.M. (1980) Phylogenetic relationships between the families of recent Thysanoptera (Insecta). *Zool. J. Linn. Soc.*, **69**, 111–41.

Mueller, L.D. and Ayala, F.J. (1982) Estimation and interpretation of genetic distance in empirical studies. *Genet. Res.*, **40**, 127–37.

Müller, F.P. (1985) Biotype formation and sympatric speciation in aphids, in *Evolution and Biosystematics of Aphids*, (ed H. Fzelegiewicz), Ossolineum, Wroclaw, pp. 135–66.

Muma, M.H. (1982) Solpugida, in *Synopsis and Classification of Living Organisms*, (ed S.P. Parker), McGraw-Hill, New York, Vol. 2, pp. 102–4.

Munroe, E.G. (1982) Lepidoptera, in *Synopsis and Classification of Living Organisms*, (ed. S.P. Parker), McGraw-Hill, New York, Vol. 2, pp. 612–51.

Naomi, S.-I. (1985) The phylogeny and higher classification of the Staphylinidae and their allied groups (Coleoptera, Staphylinoidea). *Esakia*, **23**, 1–27.

Neave, S.A. (1939–1940, 1950) *Nomenclator Zoologicus*, The Zoological Society, London, 4 vols.

Neff, N.A. (1986) A rational basis for *a priori* weighting. *Syst. Zool.*, **35**, 110–23.

Neigel, J.E. and Avise, J.C. (1986) Phylogenetic relationships of mitochondrial DNA under various demographic models of speciation, in *Evolutionary Processes and Theory*, (eds E. Nevo and S. Karlin), Academic Press, New York, pp. 515–34.

Nelson, G.J. (1969) Gill arches and the phylogeny of fishes, with notes on the classification of vertebrates. *Bull. Am. Mus. Nat. Hist.*, **141**, 475–552.

Nelson, G.J. (1970) Outline of a theory of comparative biology. *Syst. Zool.*, **19**, 373–84.

Nelson, G.J. (1971) Paraphyly and polyphyly: Redefinitions. *Syst. Zool.*, **20**, 471–2.

Nelson, G.J. (1972) Phylogenetic relationship and classification. *Syst. Zool.*, **21**, 227–31.

Nelson, G.J. (1973) 'Monophyly again?' – A replay to P.D. Ashlock. *Syst. Zool.*, **22**, 310–12.

Nelson, G.J. (1974) Classification as an expression of phylogenetic relationships. *Syst. Zool.*, **22**, 344–59.

Nelson, G. (1978) Ontogeny, phylogeny, paleontology, and the biogenetic law. *Syst. Zool.*, **27**, 324–245.

Nelson, G.J. (1989a) Cladistics and evolutionary models. *Cladistics*, **5**, 275–89.

Nelson, G. (1989b) Species and taxa. Systematics and evolution, in *Speciation and its Consequences*, (eds D. Otte and J.A. Endler), Sinauer Associates, Sunderland, MA, pp. 60–81.

Nelson, G.J. and Platnick, N.I. (1981) *Systematics and Biogeography, Cladistics and Vicariance*. Columbia University Press, New York.

Nelson, G.J. and Platnick, N.I (1984) Systematics and evolution, in *Beyond Neo-Darwinism*, (eds M.-W. Ho and P.T. Saunders), Academic Press, London, pp. 143–58.

Nelson, J.S. (1984) *Fishes of the World*, 2nd edn, Wiley, New York.

Nevo, E. (1982) Speciation in subterranean mammals, in *Mechanisms of Speciation*, (ed C. Barigozzi), Alan R. Liss, New York, pp. 191–218.

Nevo, E. and Capranica, R.R. (1985) Evolutionary origin of ethological reproductive isolation in cricket frogs, *Acris*, in *Evolutionary Biology*, (eds M.K. Hecht, B. Wallace and G.T. Prance), Plenum Press, New York, Vol. 19, pp. 147–214.

Newman, W.A. (1987) Evolution of Cirripedes and their major groups, in *Barnacle Biology (Crustacean Issues, 5)*, (ed A.J. Southward), A.A. Balkema, Rotterdam, pp. 3–42.

Newman, W.A. (1989) Barnacle taxonomy. *Nature*, **337**, 23–4.

Newton, A.F. and Thayer, M.K. (1988a) Pselaphidae and Staphylinidae (Coleoptera): The 'missing link' discovered? *Proc. XVIII Int. Congress Entomol., Vancouver*, **50**.

Newton, A.F. Jr and Thayer, M.K. (1988b) A critique on Naomi's phylogeny and higher classification of Staphylinidae and allies (Coleoptera). *Entomol. Gener.*, **14**, 63–72.

Nielsen, C. (1971) Entoproct life-cycles and the entoproct–ectoproct relationships. *Ophelia*, **9**, 209–341.

Nielsen, C. (1977) Phylogenetic considerations, in *Biology of Bryozoans*, (eds R.M. Woollacott and R.J. Zimmer), Academic Press, New York, pp. 519–34.

Nielsen, C. (1982) Entoprocta, in *Synopsis and Classification of Living Organisms*, (ed S.P. Parker), McGraw-Hill, New York, Vol. 2, pp. 771–2.

Nielsen, C. (1985) Animal phylogeny in the light of the trochaea theory. *Biol. J. Linn. Soc.*, **25**, 243–99.

Nielsen, C. (1991) The development of the brachiopod *Crania (Neocrania) anomala* (O.F. Müller) and its phylogenetic significance. *Acta Zool. (Stockholm)*, **72**, 7–28.

Nielsen, C. and Nørrevang, A. (1985) The trochaea theory: An example of life cycle phylogeny, in *The Origins and Relationships of Lower Invertebrates. The Systematics Association Special Volume No. 28*, (eds S. Conway Morris *et al.*), Clarendon Press, Oxford, pp. 28–41.

Niemann, G.J. (1988) Distribution and evolution of the flavonoids in gymnosperms, in *The Flavonoids, Advances in Research since 1980*, (ed J.B. Harborne), Chapman & Hall, London, pp. 469–478.

Nieuwkoop, P.D. and Sutasurya, L.A. (1979) *Primordial Germ Cells in the Chordates: Embryogenesis and Phylogenesis*, Cambridge University Press, Cambridge.

Nixon, K.C. and Davis, J.I. (1991) Polymorphic taxa, missing values and cladistic analysis. *Cladistics*, **7**, 233–41.

Nixon, K.C. and Wheeler, Q.D. (1990) An amplification of the phylogenetic species concept. *Cladistics*, **6**, 211–23.

Norell, M.A. and Novacek, M.J. (1992) The fossil record and evolution: Comparing cladistic and paleontological evidence for vertebrate history. *Science*, **255**, 1690–3.

Nørrevang, A. (1970) The position of Pogonophora in the phylogenetic system. *Z. Zool. Syst. Evol.-forsch.*, **10**, 161–72.

Novacek, M.J. (1992) Mammalian phylogeny: Shaking the tree. *Nature*, **356**, 121–5.

Novacek, M.J., Wyss, A.R. and McKenna, M.C. (1988) The major groups of eutherian mammals, in *The Phylogeny and Classification of the Tetrapods. Systematics Association Special Volume No. 35B*, (ed M.J. Benton), Clarendon Press, Oxford, Vol. 2, pp. 31–71.

Novak, F. (1930) *Systematicka botanika*, Praha.

Novoa, C. and Wheeler, J. (1984) Llama and alpaca, in *Evolution of Domesticated Animals*, (ed I.L. Mason), Longman, London, pp. 116–28.

Nudds, J. and Palmer, D. (1990) Societies, organizations, journals, and collections, in *Palaeobiology. A synthesis*, (eds D.E.G. Briggs and P.R. Crowther), Blackwell Scientific Publications, Oxford, pp. 522–36.

Nylander, W. (1866) Circa novum in studio lichenum criterium chimicum. *Flora*, **49**, 198.

Ohama, T., Kumazaki, T., Hori, H. and Osawa, S. (1984) Evolution of multicellular animals as deduced from 5s rRNA sequences: A possible early emergence of the Mesozoa. *Nucl. Acid. Res.*, **12**, 5101–8.

O'Hara, R.J. (1992) Telling the tree. Narrative representation and the study of evolutionary history. *Biol. Philos.*, **7**, 135–60.

Oldroyd, D. (1990) David Hull's evolutionary model for the progress and process of science. *Biol. Philos.*, **5**, 473–87.

Oldroyd, H. (1966) The future of taxonomic entomology. *Syst. Zool.*, **15**, 253–60.

Øllgaard, B. (1990) Lycopodiaceae, in *The Families and Genera of Vascular Plants. I. Pteridophytes and Gymnosperms*, (ed K. Kubitzki), Springer-Verlag, Berlin, pp. 31–9.

Olmi, M. (1984) A revision of the Dryinidae (Hymenoptera). *Mem. Am. Entomol. Inst.*, **37**, 1913 pp.

Olsen, S.L. (1981) The museum tradition in ornithology. A response to Ricklefs. *AuK*, **98**, 193–5.

Olson, R.R. (1991) Whose larvae? *Nature*, **351**, 357–8.

Oosterbroek, P. (1987) More appropriate definitions of paraphyly and polyphyly, with a comment on the Farris 1974 model. *Syst. Zool.*, **35**, 103–8.

Ortolani, G. (1989) The Ctenophores: A review. *Acta Embr. Morph. Exp.*, **10**, 13–31.

Ostrom, J.H. (1991) The bird in the bush. *Nature*, **353**, 212.

Otte, D. (1989) Speciation in Hawaiian crickets, in *Speciation and its Consequences*, (eds D. Otte and J.A. Endler), Sinauer Associates, Sunderland, MA, pp. 482–526.

Otte, D. and Endler, J.A. (1989) *Speciation and its Consequences*, Sinauer Associates, Sunderland, MA.

Owen, R. (1843) *Lectures on the Comparative Anatomy and Physiology of the Invertebrate Animals, Delivered at the Royal College of Surgeons, in 1843*, Longman, Brown, Green, and Longmans, London.

Owen, R. (1866–8) *On the Anatomy of Vertebrates*, Longmans, Green and Co., London, 3 vols.

Pääbo, S. and Wilson, A.C. (1991) Miocene DNA sequences – a dream come true? *Curr. Biol.*, **1**, 45–6.

Pagel, M.D. and Harvey, P.H. (1988) Recent developments in the analysis of comparative data. *Quart. Rev. Biol.*, **63**, 413–40.

Palenik, B. and Haselkorn, R. (1992) Multiple evolutionary origins of prochlorophytes, the chlorophyll *b*-containing prokaryotes. *Nature*, **355**, 265–7.

Palissa, A. (1989) *Diplura, Protura, Collembola*, in *Ekursionsfauna für die Gebiete der DDR und der BRD, 2/1, Wirbellose: Insekten-Erster Teil, 8, Auflage*, (ed E. Stresemann), Volk und Wissen Volkseigener Verlag, Berlin, pp. 44–52.

Palmer, J.D. (1992) Mitochondrial DNA in plant systematics: Applications and limitations, in *Molecular Systematics of Plants*, (eds P.S. Soltis, D.E. Soltis and J.J. Doyle), Chapman & Hall, New York, pp. 36–49.

Palmer, J.D., Shields, C.R., Cohen, D.B. and Orton, T.J. (1983) Chloroplast DNA evolution and the origin of amphidiploid *Brassica* species. *Theor. Appl. Genet.*, **65**, 181–9.

Panchen, A.L. and Smithson, T.R. (1988) The relationships of the earliest tetrapods, in *The Phylogeny and Classification of the Tetrapods. Systematics Association Special Volume no. 35A* (ed M.J. Benton), Clarendon Press, Oxford, Vol. 1, pp. 1–32.

Papavero, N. (ed) (1967–) *A Catalogue of the Diptera of the Americas South of the United States*, Departamento de Zoologia, Sao Paulo, Fasc. 1–110.

Papp, L. (1988) After 35 years again: Why worship types. *Proc. XVIII Intern. Congr. Entomology, Vancouver*, Abstract 56.

Parker, S.P. (ed) (1982) *Synopsis and Classification of Living Organisms*, McGraw-Hill, New York, 2 vols.

Paterson, H.E.H. (1978) More evidence against speciation by reinforcement. *S. Afr. J. Sci.*, **74**, 369–71.

Paterson, H.E. (1980) A comment on 'Mate Recognition Systems'. *Evolution*, **34**, 330–1.

Paterson, H.E.H. (1981) The continuing search for the unknown and unknowable: A critique of contemporary ideas on speciation. *S. Afr. J. Sci.*, **77**, 113–19.

Paterson, H.E.H. (1982) Perspective on speciation by reinforcement. *S. Afr. J. Sci.*, **78**, 53–7.

Paterson, H.E.H. (1985) The recognition concept of species, in *Species and Speciation*, (ed E.S. Vrba), Transvaal Museum Monograph, Pretoria, Vol. 4, pp. 21–9.

Paterson, H.E.H. (1986) Environment and species. *S. Afr. J. Sci.*, **82**, 62–5.

Paterson, H. (1988) On defining species in terms of sterility: Problems and alternatives. *Pacific Sci.*, **42**, 65–71.

Patterson, C. (1981a) Significance of fossils in determining evolutionary relationships. *Annu. Rev. Ecol. Syst.*, **12**, 195–223.

Patterson, C. (1981b) Agassiz, Darwin, Huxley, and the fossil record of teleost fishes. *Bull. Br. Mus. Nat. Hist. (Geol.)*, **35**, 213–24.

Patterson, C. (1982) Morphological characters and homology, in *Problems of Phylogenetic Reconstruction. The Systematics Association Special Volume No. 21* (eds K.A. Joysey and A.E. Friday), Academic Press, London, pp. 21–74.

Patterson, C. (ed) (1987) *Molecules and Morphology in Evolution: Conflict or Compromise?* Cambridge University Press, Cambridge.

Patterson, C. (1988) The impact of evolutionary theories on systematics, in *Prospects in Systematics*, (ed D.L. Hawksworth), Clarendon Press, Oxford, pp. 59–91.

Patterson, C. and Rosen, D. (1977) Review of ichthyodectiform and other Mesozoic teleost fishes and the theory and practice of classifying fossils. *Bull. Am. Mus. Nat. Hist.*, **158**, 81–172.

Patterson, C. and Smith, A.B. (1987) Is the periodicity of extinctions a taxonomic artefact? *Nature*, **330**, 248–51.

Patterson, D.J. and Larsen, J. (1991) Nomenclatural problems with protists, in *Improving the Stability of Names: Needs and Options. Regnum Vegetabile No. 123*, (ed D.L. Hawksworth), Koeltz, Königstein, pp. 197–208.

Paulay, G. (1985) Adaptive radiation on an isolated oceanic island: The Cryptorhynchinae (Curculionidae) of Rapa revisited. *Biol. J. Linn. Soc.*, **26**, 95–187.

Pawson, D.L. (1982) Holothuroidea, in *Synopsis and Classification of Living Organisms*, (ed S.P. Parker), McGraw-Hill, New York, Vol. 2, pp. 813–18.

Peck, S.B. (1982) Onychophora, in *Synopsis and Classification of Living Organisms*, (ed S.P. Parker), McGraw-Hill, New York, Vol. 2, pp. 729–30.

Peel, J.S. (1991) Functional morphology of the Class Helcionelloida nov., and the early evolution of the Mollusca, in *The Early Evolution of Metazoa and the Significance of Problematic Taxa*, (eds A.M. Simonetta and S. Conway Morris), Cambridge University Press, Cambridge, pp. 157–77.

Penny, D. (1989) What, if anything, is *Prochloron*? *Nature*, **337**, 304–5.

Penny, D. and O'Kelly, C.J. (1991) Seeds of a universal tree. *Nature*, **350**, 106–7.

Peters, J.L. (1931) *Check-list of Birds of the World*, Harvard University Press, Cambridge, MA, Vol. 1, 345 pp.

Peterson, A.T. and Lanyon, S.M. (1992) New bird species, DNA studies and type specimens. *Trends Ecol. Evol.*, **7**, 167–8.

Pettibone, M.H. (1982) Annelida (p.p.), in *Synopsis and Classification of Living Organisms*, (ed S.P. Parker), McGraw-Hill, New York, Vol. 2, pp. 1–43.

Pettigrew, J.D. (1986) Flying primates? Megabats have the advanced pathway from eye to midbrain. *Science*, **231**, 1304–6.

Pettigrew, J.D. (1991a) Wings or brain? Convergent evolution in the origin of bats. *Syst. Zool.*, **40**, 199–216.

Pettigrew, J.D. (1991b) A fruitful wrong hypothesis? Response to Baker, Novacek, and Simmons. *Syst. Zool.*, **40**, 231–9.

Pettigrew, J.D., Jamieson, G.M. Robson, S.K. *et al.* (1989) Phylogenetic relations between microbats, megabats and primates (Mammalia, Chiroptera and Primates). *Phil. Trans. R. Soc. London, B, Biol. Sci.*, **325**, 489–559.

Pichi Sermolli, R.E.G. (1977) Tentamen pteridophytorum genera in taxonomicum ordinem redigendi. *Webbia*, **31**, 313–512.

Pickersgill, B. (1986) Evolution of hierarchical variation patterns under domestication and their taxonomic treatment, in *Infraspecific Classification of Wild and Cultivated Plants. The Systematics Association Special Volume No. 29*, (ed B.T. Styles), Clarendon Press, Oxford, pp. 191–209.

Pimentel, R.A. and Riggins, R. (1987) The nature of cladistic data. *Cladistis*, **3**, 201–9.

Plate, L. (1914) Kultur der Systematik mit besonderer Berücksichtigung des Systems der Tiere. *Kultur der Gegenwart*, **3**, 119–59.

Platnick, N.I. (1977a) Paraphyletic and polyphyletic groups. *Syst. Zool.*, **26**, 195–200.

Platnick, N.I. (1977b) Monotypy and the origin of higher taxa: A reply to E.O. Wiley. *Syst. Zool.*, **26**, 355–7.

Platnick, N.I. (1980) Philosophy and the transformation of cladistics. *Syst. Zool.*, **28**, 537–46.

Platnick, N.I. (1989) *Advances in Spider Taxonomy*, Manchester University Press, Manchester.

Platnick, N.I., Griswold, C.E. and Coddington, J.A. (1992) On missing entries in cladistic analysis. *Cladistics*, **7**, 337–43.

Platt, H.M. and Warwick, R.M. (1983) *Free-living Marine Nematodes. Part 1. British Enoplids. Synopses of the British Fauna (New Series) No. 28*, Cambridge University Press, Cambridge.

Poche, F. (1938) Supplement zu C.D. Sherborn's *Index Animalium. Festschrift 60. Geburtstage E. Strand*, **5**, 477–615.

Powell, J.R. (1982) Genetic and non genetic mechanisms of speciation, in *Mechanisms of Speciation*, (ed C. Barigozzi), Alan R. Liss, New York, pp. 67–74.

Powell, J.R. (1991) Molecular approaches to systematics and phylogeny reconstruction. *Bull. Zool.*, **58**, 295–8.

Prance, G.T. and Campbell, D.G. (1988) The present state of tropical floristics. *Taxon*, **37**, 519–48.

Prentice, H.C. (1986) Continuous variation and classification, in *Infraspecific Classification of Wild and Cultivated Plants. The Systematics Association Special Volume No. 29*, (ed B.T. Styles), Clarendon Press, Oxford, pp. 21–32.

Prothero, D.R., Manning, E.M. and Fischer, M. (1988) The phylogeny of the ungulates, in *The Phylogeny and Classification of the Tetrapods. Systematics Association Special Volume No. 35B*, (ed M.J. Benton), Clarendon Press, Oxford, Vol. 2, 201–34.

Provine, W.B. (1989) Founder effects and genetic revolutions in microevolution: An historical perspective, in *Genetics, Speciation, and the Founder Effect*, (eds L.V. Giddings, K.Y. Kaneshiro and W.W. Anderson), Oxford University Press, New York, pp. 43–76.

Pulawski, W.J. (1988) Revision of North American *Tachysphex* wasps including Central American and Caribbean species (Hymenoptera Sphecidae). *Mem. Calif. Acad. Sci.*, **10**, 1–211.

Puthz, V. (1981) Was ist *Dianous* Leach, 1819, was ist *Stenus* Latreille, 1796? Oder: Die Aporie des Stenologen und ihre taxonomischen Konsequenzen (Coleoptera, Staphylinidae). 180. Beitrag zur Kenntnis der Steninen. *Ent. Abh. Staat. Mus. Tierk. Dresden*, **44**, 87–132.

De Puytorac, P., Grain, P., Legendre, P. and Devaux, J. (1984) Essai d'application de l'analyse phenetique à la classification du phylum des Ciliophora. *J. Protozool.*, **31**, 496–507.

De Puytorac, P., Grain, J. and Mignot, J.-P. (1987) *Précis de Protistologie*, Société Nouvelle des Editions Boubée, Paris.

Qu, L.H., Perasso, R., Baroin, A. *et al.* (1988) Molecular evolution of the 5'-terminal domain of large subunit rRNA from lower eukaryotes. A broad phylogeny covering photosynthetic protists. *BioSystems*, **21**, 203–8.

De Queiroz, K. (1985) The ontogenetic method for determining character polarity and its relevance to phylogenetic systematics. *Syst. Zool.*, **34**, 280–99.

De Queiroz, K. (1988) Systematics and the Darwinian revolution. *Phil. Sci.*, **55**, 238–59.

De Queiroz, K. (1992) Phylogenetic definitions and taxonomic philosophy. *Biol. Philos.*, **7**, 295–313.

De Queiroz, K. and Donoghue, M.J. (1988) Phylogenetic systematics and the species problem. *Cladistics*, **4**, 317–38.

De Queiroz, K. and Donoghue, M.J. (1990a) Phylogenetic systematics or Nelson's version of cladistics? *Cladistics*, **6**, 61–75.

De Queiroz, K. and Donoghue, M.J. (1990b) Phylogenetic systematics and species revisited. *Cladistics*, **6**, 83–90.

De Queiroz, K. and Gauthier, J. (1990) Phylogeny as a central principle in taxonomy: Phylogenetic definitions of taxon names. *Syst. Zool.*, **39**, 307–22.

Raff, R.A. and Kaufman, Th.C. (1991) *Embryos, Genes, and Evolution* ('Indiana Edition'). Indiana University Press, Bloomington.

Ragan, M.A. and Lee, A.R. III (1991) Making phylogenetic sense of biochemical and morphological diversity among the protists, in *The Unity of Evolutionary Biology. Proceedings of the Fourth International Congress of Systematic and Evolutionary Biology*, (ed E.C. Dudley), Dioscorides Press, Portland, OR, Vol. 1, pp. 432–41.

Raikow, R.J. (1986) Why are there so many kinds of passerine birds? *Syst. Zool.*, **35**, 255–9.

Raikow, R.J. (1988) The analysis of evolutionary success. *Syst. Zool.*, **37**, 76–9.

Ramazzotti, G. and Maucci, W. (1983) Il phylum Tardigrada. *Mem. Ist. Ital. Idrobiol.*, **41**.

Ramsköld, L. and Edgecombe, G.D. (1991) Trilobite monophyly revisited. *Historical Biol.*, **4**, 267–83.

Ramsköld, L. and Hou, X. (1991) New early Cambrian animal and onychophoran affinities of enigmatic metazoans. *Nature*, **351**, 225–8.

Raubeson, L.A. and Jansen, R.K. (1992) Chloroplast DNA evidence on the ancient evolutionary split in vascular land plants. *Science*, **255**, 1697–9.

Raup, D.M. (1976) Species diversity in the Phanerozoic: A tabulation. *Paleobiology*, **2**, 279–88.

Raup, D.M. (1987) Major features of the fossil record and their implication for evolutionary rate studies, in *Rates of Evolution*, (eds K.S.W. Campbell and M.F. Day), Allen and Unwin, London, pp. 1–14.

Raven, C.E. (1942) *John Ray: Naturalist*, Cambridge University Press, Cambridge.

Raven, P.H. (1986) Moderns aspects of the biological species in plants, in *Modern Aspects of Species*, (eds K. Iwatsuki, P.H. Raven and W.J. Bock), University of Tokyo Press, Tokyo.

Raven, P.H. (1988) Tropical floristics tomorrow. *Taxon*, **37**, 549–60.

Raven, P.H. Berlin, B. and Breedlove, D.E. (1971) The origins of taxonomy. *Science*, **174**, 1210–13.

Raven, P.H., Dietrich, W. and Stubbe, W. (1979) An outline of the systematics of *Oenothera* subsect. *Euoenothera* (Onagraceae). *Syst. Bot.*, **4**, 242–52.

Ray, J. (1690) *Synopsis Methodica Stirpium Britannicarum*, Sam. Smith, London.

Reeck, G.R.C., De Haen, C., Doolittle, D.C. *et al.* (1987) 'Homology' in proteins and nucleic acids: A terminology muddle and a way out of it. *Cell*, **50**, 667.

Reif, W.F. (1984) Artabgrenzung und das Konzept der evolutionärer Art in der Paläontologie. *Zeitschr. Zool. Syst. Evol.–forsch.*, **22**, 263–86.

Reig, O.A., Aguilera, M., Barros, M.A. and Useche, M. (1980) Chromosomal speciation in a Rassenkreis of Venezuelan spiny rats (genus *Proechimys*) (Rodentia, Echimyidae). *Genetica*, **52/53**, 291–312.

Reisinger, E. (1960) Was ist *Xenoturbella*? *Z. Wiss. Zool.*, **164**, 188–98.

Reiswig, H.M. and Mackie, G.O. (1983) Studies on hexactinellid sponges. III. The taxonomic status of Hexactinellida within the Porifera. *Phil. Trans. R. Soc.*, **B 301**, 419–28.

Remane, A. (1952) *Die Grundlagen des natürlichen Systems, der vergleichenden Anatomie und der Phylogenetik*, Akademische Verlagsgesellschaft Geest and Portig, Leipzig.

Remington, C.L. (1968) Suture-zones of hybrid interaction between recently joined biotas. *Evol. Biol.*, **2**, 321–428.

Renno, J.-F., Berrebi, P., Boujard, T. *et al.* (1990) Intraspecific genetic differentiation of *Leporinus friderici* (Anostomidae, Pisces) in French Guiana and Brazil: A genetic approach to the refuge theory. *J. Fish Biol.*, **36**, 85–95.

Rensch, B. (1928) Grenzfälle von Rasse und Art. *J. Ornithol.*, **76**, 222–31.

Rensch, B. (1929) *Ds Prinzip geographischer Rassenkreise und das Problem der Artbildung*. Bornträger, Berlin.

Reyes, J.C., Mead, J.G. and Van Waerebeek, K.V. (1991) A new species of beaked whale *Mesoplodon peruvianus* sp. n. (Cetacea: Ziphiidae) from Peru. *Marine Mammal Sci.*, **7**, 1–24.

Reynolds, D.R. (1989) Consensus mycology. *Taxon*, **38**, 62–8.

Reynolds, D.R. and Taylor, J.W. (1991) DNA specimens and the 'International code of botanical nomenclature'. *Taxon*, **40**, 311–15.

Reynolds, J.W. and Cook, D.G. (1976) *Nomenclatura Oligochaetologica. A Catalogue of Names, Descriptions and Type Specimens of the Oligochaeta*, The University of New Brunswick, Fredericton, Canada.

Reynolds, J.W. and Cook, D.G. (1981) *Nomenclatura Oligochaetologica. Supplementum*

Primum. A Catalogue of Names, Desciptions and Type Specimens of the Oligochaeta, The University of Brunswick, Fredericton, Canada.

Rice, M.E. (1982) Sipuncula, in *Synopsis and Classification of Living Organisms*, (ed S.P. Parker), McGraw-Hill, New York, Vol. 2, pp. 67–9.

Richards, A.J. (1986) *Plant Breeding Systems*, Allen and Unwin, London.

Ride, W.D.L. (1970) *A Guide to the Native Mammals of Australia*. Oxford University Press, Melbourne.

Ride, W.D.L. (1988) Towards a unified system of biological nomenclature, in *Prospects in Systematics*, (ed D.L. Hawksworth), Clarendon Press, Oxford, pp. 332–53.

Ride, W.D.L. (1991) Justice for the living: A review of bacteriological and zoological initiatives in nomenclature, in *Improving the Stability of Names: Needs and Options. Regnum Vegetabile No. 123*, (ed D.L. Hawksworth), Koeltz, Königstein, pp. 105–22.

Ride, D.L. and Younès, T. (eds) (1986) *Biological Nomenclature Today. A Review of the Present State and Current Issues of Biological Nomenclature of Animals, Plants, Bacteria and Viruses. IUBS Monograph series No. 2*, IRL Press, Eynsham, Oxford.

Ridley, M. (1986) *Evolution and Classification. The Reformation of Cladism*, Longman, London.

Ridley, M. (1989) The cladistic solution to the species problem. *Biol. Philos.*, **4**, 1–16.

Rieger, R.M. (1980) A new group of interstitial worms, Lobatocerebridae nov. fam. (Annelida) and its significance for metazoan phylogeny. *Zoomorphologie*, **95**, 41–84.

Rieger, R.M. (1985) The phylogenetic status of the acoelomate organization within the Bilateria: A histological perspective, in *The Origins and Relationships of Lower Invertebrates. The Systematics Association Special Volume No. 28*, (eds S. Conway Morris *et al.*), Clarendon Press, Oxford, pp. 100–22.

Rieppel, O. (1979) Ontogeny and the recognition of primitive character states. *Z. Zool. Syst. Evol.-forsch.*, **17**, 57–61.

Rieppel, O.C. (1988a) *Fundamentals of Comparative Biology*, Birkhäuser, Basel.

Rieppel, O. (1988b) A review of the origin of snakes, in *Evolutionary Biology*, (eds M.K. Hecht, B. Wallace and G.T. Prance), Plenum Press, New York, Vol. 22, pp. 37–130.

Rieppel, O. (1988c) The classification of the Squamata, in *The Phylogeny and Classification of the Tetrapods. Systematics Association Special Volume No. 35A*, (ed M.J. Benton), Clarendon Press, Oxford, Vol. 1, pp. 261–93.

Rieppel, O. (1991) Things, taxa and relationships. *Cladistics*, **7**, 93–100.

Rieseberg, L.H. and Brunsfeld, S.J. (1992) Molecular evidence and plant introgression, in *Molecular Systematics of Plants*, (eds P.S. Soltis, D.E. Soltis and J.J. Doyle), Chapman & Hall, New York, pp. 151–76.

Riley, J., Banaja, A.A. and James, J.L. (1978) The phylogenetic relationships of the Pentastomida: The case for their inclusion within the Crustacea. *Int. J. Parasitol.*, **8**, 245–54.

Ringuelet, R.A. (1954) La clasificacion de los Hirudineos. *Notas Mus. Ciud. Eva Peron*, **17**, 1–15.

Rippka, R., Deruelles, J., Waterbury, J.B. *et al.* (1979) Generic assignments, strain histories and properties of pure cultures of Cyanobacteria. *J. Gen. Microbiol.*, **111**, 1–61.

Rivera, M.C. and Lake, J.A. (1992) Evidence that eukaryotes and eocyte prokaryotes are immediate relatives. *Science*, **257**, 74–6.

Robinson, H. and King, R.M. (1977) Eupatorieae – systematic review, in *The Biology and Chemistry of the Compositae*, (eds V.H. Heywood, J.B. Harborne and B.L. Turner), Academic Press, London, pp. 437–85.

Robinson, H. and King, R.M. (1987) Comments on the generic concepts in the Eupatorieae. *Taxon*, **34**, 11–16.

Robinson, R.A. (1991) Middle Cambrian biotic diversity: Examples from four Utah Lagerstätten, in *The Early Evolution of Metazoa and the Significance of Problematic Taxa*, (eds A.M. Simonetta and S. Conway Morris), Cambridge University Press, Cambridge, pp. 77–98.

Roewer, C.F. (1923) *Die Weberknechte der Erde*, Fischer, Jena.

Roewer, C.F.R. (1934) Solifugae, Palpigrada, in *H.G. Bronn's Klassen und Ordungen des Tierreichs*, Akademische Verlagsgesellschaft, Leipzig, **5**:4:4.

Roewer, C.F. (1942, 1954) *Katalog der Araneae von 1758 bis 1940 (bzw. 1954)*, Bremen: Natura; Bruxelles: Inst. R. Sci. Nat.

Rong, R., Chandley, A.C., Song, J. *et al.* (1988) A fertile mule and hinny in China. *Cytogenet. Cell. Genet.*, **47**, 134–9.

Rosen, D.E., Forey, P.L., Gardiner, B.G. and Patterson, C. (1981) Lungfishes, tetrapods, paleontology, and plesiomorphy. *Bull. Am. Mus. Nat. Hist.*, **167**, 159–276.

Rosenberg, G., Davis, G.M. and Kuncio, G.S. (1992) Extraordinary variation in conservation of D6 28S ribosomal RNA sequences in mollusks: Implications for phylogenetic analysis, in *Abstracts of the 11th International Congress of Malacology, Siena 1992*, (eds F. Giusti and G. Manganelli), University of Siena, Siena, pp. 221–2.

Ross, H.H. (1974) *Biological Systematics*, Addison-Wesley, Reading.

Ross, R. (1982) Bacillariophyceae, in *Synopsis and Classification of Living Organisms*, (ed S.P. Parker), McGraw-Hill, New York, Vol. 1, pp. 95–101.

Roth, L.M. (1982) Blattaria, in *Synopsis and Classification of Living Organisms*, (ed S.P. Parker), McGraw-Hill, New York, Vol. 2, pp. 346–7.

Roth, V.L. (1984) On homology. *Biol. J. Linn. Soc.*, **22**, 13–29.

Roth, V.L. (1988) The biological basis of homology, in *Ontogeny and Systematics*, (ed C.J. Humphries), British Museum (Natural History), London, pp. 1–26.

Roth, V.L. (1991) Homology and hierarchies: Problems solved and unresolved. *J. Evol. Biol.*, **4**, 167–94.

Rothschild, L.J. and Heywood, P. (1987) Protistan phylogeny and chloroplast evolution: Conflicts and congruence. *Prog. Protistol.*, **2**, 1–68.

Rothschild, L.J. and Heywood, P. (1988) 'Protistan' nomenclature: Analysis and refutation of some potential objections. *BioSystems*, **21**, 197–202.

Rouch, R. (1986) Copepoda: Les Harpacticoides souterrains des eaux douces continentales, in *Stygofauna Mundi*, (ed L. Botosaneanu), W.J. Brill/ Dr. W. Backhuys, Leiden, pp. 321–55.

Round, F.E., Crawford, R.M. and Mann, D.G. (1990) *The Diatoms. Biology and Morphology of the Genera*, Cambridge University Press, Cambridge.

Runnegar, B. (1987) Rates and modes of evolution in the Mollusca, in *Rates of Evolution*, (eds K.S.W. Campbell and M.F. Day), Allen and Unwin, London, pp. 39–60.

Saether, O.A. (1977) Female genitalia in Chironomidae and other Nematocera: Morphology, phylogenies, keys. *Bull. Fish. Res. Board Can.*, **197**, 1–211.

Saether, O.A. (1979a) Hierarchy of the Chironomidae with special emphasis on the female genitalia. *Ent. Scand. Suppl.*, **10**, 17–26.

Saether, O.A. (1979b) Underlying synapomorphies and anagenetic analysis. *Zool. Scr.*, **8**, 305–12.

Saether, O.A. (1979c) Underliggende synapomorfi og enestaende innvendig parallelisme belyst ved eksempler fra Chironomidae og Chaoboridae (Diptera). *Ent. Tidsskr.*, **100**, 173–80.

Saether, O.A. (1983) The canalized evolutionary potential: Inconsistencies in phylogenetic reasoning. *Syst. Zool.*, **32**, 343–59.

Saether, O.A. (1986) The myth of objectivity – Post-Hennigian deviations. *Cladistics*, **2**, 1–13.

Säve-Söderbergh, G. (1945) Notes on the trigeminal musculature in non-mammalian tetrapods. *Nova Acta Regiae Soc. Sci. Ups.*, **13**(4), 1–59.

Sakai, S. (1970–88) *Dermapterorum Catalogus*, Tokyo, **I–XXI**.

Salmon, J.T. (1964–1965) An index to the Collembola. *R. Soc. N. Z. Bull.*, **7**, 1–651.

Salvini-Plawen, L. (1968) Die Functions-Coelomtheorie in der Evolution der Mollusken. *Syst. Zool.*, **17**, 192–208.

Salvini-Plawen, L. (1978) On the origin and evolution of the lower Metazoa. *Z. Zool. Syst. Evol.-forsch.*, **16**, 40–87.

Salvini-Plawen, L. (1980) A reconsideration of systematics in the Mollusca (Phylogeny and higher classification). *Malacologia*, **19**, 249–78.

Salvini-Plawen, L. (1986) Systematic notes on *Spadella* and on the Chaetognatha in general. *Z. Zool. Syst. Evol.-forsch.*, **24**, 122–8.

Salvini-Plawen, L. (1988) Annelida and Mollusca – A prospectus, in *The Ultrastructure of Polychaeta*, (eds W. Westheide and C.O. Hermans), Gustav Fischer, Stuttgart, *Microfauna Marina*, **4**, 383–96.

Sander, K. (1989) Zum Geleit: Homologie und Ontogenese. *Zool. Beitr.*, **32**, 323–6.

Sanders, H.L., Hessler, R.R. and Garner, S.P. (1985) *Hirsutia bathyalis*, a new unusual deep-sea benthic peracaridan crustacean from the tropical Atlantic. *J. Crust. Biol.*, **5**, 30–57.

Sanderson, M.J. (1989) Confidence limits on phylogenies: The bootstrap revisited. *Cladistics*, **5**, 113–29.

Sanderson, M.J. (1991) Phylogenetic relationships within North American *Astragalus* L. (Fabaceae). *Syst. Bot.*, **16**, 414–30.

Sanderson, M.J. and Donoghue, M.J. (1989) Patterns of variation in levels of homoplasy. *Evolution*, **43**, 1781–95.

Sansom, I.J., Smith, M.P., Armstrong, H.A. and Smith, M.M. (1992) Presence of the earliest vertebrate hard tissues in conodonts. *Science*, **256**, 1308–11.

Sarich, V.M., Schmid, C.W. and Marks, J. (1989) DNA hybridization as a guide to phylogenies: A critical analysis. *Cladistics*, **5**, 3–32.

Sattler, R. (1984) Homology. A continuing challenge. *Syst. Bot.*, **9**, 382–94.

Savage, J.M. (1990) Meetings of the International Commission on Zoological Nomenclature. *Syst. Zool.*, **39**, 424–5.

Sawyer, R.T. (1984) Arthropodization in the Hirudinea: evidence for a phylogenetic link with insects and other Uniramia? *Zool. J. Linn. Soc.*, **80**, 303–22.

Sawyer, R.T. (1986) *Leech Biology and Behaviour*, Clarendon Press, Oxford, 3 vols.

Schaefer, C.W. (1988) The food plants of some 'primitive' Pentatomoidea (Hemiptera: Heteroptera). *Phytophaga*, **2**, 19–45.

Schaefer, C.W. and Ahmad, I. (1987) The food plants of four pentatomoid families (Hemiptera: Acanthosomatidae, Tessaratomidae, Urostylidae, and Dinidoridae). *Phytophaga*, **1**, 21–34.

Schaefer, C.W. and Mitchell, P.L. (1983) Food plants of the Coreoidea (Hemiptera: Heteroptera). *Ann. Entomol. Soc. America*, **76**, 591–615.

Schaefer, S.A. and Lauder, G.V. (1986) Historical transformation of functional design: Evolutionary morphology of feeding mechanisms of loricarioid catfishes. *Syst. Zool.*, **35**, 489–508.

Schaeffer, B. (1987) Deuterostome monophyly and phylogeny, in *Evolutionary Biology*, (eds M.K. Hecht, B. Wallace and G.T. Prance), Plenum Press, New York, **21**, 179–235.

Schaeffer, B., Hecht, M.K. and Eldredge, N. (1972) Palaeontology and Phylogeny. *Evol. Biol.*, **6**, 31–44.

Scheel, J. (1968) *Rivulius of the Old World*, T.F.H. Publications, Jersey City.

Scheller, U. (1982a) Symphyla, in *Synopsis and Classification of Living Organisms*, (ed S.P. Parker), McGraw-Hill, New York, Vol. 2, pp. 688–9.

Scheller, U. (1982b) Pauropoda, in *Synopsis and Classification of Living Organisms*, (ed S.P. Parker), McGraw-Hill, New York, Vol. 2, pp. 724–6.

Schilder, F.A. (1962) Das geographische Prinzip in der Taxonomie. *XI Intern. Kongr. Entomol.*, **3**, 329–33.

Schmidt, G.D. (1986) *Handbook of Tapeworm Identification*, CRC Press, Boca Raton.

Schmidt, T.M., De Long, E.F. and Pace, N.R. (1991) Analysis of a marine picoplankton community by 16S rRNA gene cloning and sequencing. *J. Bacteriol.*, **173**, 4371–8.

Schminke, H.K. (1982) Syncarida, in *Synopsis and Classification of Living Organisms*, (ed S.P. Parker), McGraw-Hill, New York, Vol. 2, pp. 233–7.

Schopf, T.J.M., Raup, D.M., Gould, S.J. and Simberloff, D.S. (1975) Genomic versus morphologic rates of evolution: Influence of morphologic complexity. *Paleobiology*, **1**, 63–70.

Schram, F.R. (1983a) Remipedia and crustacean phylogeny, in *Crustacean Phylogeny (Crustacean Issues, 1)*, (ed F.R. Schram), A.A. Balkema, Rotterdam, pp. 23–8.

Schram, F.R. (1983b) Introduction, in *Crustacean Phylogeny (Crustacean Issues, 1)*, (ed F.R. Schram), A.A. Balkema, Rotterdam, pp. IX–XI.

Schram, F.R. (1986) *Crustacea*, Oxford University Press, New York.

Schram, F.R. (1991) Cladistic analysis of metazoan phyla and the placement of fossil problematica, in *The Early Evolution of Metazoa and the Significance of Problematic Taxa* (eds A.M. Simonetta and S. Conway Morris), Cambridge University Press, Cambridge, pp. 35–46.

Schram, F.R. and Emerson, J.M. (1991) Arthropod pattern theory: A new approach to arthropod phylogeny. *Mem. Queensland Mus.*, **31**, 1–18.

Schultz, R.J. (1961) Reproductive mechanism of unisexual and bisexual strains of the viviparous fish *Poeciliopsis*. *Am. Nat.*, **15**, 302–25.

Schulze, F.E. (1883) *Trichoplax adhaerens* n. g. n. sp. *Zool. Anz.*, **6**, 92–7.

Schulze, F.E., Kükenthal, W. and Heider, K. (1926–1954) *Nomenclator animalium generum et subgenerum*, Akademische Verlagsgesellschaft, Berlin, 5 vols.

Schuster, R.M. (1966) Studies on Hepaticae. XV, Calobryales. *Nova Hedwigia*, **13**, 1–63.

Scoble, M.J. (1986) The structure and affinities of the Hedyloidea: a new concept of the butterflies. *Bull. Br. Mus. (Nat. Hist.)*, *Entomol.*, **53**, 251–86.

Scott-Ram, N.R. (1990) *Transformed Cladistics, Taxonomy and Evolution*, Cambridge University Press, Cambridge.

Seaman, F.C. (1982) Sesquiterpene lactones as taxonomic characters in the Asteraceae. *Bot. Rev.*, **48**, 121–595.

Searcy, D.G. (1987) Phylogenetic and phenotype relationships between the eukaryotic nucleocytoplasm and thermophilic archaebacteria. *Ann. N. Y. Acad. Sci.*, **503**, 168–70.

Searle, J.B. (1991) A hybrid zone comprising staggered chromosomal clines in the house mouse (*Mus musculus domesticus*). *Proc. R. Soc. London*, B, **246**, 47–52.

Seeno, T. and Wilcox, J.A. (1982) Leaf-beetle genera. *Entomography*, **1**, 1–221.

Self, J.T. (1982) Pentastomida, in *Synopsis and Classification of Living Organisms*, (ed S.P. Parker), McGraw-Hill, New York, Vol. 2, pp. 726–8.

Serres, E. (1827) Recherches d'anatomie trascendante, sur les lois de l'organogénie appliquées à l'anatomie pathologique. *Ann. Sci. Nat., Paris*, **11**, 47–70.

Shaw, A.B. (1969) Adam and Eve, paleontology, and the non-objective arts. *J. Paleontol.*, **43**, 1085–98.

Shear, W.A. (1992) End of the 'Uniramia' taxon. *Nature*, **359**, 477–8.

Sherborn, C.D. (1902; 1922–33) *Index Animalium sive Index Nominum quae ab A.D. MDCCLVIII Generibus et Speciebus Animalium Imposita Sunt*, Sectio prima (1758–1800) (1902); Sectio secunda (1901–1850) (1922–33). Cantabrigiae et Londinii.

Shubin, N.H. and Alberch, P. (1986) A morphogenetic approach to origin and basic organization of the tetrapod limb, in *Evolutionary Biology*, (eds M.K. Hecht, B. Wallace and G.T. Prance), Plenum Press, New York, Vol. 20, pp. 319–87.

Shultz, J.W. (1990) Evolutionary morphology and phylogeny of Arachnida. *Cladistics*, **6**, 1–38.

Sibley, C.G. (1985) DNA and proteins as sources of taxonomic data, in *A Dictionary of Birds*, (eds B. Campbell and E. Lack), The British Ornithologists' Union, pp. 150–1.

Sibley, C.G. and Ahlquist, J.E. (1973) The relationships of the hoatzin. *Auk*, **90**, 1–13.

Sibley, C.G. and Ahlquist, J.E. (1983) Phylogeny and classification of birds based on the data of DNA–DNA hybridization, in *Current Ornithology*, (ed R.F. Johnston), Plenum Press, New York, Vol. 1, pp. 245–92.

Sibley, C.G. and Ahlquist J.F. (1984) The phylogeny of hominoid primates, as indicated by DNA–DNA hybridization. *J. Mol. Evol.*, **20**, 2–15.

Sibley, Ch.G. and Ahlquist, J.E. (1987) Avian phylogeny reconstructed from comparisons of the genetic material, DNA, in *Molecules and Morphology in Evolution: Conflict or Compromise?* (ed C. Patterson), Cambridge University Press, Cambridge, pp. 95–121.

Sibley, Ch.G. and Ahlquist, J.E. (1990) *Phylogeny and Classification of Birds*. Yale University Press, New Haven.

Sibley, Ch.G., Ahlquist, J.E. and Monroe, B.L. Jr. (1988) A classification of the living birds of the world based on DNA–DNA hybridization studies. *Auk*, **105**, 409–23.

Sibley, Ch.G., Ahlquist, J.E. and Sheldon, F.H. (1987) DNA hybridization and avian phylogenetics. Reply to Cracraft, in *Evolutionary Biology*, (eds M.K. Hecht, B. Wallace and G.T. Prance), Plenum Press, New York, Vol. 21, pp. 97–125.

Sibley, Ch.G. and Monroe, B.L. Jr (1990) *Distribution and Taxonomy of Birds of the World*, Yale University Press, New Haven.

Siddiqi, M.R. (1986) *Tylenchida, parasites of plants and insects*, CAB, London.

Sidow, A. and Bowman, B.H. (1991) Molecular phylogeny. *Curr. Op. Genet. Dev.*, **1**, 451–6.

Siewing, R. (1977) Mesoderm bei Ctenophoren. *Z. Zool. Syst. Evol.-forsch.*, **15**, 1–8.

Silberman, J.D. and Walsh, P.J. (1992) Species identification of spiny lobster phyllosome larvae via ribosomal DNA analysis. *Mol. Mar. Biol. Biotechnol.*, **1**, 195–205.

Simmons, N.B., Novacek, M.J. and Baker, R.J. (1991) Approaches, methods, and the future of the Chiropteran monopyly controversy: A reply to J.D. Pettigrew. *Syst. Zool.*, **40**, 239–43.

Simpson, G.G. (1945) The principles of classification and a classification of mammals. *Bull. Am. Mus. Nat. Hist.*, **85**, 1–350.

Simpson, G.G. (1961) *Principles of Animal Taxonomy*, Columbia University Press, New York.

Sims, R.W. and Gerard, B.M. (1985) *Earthworms. Synopses of the British Fauna (New Series) No. 31,* E.J. Brill/Dr. W. Backhuys, London.

Skerman, V.B.D., McGowan, V. and Sneath, P.H.A. (eds) (1980) Approved lists of bacterial names. *Int. J. Syst. Bacteriol.,* **30,** 225–420.

Slater, J.A. (1964) *A Catalogue of the Lygaeidae of the World,* University of Connecticut, Storrs.

Sluiman, H.J. (1985) A cladistic evaluation of the lower and higher green plants (Viridiplantae). *Plant Syst. Evol.,* **149,** 217–32.

Sluys, R. (1989) Rampant parallelism: an appraisal of the use of nonuniversal derived character states in phylogenetic reconstruction. *Syst. Zool.,* **38,** 350–70.

Small, E.B. and Lynn, D.H. (1981) A new macrosystem for the phylum Ciliophora Doflein, 1901. *BioSystems,* **14,** 387–401.

Small, E.B. and Lynn, D.H. (1985) Phylum Ciliophora, in *Illustrated Guide to the Protozoa,* (eds J.J. Lee, S.H. Hutner and E.C. Bovee), Society of Protozoologists, Lawrence, Kansas, pp. 393–575.

Smit, F.G.A.M. (1982) Siphonaptera, in *Synopsis and Classification of Living Organisms,* (ed S.P. Parker), McGraw-Hill, New York, Vol. 2, pp. 557–63.

Smith, A. (1986) Bryophyte phylogeny: fact or fiction? *J. Bryol. ,* **14,** 83–9.

Smith, A.B. (1992) Echinoderm phylogeny: Morphology and molecules approach accord. *Trends Ecol. Evol.,* **7,** 224–8.

Smith, A.B. and Patterson, C. (1988) The influence of taxonomic method on the perception of patterns of evolution. *Evol. Biol.,* **23,** 127–216.

Smith, E.F.G., Arctander, P., Fieldsa, J. and Amir, O.G. (1991) A new species of shrike (Laniidae: *Laniarius*) from Somalia, verified by DNA sequence data from the only known individual. *Ibis,* **133,** 227–35.

Smith, D.C. (1988) Heritable divergence in *Rhagoletis pomonella* host races by seasonal asynchrony. *Nature,* **336,** 66–7.

Smith, H.M. (1965) More evolutionary terms. *Syst. Zool.,* **14,** 57–8.

Smith, H.M. (1969) Parapatry: Sympatry or allopatry? *Syst. Zool.,* **18,** 254–5.

Smithers, C.N. (1967) A catalogue of the Psocoptera of the world. *Aust. Zool.,* **14,** 1–145.

Smithies, O. (1955) Zone electrophoresis in starch gels: Group variation in the serum proteins of normal human adults. *Biochem. J.,* **61,** 629–41.

Smouse, P.E., Dowling, T.E., Tworek, J.A. *et al.* (1991) Effects of intraspecific variation on phylogenetic inference: A likelihood analysis of mtDNA restriction site data in cyprinid fishes. *Syst. Zool.,* **40,** 393–409.

Sneath, P.H.A. (1962) The construction of taxonomic groups, in *Microbial Classification,* (eds G.C. Ainsworth and P.H.A. Sneath), Cambridge University Press, Cambridge, pp. 289–332.

Sneath, P.H.A. (1986) Nomenclature of Bacteria, in *Biological Nomenclature Today. A Review of the Present State and Current Issues of Biological Nomenclature of Animals, Plants, Bacteria and Viruses. IUBS Monograph Series No. 2,* (eds D.L. Ride and T. Younès), IRL Press, Eynsham, Oxford, pp. 36–48.

Sneath, P.H.A. (ed) (1992) *International Code of Nomenclature of Bacteria and Statutes of the International Committee on Systematic Bacteriology and Statutes of the Bacteriology and Applied Microbiology Section of the International Union of Microbiological Societies: Bacteriological Code,* American Society for Microbiology, Washington.

Sneath, P.H.A. and Sokal, R.R. (1973) *Numerical Taxonomy,* W.H. Freeman, San Francisco.

Snowden, J.D. (1936) A classification of the cultivated sorghums. *Kew Bull.,* **1935,** 221–5.

Snowden, J.D. (1954) The wild fodder sorghums of the section Eu-Sorghum. *J. Linn. Soc. Bot.*, **4**, 191–260.

Sober, E. (1983) Parsimony methods in systematics, in *Advances in Cladistics*, 2,(eds N. Platnik and V. Funk), Columbia University Press, New York, pp. 37–47.

Sober, E. (1986) Parsimony and character weighting. *Cladistics*, **2**, 28–42.

Sober, E. (1988) *Reconstructing the Past. Parsimony, Evolution, and Inference*, MIT Press, Cambridge, MA.

Sogin, M.L., (1991) Early evolution and the origin of eukaryotes. *Curr. Op. Genet. Dev.*, **1**, 457–63.

Sogin, M.L., Gunderson, J.H., Elwood, H.J. *et al.* (1989) Phylogenetic meaning of the kingdom concept: An unusual ribosomal RNA from *Giardia lamblia*. *Science*, **243**, 75–7.

Sokal, R.R. and Sneath, P.H.A. (1963) *Principles of Numerical Taxonomy*, W.H. Freeman, San Francisco.

Solignac, M. and Monnerot, M. (1986) Race formation, speciation, and introgression within *Drosophila simulans*, *D. mauritiana*, and *D. sechellia* inferred from mitochondrial DNA analysis. *Evolution*, **40**, 531–9.

Soltis, D.E., Soltis, P.S. and Milligan, B.G. (1992b) Intraspecific chloroplast DNA variation: Systematic and phylogenetic implications, in *Molecular Systematics of Plants*, (eds P.S. Soltis, D.E. Soltis and J.J. Doyle), Chapman and Hall, New York, pp. 117–50.

Soltis, P.S., Doyle, J.J. and Soltis, D.E. (1992a) Molecular data and polyploid evolution in plants, in *Molecular Systematics of Plants*, (eds P.S. Soltis, D.E. Soltis and J.J. Doyle), Chapman and Hall, New York, pp. 177–201.

Soós, A. and Papp, L. (eds) (1984) *Catalogue of Palaearctic Diptera*, Elsevier, Amsterdam.

Southward, E.C. and Southward, A.J. (1988) New observations on the biology and phylogeny of Pogonophora. *Am. Zool.*, **28**, 8A.

Spolsky, C. and Uzzell, T. (1986) Evolutionary history of the hybridogenetic frog *Rana esculenta* as deduced from mtDNA analyses. *Mol. Biol. Evol.*, **3**, 44–56.

Spolsky, C.M., Phillips, C.A. and Uzzell, T. (1992) Antiquity of clonal salamander lineages revealed by mitochondrial DNA. *Nature*, **356**, 706–8.

Sprague, G.F. Jr (1991) Genetic exchange between kingdoms. *Curr. Opin. Gen. Dev.*, **1**, 530–3.

Springer, M. and Krajewski, C. (1989) DNA hybridization in animal taxonomy: A critique from first principles. *Quart. Rev. Biol.*, **64**, 291–318.

Stace, C.A. (1986) The present and future infraspecific classification of wild plants. In *Infraspecific Classification of Wild and Cultivated Plants. The Systematics Association Special Volume No. 29*, (ed B.T. Styles), Clarendon Press, Oxford, pp. 9–20.

Stace, C.A. (1989) *Plant Taxonomy and Biosystematics*, 2nd edn, Edward Arnold, London.

Stafleu, F.A. and Cowan, R.S. (1976–1988) *Taxonomic Literature*, International Association of Plant Taxonomists, Utrecht, 7 vols (also Supplement I, by Stafleu, F.A. and Mennega, E.A., 1992).

Stanier, R.Y., Sistrom, W.R. Hansen, T.A. *et al.* (1978) Proposal to place the nomenclature of the Cyanobacteria (blue–green algae) under the rules of the International Code of Nomenclature of Bacteria. *Int. J. Syst. Bacteriol.*, **28**, 335–6.

Stanley, G.D. Jr and Stürmer, W. (1983) The first fossil ctenophore from the Lower Devonian of West Germany. *Nature*, **303**, 518–20.

Stanley, G.D. Jr and Stürmer, W. (1987) A new fossil ctenophore discovered by X-rays. *Nature*, **328**, 61–3.

Starobogatov, Y.I. (1988) Systematics of Crustacea. *J. Crust. Biol.*, **8**, 300–11.

Starobogatov, Y.I. (1991) Problems in the nomenclature of higher taxonomic categories. *Bull. Zool. Nomencl.*, **48**, 6–18.

Stearn, W.T. (1981) Bibliography in the British Museum (Natural History), in *History in the Service of Systematics. Society for the Bibliography of Natural History Special Volume Publication Number 1*, (eds A. Wheeler and J.H. Price), Society for the Bibliography of Natural History, London, pp. 1–6.

Stebbins, G.L. (1950) *Variation and Evolution in Plants*, Columbia University Press, New York.

Stebbins, G.L. (1981) Why are there so many species of flowering plants? *BioScience*, **31**, 573–7.

Stefani, R. (1956) Il problema della partenogenesi in 'Haploembia solieri' Ramb. (Embioptera, Oligotomidae). *Atti Accad. Naz. Lincei, Mem. Cl. Sci. Fis. Mat. Nat., sez. IIIa*, **5**, 127–201.

Stefani, R. (1960) L'*Artemia salina* partenogenetica a Cagliari. *Riv. Biol.*, **53**, 463–91.

Steiner, W.W.M. (1981) Parasitization and speciation in mosquitoes: A hypothesis, in *Cytogenetics and Genetics of Vectors*, (eds R. Pal, J.B. Kitzmiller and T. Kanda), Elsevier Biomedical Press, New York, pp. 91–119.

Steinmann, H. (1989) *World Catalogue of Dermaptera*, Series Entomologica, **43**.

Stephen, A.C. and Edmonds, S.J. (1972) *The Phyla Sipuncula and Echiura*, Trustees of British Museum (Natural History), London.

Sterck, A.A., Groenhart, M.C. and Mooren, J.F.A. (1983) Aspects of the ecology of some microspecies of *Taraxacum* in the Netherlands. *Acta Bot. Neerl.*, **32**, 385–415.

Sterrer, W., Mainitz, M. and Rieger, R.M. (1985) Gnathostomulida: Enigmatic as ever, in *The Origins and Relationships of Lower Invertebrates. The Systematics Association Special Volume No. 28*, (eds S. Conway Morris *et al.*), Clarendon Press, Oxford, pp. 181–99.

Stevens, P.F. (1986) Evolutionary classification in botany, 1960–1985. *J. Arnold Arboretum*, **67**, 313–39.

Stevens, P.F. (1991) Character states, morphological variation, and phylogenetic analysis: A review. *Syst. Bot.*, **16**, 553–83.

Steyskal, G.C. (1965) Trend curves of the rate of species description in zoology. *Science*, **149**, 880–2.

Steyskal, G.C. (1973) Notes on the growth of taxonomic knowledge in the Psocoptera and the grammar of the nomenclature of the order. *Proc. Entomol. Soc. Washington*, **75**, 160–4.

Steyskal, G.C. (1976) Notes on the nomenclature and taxonomic growth of the Plecoptera. *Proc. Biol. Soc. Washington*, **88**, 399–428.

Stock, D.W., Swoffords, D.L., Sharp, P.M. *et al.* (1991) Coelacanth's relationships. *Nature*, **353**, 217–19.

Stock, D.W. and Whitt, G.S. (1992) Evidence from 18S ribosomal RNA sequences that lampreys and hagfishes form a natural group. *Science*, **257**, 787–9.

Stone, A.R. and Hawksworth, D.L. (eds) (1986) *Coevolution and Systematics. The Systematics Association Special Volume No. 32*, Clarendon Press, Oxford.

Stone, A.C., Sabrosky, C.W., Wirth, W.W. *et al.* (eds) (1965) *A Catalog of the Diptera of America North of Mexico*, U.S. Dept. Agric., Agric. Res. Serv., Agric. Handbook.

Stork, N.E. (1988) Insect diversity: Facts, fiction and speculation. *Biol. J. Linn. Soc.*, **35**, 321–37.

Stuessy, T.F. (1990) *Plant Taxonomy: The Systematic Evaluation of Comparative Data* Columbia University Press, New York.

Stunkard, H.W. (1982) Mesozoa, in *Synopsis and Classification of Living Organisms*, (ed S.P. Parker), McGraw-Hill, New York, Vol. 1, pp. 853–5.

Sturmbauer, C. and Meyer, A. (1992) Genetic divergence, speciation and morphological stasis in a lineage of African cichlid fishes. *Nature*, **358**, 578–81.

Styles, B.T. (ed) (1986) *Infraspecific Classification of Wild and Cultivated Plants. The Systematics Association Special Volume No. 29*, Clarendon Press, Oxford.

Sues, H.-D. (1992) No Palaeocene 'mammal-like reptile'. *Nature*, **359**, 278.

Sugihara, G. and May, R.M. (1990) Applications of fractals in ecology. *Trends Ecol. Evol.*, **5**, 79–86.

Swanson, C.J. (1982) Nematomorpha, in *Synopsis and Classification of Living Organisms*, (ed S.P. Parker), McGraw-Hill, New York, Vol. 1, pp. 931–2.

Swofford, D.L. and Olsen, G.J. (1990) Phylogeny reconstruction, in *Molecular Systematics*, (eds D.M. Hillis and C. Moritz), Sinauer Associates, Sunderland, MA, pp. 411–501.

Sytsma, K.J. (1990) DNA and morphology: Inference of plant phylogeny. *Trends Ecol. Evol.*, **5**, 104–10.

Sytsma, K. and Schaal, B. (1991) Ribosomal DNA variation within and among individuals of *Lisianthus* (Gentianaceae) populations. *Plant Syst. Evol.*, **170**, 97–106.

Szmidt, A.E., Alden, T. and Hallgren, J.-E. (1987) Paternal inheritance of chloroplast DNA in *Larix*. *Plant Mol. Biol.*, **9**, 59–64.

Szymura, J.M., Spolsky, C. and Uzzell, T. (1985) Concordant change in mitochondrial and nuclear genes in a hybrid zone between two frog species (genus *Bombina*). *Experientia*, **41**, 1464–70.

Takhtajan, A. (1969) *Flowering Plants. Origin and Dispersal*, Oliver and Boyd, Edinburgh.

Takhtajan, A.L. (1980) Outline of the classification of flowering plants (Magnoliophyta). *Bot. Rev.*, **46**, 225–359.

Takhtajan, A. (1983) The systematic arrangement of dicotyledonous familes, in Metcalfe, C.R. and Chalk, L. (eds), *Anatomy of Dicotyledons*, (eds C.R. Metcalfe and L. Chalk), 2nd edn, Clarendon Press, Oxford, Vol. 2, pp. 180–201.

Takhtajan, A. (1987) *Sistema Magnoliophytov*, Izdatel'stvo 'Nauko', Leningrad.

Tassy, P. (1988) The classification of Proboscidea. *Cladistics*, **4**, 43–57.

Tassy, P. and Shoshani, J. (1988) The Tethytheria: Elephants and their relatives, in *The Phylogeny and Classification of the Tetrapods. Systematics Association Special Volume No. 35B*, (ed M.J. Benton), Clarendon Press, Oxford, Vol. 2, pp. 283–315.

Taylor, F.J.R., Sarjeant, W.A.S., Fensome, R.A. and Williams, G.L. (1987) Standardisation of nomenclature in flagellate groups treated by both the botanical and zoological codes of nomenclature. *Syst. Zool.*, **36**, 79–85.

Tehler, A. (1988) A cladistic outline of the Eumycota. *Cladistics*, **4**, 227–77.

Templeton, A.R. (1982) Genetic architectures of speciation, *in Mechanisms of Speciation*, (ed C. Barigozzi), Alan R. Liss, New York, pp. 105–21.

Templeton, A.R. (1987) Genetic systems and evolutionary rates, in *Rates of Evolution*, (eds K.S.W. Campbell and M.F. Day), Allen and Unwin, London, pp. 218–34.

Templeton, A.R. (1989) The meaning of species and speciation: A genetic perspective, in *Speciation and its Consequences*, (eds D. Otte and J.A. Endler), Sinauer Associates, Sunderland, MA, pp. 3–27.

Thomas, C.D. (1990) Fewer species. *Nature*, **347**, 237.

Thompson, J.N. (1987) Symbiont induced speciation. *Biol. J. Linn. Soc.*, **32**, 385–92.

Thorne, A.G. and Wolpoff, M.H. (1992) The multiregional evolution of humans. *Sci. Am.*, **266**(4), 28–33.

Thorne, R.F. (1968) Synopsis of a putatively phylogenetic classification of the flowering plants. *Aliso*, **6**(4), 57–66.

Thorne, R.F. (1976) A phylogenetic classification of the Angiospermae, in *Evolutionary Biology*, (eds M.K. Hecht, W.C. Steere and B. Wallace), Plenum Press, New York, Vol. 9, pp. 35–106.

Thorne, R.F. (1983) Proposed new realignments in the angiosperms. *Nordic J. Bot.*, **3**, 85–117.

Thorpe, W.H. (1929) Biological races in *Hyponomeuta padella* L. *Zool. J. Linn. Soc.*, **36**, 621–34.

Thorpe, W.H. (1930) Biological races in insects and allied groups. *Biol. Rev.*, **5**, 177–212.

Tillier, S., Masselot, M., Guerdoux, J. and Tillier, A. (1992a) Large subunit ribosomal RNA phylogeny of Gastropoda, in *Abstracts of the 11th International Malacological Congress, Siena 1992*, (eds F. Giusti and G. Manganelli), University of Siena, Siena, pp. 225–6.

Tillier, S., Masselot, M., Philippe, H. and Tiller, A. (1992b) Phylogénie moléculaire des Gastropoda (Mollusca) fondé sur le séquençage partial de l'ARN ribosomique 28S. *C. R. Acad. Sci. Paris, Sér. 3*, **314**, 79–85.

Townes, H. (1969–1971) The genera of Ichneumonidae. *Mem. Am. Entomol. Inst.*, **11**, 1–300; **12**, 1–537; **13**, 1–307; **17**, 1–372.

Townes, H. and Townes, M. (1981) A revision of the Serphidae (Hymenoptera). *Mem. Am. Entomol. Inst.*, **32**, 1–541.

Traverse, A. (1988) Plant evolution dances to a different beat. Plant and animal evolutionary mechanisms compared. *Historical Biol.*, **1**, 277–301.

Tunner, H.G. (1970) Das Serumeiweissbild einheimischer Wasserfrösche und der Hybridcharakter von *Rana esculenta*. *Verh. Dtsch. Zool. Ges.*, **64**, 352–8.

Turesson, G. (1922) The species and the variety as ecological units. *Hereditas*, **3**, 100–13.

Turner, S., Burger-Wiersma, T., Giovannoni, S.J. *et al.* (1989) The relationship of a prochlorophyte *Prochlorothrix hollandica* to green chloroplasts. *Nature*, **337**, 380–2.

Tutin, T.G., Heywood, V.H., Burges, N.A. *et al.* (eds) (1964–1980) *Flora Europea*, Cambridge University Press, Cambridge, **1** (1964), 464 pp.; **2** (1968), 455 pp.; **3** (1972), 370 pp.; **4** (1976), 505 pp.; **5** (1980), 452 pp.

Udayagiri, S. and Wadhii, S.R. (1989) *Catalogue of Bruchidae*, Am. Entomol. Inst. Mem. No. 45.

Urbach, E., Robertson, D.L. and Chisholm, S.W. (1992) Multiple evolutionary origins of prochlorophytes within the cyanobacterial radiation. *Nature*, **355**, 267–9.

Uzzell, T. and Berger, L. (1975) Electrophoretic phenotypes of *Rana ridibunda, Rana lessonae*, and their hybridogenetic associate, *Rana esculenta*. *Proc. Acad. Nat. Sci. Philadelphia*, **127**, 13–23.

Vacelet, J. (1985) Coralline sponges and the evolution of Porifera, in *The Origins and Relationships of Lower Invertebrates. The Systematics Association Special Volume No. 28*, (eds S. Conway Morris *et al.*), Clarendon Press, Oxford, pp. 1–13.

Van Beneden, E. (1882) Contribution a l'histoire des Dicyemides. *Arch. Biol.*, **3**, 195–228.

Van der Vecht, J.L. and Ferrière, C. (eds) (1965–) *Hymenopterorum Catalogus*, W. Junk, Gravenhage.

Vane-Wright, R.I., Schulz, S. and Boppré, M. (1992) The cladistics of *Amauris* butterflies: Congruence, consensus and total evidence. *Cladistics*, **8**, 125–38.

Van Steenis, C.G.G. (1969) Plant speciation in Malesia, with special reference to the theory of non-adaptive saltatory evolution. *Biol. J. Linn. Soc.*, **1**, 97–133.

Van Valen, L. (1973) Are categories in different phyla comparable? *Taxon*, **22**, 333–73.

Van Valen, L. (1976) Ecological species, multispecies, and oaks. *Taxon*, **25**, 233–9.

Van Valen, L.M. (1978) Arborescent animals and other colonoids. *Nature*, **276**, 318.

Van Valen, L. (1982) Homology and causes. *J. Morphol.*, **173**, 305–12.

Van Valen, L.M. (1988) Species, sets and the derivative nature of philosophy. *Biol. Philos.*, **3**, 49–66.

Van Z. and Engelbrecht, D. (1969) The annelid ancestry of the Chordates and the origin of the chordate central nervous system and the notochord. *Z. Zool. Syst. Evol.-forsch.*, **7**, 18–30.

Verity, R. (1925) Remarks on the evolution of the *Zygaenae* and an attempt to analyse and classify the variation of *Z. lonicerae* Sch. and of *Z. trifolii* Esp. *Ent. Rec.*, **37**, 101–4, 154–8.

Verity, R. (1926) The geographical and seasonal variations of *Coenonympha pamphilus*. *Ztschr. Wiss. Insektenbiol.*, **21**, 191–208.

Vermeij, G.J. (1988) The evolutionary success of passerines: A question of semantics? *Syst. Zool.*, **37**, 69–71.

Via, S. (1984a) The quantitative genetics of polyphagy in an insect herbivore: 1. Genotype–environment interaction in larval performance on different host plant species. *Evolution*, **38**, 881–95.

Via, S. (1984b) The quantitative genetics of polyphagy in an insect herbivore: 2. Genetic correlations in larval performance within and among plants. *Evolution*, **38**, 896–905.

Via, S. (1986) Genetic covariance between oviposition preference and larval performance in an insect herbivore. *Evolution*, **40**, 778–85.

Vickerman, K. (1982) Zoomastigophora, in *Synopsis and Classification of Living Organisms*, (ed S.P. Parker), McGraw-Hill, New York, Vol. 2, pp. 496–508.

Vicq d'Azyr, F. (1786) *Traité d'Anatomie et de Physiologie*, **1**. Paris.

Vigna Taglianti, A. (1982) Carabidae. Introduzione, in *Fauna d'Italia, 18. Coleoptera Carabidae. I. Introduzione, Paussinae, Carabinae*, (eds A. Casale, M. Sturani and A. Vigna Taglianti), Calderini, Bologna, pp. 1–51.

Vives Moreno, A. (1988) Catalogo mundial sistematico y de distribución de la familia Coleophoridae Hübner (1825) (Insecta, Lepidoptera). *Bol. Sanidad Vegetal*, **12**, 196 pp.

Vossbrinck, C.R., Maddox, J.V., Friedman, S. *et al.* (1987) Ribosomal RNA sequence suggests Microsporidia are extremely ancient eukaryotes. *Nature*, **326**, 411–14.

Vossbrinck, C.R. and Freidman, S. (1989) A 28s ribosomal RNA phylogeny of certain cyclorrhaphous Diptera based upon a hypervariable region. *Syst. Entomol.*, **14**, 417–31.

Vrana, P. and Wheeler, W. (1992) Individual organisms as terminal entities: Laying the species problem to rest. *Cladistics*, **8**, 67–72.

Vrba, E.S. (ed), (1985) *Species and Speciation*, Transvaal Museum Monograph 4.

Vrijenhoek, F.C., Dawley, R.M., Cole, C.J. *et al.* (1989) A list of known unisexual vertebrates, in *Evolution and Ecology of Unisexual Vertebrates*, (eds R.M. Dawley and J.P. Bogart), State University of New York Press, Albany.

Waddington, C. (1957) *The Strategy of the Genes*, Macmillan, New York.

Wägele, J.-W. (1989) Evolution und phylogenetisches System der Isopoda. *Zoologica*, **140**, 1–262.

Wagner, D.F., Furnier, G.R., Saghai-Maroof, M.A. *et al.* (1987) Chloroplast DNA polymorphism in lodgepole and jack pines and their hybrids. *Proc. Natl. Acad. Sci. USA*, **84**, 2097–100.

Wagner, G.P. (1989a) The biological homology concept. *Annu. Rev. Ecol. Syst.*, **20**, 51–69.

Wagner, G.P. (1989b) The origin of morphological characters and the biological basis of homology. *Evolution*, **43**, 1157–71.

Wagner, W.H. Jr (1970) Biosystematics and evolutionary noise. *Taxon*, **19**, 146–51.

Walker, J.M. (1986) The taxonomy of parthenogenetic species of hybrid origin: Cloned hybrid populations of *Cnemidophorus* (Sauria, Teiidae). *Syst. Zool.*, **35**, 427–40.

Walker, W.F. (1985) 5S and 5.8S ribosomal RNA sequences and protist phylogenetics. *BioSystems*, **18**, 269–78.

Walter, M.R. (1987) The timing of major evolutionary innovations from the origin of life to the origins of the Metaphyta and Metazoa: The geological evidence, in *Rates of Evolution*, (eds K.S.W. Campbell and M.F. Day), Allen and Unwin, London, pp. 15–38.

Walters, S.M. (1961) The shaping of angiosperm taxonomy. *New Phytol.* , **60**, 74–84.

Walters, S.M. (1986) The name of the rose: A review of ideas on the European bias in angiosperm classification. *New Phytol.*, **104**, 527–46.

Ward, D.M., Weller, R. and Bateson, M.M. (1990) 16S rRNA sequences reveal numerous uncultured microorganisms in a natural community. *Nature*, **345**, 63–5.

Watling, L. (1982) Cumacea, in *Synopsis and Classification of Living Organisms*, (ed S.P. Parker), McGraw-Hill, New York, Vol. 2, pp. 243–5.

Watling, L. (1983) Peracaridan disunity and its bearing on eumalacostracan phylogeny with a redefinition of eumalacostracan superorders, in *Crustacean Phylogeny (Crustacean Issues, 1)*, (ed F.R. Schram), A.A. Balkema, Rotterdam, pp. 213–28.

Watson, L. (1971) Basic taxonomy data: The need for organisation over presentation and accumulation. *Taxon*, **20**, 131–6.

Wayne, R.K., George, S.B., Gilbert, D. *et al.* (1991) A morphologic and genetic study of the Island Fox, *Urocyon littoralis*. *Evolution*, **45**, 1849–68.

Webb, M. (1969) *Lamellibrachia barhami* gen. nov., sp. nov. (Pogonophora) from the northeast Pacific. *Bull. Mar. Sci.*, **19**, 18–47.

Weber, H.E. (1981) *Sonderb. Naturwiss. Ver. Hamburg*, **4**, 1–229.

Wendel, J.F. and Albert, V.A. (1992) Phylogenetics of the cotton genus (*Gossypium*): Character-state weighted parsimony analysis of chloroplast-DNA restriction site data and its systematic and biogeographic implications. *Syst. Bot.*, **17**, 115–43.

Wenzel, W. and Hemleben, V. (1982) A comparative study of genomes in angiosperms. *Plant. Syst. Evol.*, **139**, 209–27.

Werman, S.D., Springer, M.S. and Britten, R.J. (1990) Nucleic acids I: DNA–DNA hybridization, in *Molecular Systematics*, (eds D.M. Hillis and C. Moritz), Sinauer Associates, Sunderland, MA, pp. 204–49.

Werner, B. (1973) New investigations on systematics and evolution of the class Scyphozoa and the phylum Cnidaria. *Publ. Seto Mar. Biol. Lab.*, **20**, 35–61.

Werner, B. (1975) Bau und Lebensgeschichte des Polypen von *Tripedalia cystophora* (Cubozoa, class. nov., Carybdeidae) und seine Bedeutung für die Evolution der Cnidaria. *Helgoländer Wiss. Meeresunters.*, **27**, 461–504.

Werner, B. (1984) 4. Stamm Cnidaria. – 5.Stamm Ctenophora, in *Kaestner's Lehrburch der speziellen Zoologie*, (ed H.E. Gruner), Gustav Fischer, Stuttgart, **1**(2), 11–335.

Wernham, H.F. (1912). Floral evolution: with particular reference to the sympetalous dicotyledons. IX.– Summary and conclusions. Evolutionary genealogy; and some principles of classification. *New Phytol.*, **11**, 373–97.

Werth, C.R., Guttman, S.I. and Esbaugh, W.H. (1985) Recurring origins of allopolyploid species in *Asplenium*. *Science*, **228**, 731–3.

West, J.G., Conn, B.J., Jarzembowski, E.A. *et al.* (1990) In defence of taxonomy. *Nature*, **347**, 222–4.

Westblad, E. (1949) *Xenoturbella bocki* n. g., n. sp., a peculiar, primitive turbellarian type. *Arkiv f. Zool.*, **1**, 11–29.

Westheide, W. (1985) The systematic position of the Dinophilidae and the archiannelid problem, in *The Origins and Relationships of Lower Invertebrates. The Systematics Association Special Volume No. 28*, (eds S. Conway Morris *et al.*), Clarendon Press, Oxford, pp. 310–26.

De Wet, J.M.J. (1978) Systematics and evolution of *Sorghum* sect. *Sorghum* (Gramineae). *Am. J. Bot.*, **65**, 477–84.

De Wet, J.M.J., Harlan, J.R. and Brink, D.E. (1986) Reality of infraspecific taxonomic units in domesticated cereals, in *Infraspecific Classification of Wild and Cultivated Plants. The Systematics Association Special Volume No. 29*, (ed B.T. Styles), Clarendon Press, Oxford, pp. 211–22.

Wetzel, R.M., Dubos, R.E., Martin, R.L. and Myers, P. (1975) *Catagonus*, an 'extinct' peccary, alive in Paraguay. *Science*, **189**, 379–81.

Weygoldt, P. and Paulus, H.P. (1979) Untersuchung zur Morphologie, Taxonomie und Phylogenie der Chelicerata. II. Cladogramme und die Entfaltung der Chelicerata. *Z. Zool. Syst. Evol.-forsch.*, **17**, 177–200.

Wheatley, B.P. (1980) Malaria as a possible selective factor in the speciation of macaques. *J. Mammalogy*, **61**, 307–11.

Wheeler, Q.D. (1986) Character weighting and cladistic analysis. *Syst. Zool.*, **35**, 102–9.

Wheeler, Q.D. (1990) Ontogeny and character phylogeny. *Cladistics*, **6**, 225–68.

White, M.J.D. (1968) Models of speciation. *Science*, **159**, 1065–70.

White, M.J.D. (1978) *Modes of Speciation*, Freeman, San Francisco.

White, M.J.D. (1982) Rectangularity, speciation, and chromosome architecture, in *Mechanisms of Speciation*, (ed C. Barigozzi), Alan R. Liss, New York, pp. 75–103.

White, M.J.D. (1985) Types of hybrid zones. *Boll. Zool.*, **52**, 1–20.

White, R.E. (1975) Trend curves of the rate of species description for certain north American Coleoptera. *The Coleopterists Bull.*, **29**, 281–95.

Whittaker, R.H. (1959) On the broad classification of organisms. *Quart. Rev. Biol.*, **34**, 210–26.

Whittaker, R.H. (1969) New concepts of kingdoms of organisms. *Science*, **163**, 150–60.

Whittaker, R.H. and Margulis, L. (1978) Protist classification and the kingdoms of organisms. *BioSystems*, **10**, 3–18.

Whittemore, A. and Schaal, B. (1991) Interspecific gene flow in sympatric oaks. *Proc. Natl. Acad. Sci. USA*, **88**, 2540–4.

Wible, J.R. and Novacek, M.J. (1988) Cranial evidence for the monophyletic origin of bats. *Am. Mus. Novit.*, **2911**, 1–19.

Wiggins, G.B. (1982) Trichoptera, in *Synopsis and Classification of Living Organisms*, (ed S.P. Parker), McGraw-Hill, New York, Vol. 2, pp. 599–612.

Wijk, R. van der, Margadant, W.D. and Florschutz, P.A. (1959–1969) *Index Muscorum*, The International Association for Plant Taxonomy, Utrecht, 5 vols.

Wilcox, J.A. (1973) Chrysomelidae: Galerucinae. Luperini: Luperina. *Coleopterorum Catalogus*, **78**(3) (suppl.), 433–663.

Wiley, E.O. (1978) The evolutionary species concept reconsidered. *Syst. Zool.*, **28**, 88–92.

Wiley, E.O. (1981) *Phylogenetics: The Theory and Practice of Phylogenetic Systematics*, Wiley, New York.

Williams, C.A. and Harborne, J.B. (1988) Distribution and evolution of flavonoids in the monocotyledons, in *The Flavonoids, Advances in Research Since 1980*, (ed J.B. Harborne), Chapman & Hall, London, pp. 505–24.

Williams, C.B. (1947) The logarithmic series and its application to biological problems. *J. Ecol.*, **34**, 253–72.

Williams, C.B. (1964) *Patterns in the Balance of Nature*, Academic Press, London.

Williams, D.M. (1991) Phylogenetic relationships among the Chromista: A review and preliminary analysis. *Cladistics*, **7**, 141–56.

Williams, P.H., Humphries, C.J. and Vane-Wright, R.I. (1991) Measuring biodiversity: Taxonomic relatedness for conservation priorities. *Aust. Syst. Bot.*, **4**, 665–79.

Willis, J.C. (1922) *Age and Area*, Cambridge University Press, Cambridge.

Willis, J.C. (1973) *A Dictionary of the Flowering Plants*, (revised by H.K. Airy Shaw), 8th edn, Cambridge University Press, Cambridge.

Willis, J.C. and Yule, G.U. (1922) Some statistics of evolution and geographical distribution in plants and animals, and their significance. *Nature*, **109**, 177–9.

Willmann, R. (1983) Biospecies und phylogenetische Systematik. *Z. Zool. Syst. Evol.-forsch.*, **21**, 241–9.

Willmann, R. (1985) *Die Art in Raum und Zeit. Das Artkonzept in der Biologie und Paläontologie*, Paul Parey, Berlin.

Willmann, R. (1986) Reproductive isolation and the limits of the species in time. *Cladistics*, **2**, 356–8.

Willamnn, R. (1987) Missverständnisse um das biologische Artkonzept. *Paläont. Zeitschr.*, **61**, 3–15.

Willmer, P. (1990) *Invertebrate Relationships. Patterns in Animal Evolution*, Cambridge University Press, Cambridge.

Wilson, A.C. and Cann, R.L. (1992) The recent African genesis of humans. *Sci. Am.*, **266**(4), 22–7.

Wilson, A.C., Cann, R.L., Carr, S.M. *et al.* (1985) Mitochondrial DNA and two perspectives on evolutionary genetics. *Biol. J. Linn. Soc.*, **26**, 375–400.

Wilson, A.C., Ochman, H. and Praeger, E.M. (1987) Molecular time scale for evolution. *Trends Genet.*, **3**, 241–7.

Wilson, D.F. and Hessler, R.R. (1987) Speciation in the deep sea. *Annu. Rev. Ecol. Syst.*, **18**, 185–207.

Wilson, E.O. (ed) (1988a) *Biodiversity*, National Academy Press, Washington.

Wilson, E.O. (1988b) The current state of biological diversity, in *Biodiversity*, (ed E.O. Wilson), National Academy Press, Washington, pp. 3–18.

Wingstrand, K.G. (1972) Comparative spermatology of a pentastomid, *Raillietiella hemidactyli*, and a brachiuran crustacean, *Argulus foliaceus*, with a discussion of pentastomid relationships. *Biol. Skr. Danske Videnskab. Selskab*, **19**(4), 1–72.

Wingstrand, K.G. (1985) On the anatomy and relationships of recent Monoplacophora. *Galathea Rep.*, **16**, 7–94.

Woese, C.R. (1978) Macroevolution in the microscopic world, in *Molecules and Morphology in Evolution: Conflict or Compromise?* (ed C. Patterson), Cambridge University Press, Cambridge, pp. 177–202.

Woese, C.R. and Fox, G.E. (1977) Phylogenetic structure of the prokaryotic domain: The primary kingdoms. *Proc. Natl. Acad. Sci. USA*, **74**, 5088–90.

Woese, C.R., Kandler, O. and Wheelis, M.L. (1990) Towards a natural system of organisms: Proposal for the domains Archaea, Bacteria, and Eucarya. *Proc. Natl. Acad. Sci. USA*, **87**, 4576–9.

Woese, C.R., Kandler, O. and Wheelis, M.L. (1991) A natural classification. *Nature*, **351**, 528–9.

Wolf, K. and Markiw, M.E. (1984) Biology contravenes taxonomy in the Myxozoa: New discoveries show alternation of invertebrate and vertebrate hosts. *Science*, **225**, 1449–52.

Wolpert, L. (1969) Positional information and the spatial pattern of cellular differentiation. *J. Theor. Biol.*, **25**, 1–47.

Wolstenholme, D.R., MacFarlane, J.L., Okimoto, R. *et al.* (1987) Bizarre tRNAs inferred from DNA sequences of mitochondrial genomes of nematode worms. *Proc. Natl. Acad. Sci. USA*, **84**, 1324–8.

Wolters, J. (1991) The troublesome parasites: Molecular evidence that apicomplexa belong to the dinoflagellate–ciliate clade. *BioSystems*, **25**, 75–83.

Wolters, J. and Erdmann, V.A. (1988) Cladistic analysis of ribosomal RNAS – the phylogeny of eukaryotes with respect to the endosymbiotic theory. *BioSystems*, **21**, 209–14.

Wood, D.M. and Borkent, A. (1989) Phylogeny and classification of the Nematocera, in *Manual of Nearctic Diptera. Volume 3. Research Branch Agriculture Canada Monograph No. 32*, (ed J.F. McAlpine), pp. 1333–70.

Wood, T.K. (1987) Host plant shifts and speciation in the *Enchenopa binotata* Say complex, in *Proceedings of the II International Workshop on Leafhoppers and Planthoppers of Economic Importance*, (eds M.R. Wilson and L.R. Nault), CAB Commonwealth Institute of Entomology, London, pp. 361–8.

Wood, T.K. and Guttman, S.I. (1983) *Enchenopa binotata* complex: Sympatric speciation? *Science*, **220**, 310–2.

Woodley, N.E. (1989) Phylogeny and classification of the 'orthorrhaphous' Brachycera, in *Manual of Nearctic Diptera. Volume 3. Research Branch Agriculture Canada Monograph No. 32*, (ed J.F. McAlpine), pp. 1371–95.

Wright, J.W., Spolsky, C. and Brown, W.M. (1983) The origin of the parthenogenetic lizard *Cnemidophorus laredoensis* inferred from mitochondrial DNA analysis. *Herpetologica*, **39**, 410–6.

Wyss, A.R. (1988) Evidence from flipper structure for a single origin of pinnipeds. *Nature*, **334**, 427–8.

Yaeger, J. (1981) Remipedia, a new class of Crustacea from a marine cave in the Bahamas. *J. Crust. Biol.*, **1**, 328–33.

Yen, J.H. and Barr, A.R. (1973) The etiological agent of cytoplasmic incompatibility in *Culex pipiens*. *J. Invert. Pathol.*, **22**, 242–50.

Young, J.P.W. (1988) The estimation of protein and nucleic acid homologies, in *Prospects in Systematics*, (ed D.L. Hawksworth), Clarendon Press, Oxford, pp. 169–83.

Young, Y.Z. (1981) *The Life of Vertebrates*, 3rd edn, Clarendon Press, Oxford.

Zhang, Y. -P. and Shi, L.-M. (1991) Riddle of the giant panda. *Nature*, **352**, 573.

Zimmermann, E.A.W. (1778) *Geographische Geschichte des Menschen und der allgemein verbreiteten vierfüssigen Thiere, nebst einer hierher gehörigen Weltcharte*, 1. Leipzig.

Zouros, E., Freeman, K.R., Ball, A.O. and Pogson, G.H. (1992) Direct evidence for extensive paternal mitochondrial DNA inheritance in the marine mussel *Mytilus*. *Nature*, **359**, 412–4.

Zuckerkandl, E. (1987) On the molecular evolutionary clock. *J. Mol. Evol.*, **26**, 34–46.

Zuckerkandl, E. and Pauling, L. (1965) Evolutionary divergence and convergence in proteins, in *Evolving Genes and Proteins*, (eds V. Bryson and H.J. Vogel), Academic Press, New York, pp. 97–165.

Zwick, P. (1973) Insecta, Plecoptera. Phylogenetisches System und Katalog. *Das Tierreich*, **94**, 465 pp.

Author index

Subject index

Taxonomic index

Vernacular terms which are very close to scientific names already in the index have been generally omitted; the corresponding entries are entered under the Latin names.

Page numbers in *italic* refer to Appendices 2–23